Internet of Things

Technology, Communications and Computing

Series Editors

Giancarlo Fortino, Rende (CS), Italy

Antonio Liotta
Edinburgh Napier University, School of Computing
Edinburgh, UK

The series Internet of Things - Technologies, Communications and Computing publishes new developments and advances in the various areas of the different facets of the Internet of Things. The intent is to cover technology (smart devices, wireless sensors, systems), communications (networks and protocols) and computing (theory, middleware and applications) of the Internet of Things, as embedded in the fields of engineering, computer science, life sciences, as well as the methodologies behind them. The series contains monographs, lecture notes and edited volumes in the Internet of Things research and development area, spanning the areas of wireless sensor networks, autonomic networking, network protocol, agent-based computing, artificial intelligence, self organizing systems, multi-sensor data fusion, smart objects, and hybrid intelligent systems.

** Indexing: *Internet of Things* is covered by Scopus and Ei-Compendex **

More information about this series at http://www.springer.com/series/11636

Deepak Gupta
Victor Hugo C. de Albuquerque
Ashish Khanna • Purnima Lala Mehta
Editors

Smart Sensors for Industrial Internet of Things

Challenges, Solutions and Applications

Springer

Editors
Deepak Gupta
Department of Computer Science and Engineering
Maharaja Agrasen Institute of Technology
GGSIP University
Delhi, India

Ashish Khanna
Department of Computer Science and Engineering
Maharaja Agrasen Institute of Technology
GGSIP University
Delhi, India

Victor Hugo C. de Albuquerque
University of Fortaleza, UNIFOR
Fortaleza, Brazil

Purnima Lala Mehta
IILM Academy of Higher Learning-College of Engineering & Technology
Greater Noida, India

ISSN 2199-1073 ISSN 2199-1081 (electronic)
Internet of Things
ISBN 978-3-030-52623-8 ISBN 978-3-030-52624-5 (eBook)
https://doi.org/10.1007/978-3-030-52624-5

This Springer imprint is published by the registered company Springer Nature Switzerland AG
The registered company address is: Gewerbestrasse 11, 6330 Cham, Switzerland

Dr. Deepak Gupta would like to dedicate this book to his father Sh. R. K. Gupta and his mother Smt. Geeta Gupta for their constant encouragement, his family members including his wife, brothers, sisters, kids, and to his students close to his heart.

Dr. Ashish Khanna would like to dedicate this book to his mentors Dr. A. K. Singh and Dr. Abhishek Swaroop for their constant encouragement and guidance and his family members including his mother, wife, and kids. He would also like to dedicate this work to his (Late) father Sh. R. C. Khanna with folded hands for his constant blessings.

Dr. Purnima Lala Mehta would like to dedicate this book to the almighty, her teachers, family, and friends, and to her dear students.

Foreword

I am very much delighted to have been invited to write the foreword for this edited book on *Smart Sensors for Industrial Internet of Things*. The title of the book itself is self-explanatory and very interesting. This book highlights very important concepts related to the theory, design, and applications of the industrial smart sensor technology, which have advanced the fields of sensors and the industrial Internet of Things. This book is multidisciplinary in nature that includes different applications of the monitoring system.

With the rapid growth of connected technologies, the industrial world is transforming in a trend that conforms to several headlined names including the fourth industrial revolution, smart manufacturing, and industrial Internet of Things (IIoT). Industry 4.0 is a hot topic since it was first introduced, and which primarily focuses on the automation of factories and the implementation of IoT in industries. It enables industrial advancements with the help of advanced computing, analytics, low-cost sensing, and new levels of connectivity enabled through the Internet. Some of the technologies supporting this new form of industrial revolution are cloud services, big data analytics, and pervasive, intelligent, sensing technologies.

In modern industry, productivity, quality, reliability, and safety heavily depend upon the performance of the sensors employed. They form an interface between the production equipment and the surrounding environment providing feedback based on the results of the executed operations. The significant benefits of using intelligent sensing technology in industries are accuracy and consistency, which enable functions such as picking, placing, labeling, and printing to be performed at higher production rates, leading to low wastage, minimal downtime, and better quality control. These capabilities have made industrial smart sensors capable of more complex data processing enabled within the sensor unit while being independent of PLC. Given these abilities, it is quite certain that the manufacturing industry majorly depends on smart sensor devices in ensuring the accuracy and efficiency of the source data and eventually will hinge on the reliability of the information for the process chain. These days, two trends are extremely popular in researching and developing sensors. First is developing integrated sensors, which is inclined to supplying an advanced level of information by directly estimating the sensed data. Second is the augmentation of multi-sensor systems, which allows huge quantities of data to be acquired in the system.

This book provides a window to the research and development in the field of smart sensors and the industrial Internet of Things in a comprehensive way. The advances and challenges are discussed with a focus on successes, failures, and lessons learned, open issues, unmet challenges, and future directions. The contributions in this book cover a wide range of interdisciplinary areas including Internet of Things and its applications in smart sensors, industry 4.0, autonomous vehicles, future of mobility, intelligent transport systems, auto-braking accidental detection system, healthcare, artificial intelligence, cloud teleophthalmology, agriculture, monitoring systems, renewable energy, security, smart homes, real-time data collection, data management, systems design/analysis, web services, and other industrial applications. This book contribution emphasizes several topics in the area of smart sensors in industrial real-world applications.

I highly recommend this book to a variety of audiences, including numerous researchers working in the industrial smart sensor technology. This book is primarily intended for researchers from academia, and industry, who are working in the research areas such as sensor technology, system design, computer science, electronics engineering, wireless communication, IoT, big data, monitoring systems design, and information technology. It is my hope and expectation that this book will provide an effective learning experience, a contemporary update, and a practical reference for researchers, professionals, and students that are interested in the advances of sensor technology and its integration to the industrial Internet of Things.

Joel J. P. C. Rodrigues
Federal University of Piauí (UFPI)
Teresina–PI, Brazil
Instituto de Telecomunicações
Lisbon, Portugal

Preface

We are delighted to launch our book entitled *Smart Sensors for Industrial Internet of Things* that aims to attract a number of amateur researchers, engineering practitioners, academicians, scientists, scholars, and industry delegates with the respective book chapters. The book received plentiful abstracts from different parts of the world, and assured selection procedure was conducted to maintain the research standards. The book is composed of 18 full-length chapters in this volume. All the chapters submitted were peer reviewed by at least two independent reviewers, who were provided with a detailed review pro forma for evaluating each chapter. The valuable comments and suggestions from the reviewers were communicated to the authors and provided with their revised manuscripts. The exhaustiveness of the review process is evident, given the large number of articles received addressing a wide range of research areas. The stringent review process ensured that each published chapter met the rigorous academic and scientific standards.

We would also like to thank the authors of the published chapters for adhering to the time schedule and for incorporating the review comments. We wish to extend my heartfelt acknowledgment to the authors, peer reviewers, committee members, and production staff whose diligent work put shape to this volume. We especially want to thank our dedicated team of peer reviewers who volunteered for the arduous and tedious step of quality checking and critique on the submitted chapters.

Lastly, we would like to thank Springer for accepting our proposal for publishing the volume titled *Smart Sensors for Industrial Internet of Things*.

Delhi, India — Deepak Gupta
Fortaleza, Brazil — Victor Hugo C. de Albuquerque
Delhi, India — Ashish Khanna
Greater Noida, India — Purnima Lala Mehta

About the Book

Smart Sensors for Industrial Internet of Things brings together the latest research in smart sensors technology and exposes the reader to myriad industrial applications that this technology has enabled. The book emphasizes several topics in the area of smart sensors in industrial real-world applications. The contributions in this book give a broader view on the usage of smart sensor devices covering a wide range of interdisciplinary areas like Intelligent Transport Systems, Healthcare, Agriculture, Drone communications, and Security. This book series will demonstrate that this new domain is an outstanding and significant domain that has a brilliant future.

The book reflects specific topics like smart sensors for industrial IoT, intelligent connected vehicles, green IoT, heart disease prediction, medical image classification, smart fitness diagnosis, Internet of drones, and so on.

By presenting an insight into smart sensors for industrial IoT, this book directs the reader to explore the utility and advancements in smart sensors and their applications into numerous research fields. Lastly, the book aims to reach through a mass number of amateur researchers, industry experts, researchers, scientists, engineers, and practitioners and help them guide and evolve to advance research practices.

Contents

About the Editors

Deepak Gupta is an eminent academician and plays versatile roles and responsibilities juggling between lectures, research, publications, consultancy, community service, PhD and postdoctorate supervision, etc. With 12 years of rich expertise in teaching and two years in industry, he focuses on rational and practical learning. He has contributed massive literature in the fields of Human-Computer Interaction, Intelligent Data Analysis, Nature-Inspired Computing, Machine Learning, and Soft Computing. He has served as Editor-in-Chief, Guest Editor, and Associate Editor in SCI and various other reputed journals. He has completed his postdoc from Inatel, Brazil, and PhD from Dr. APJ Abdul Kalam Technical University. He has authored/ edited 33 books with national/international level publisher (Elsevier, Springer, Wiley, Katson). He has published 118 scientific research publications in reputed international journals and conferences including 56 SCI indexed journals of IEEE, Elsevier, Springer, Wiley, and many more. He is the convener and organizer of "ICICC" Springer conference series.

Victor Hugo C. de Albuquerque [M′17, SM′19] is a professor and senior researcher at the University of Fortaleza, UNIFOR, Brazil, and Data Science Director at the Superintendency for Research and Public Safety Strategy of Ceará State (SUPESP/CE), Brazil. He has a PhD in Mechanical Engineering from the Federal University of Paraíba, an MSc in Teleinformatics Engineering from the Federal University of Ceará, and graduated in Mechatronics Engineering at the Federal Center of Technological Education of Ceará. He is currently an Associate Professor of the Graduate Program in Applied Informatics of UNIFOR and leader of the Industrial Informatics, Electronics and Health Research Group (CNPq). He is a specialist, mainly, in IoT, Machine/Deep Learning, Pattern Recognition, and Robotics.

Ashish Khanna has 16 years of expertise in Teaching, Entrepreneurship, and Research and Development He received his PhD from National Institute of Technology, Kurukshetra. He has completed his M. Tech. and B. Tech. from GGSIPU, Delhi. He has completed his postdoc from Internet of Things Lab at

Inatel, Brazil, and University of Valladolid, Spain. He has published around 45 SCI indexed papers in IEEE Transaction, Springer, Elsevier, Wiley, and many more reputed journals with cumulative impact factor of above 100. He has around 100 research articles in top SCI/Scopus journals, conferences, and book chapters. He is coauthor of around 20 edited books. His research interest includes Distributed Systems, MANET, FANET, VANET, IoT, Machine Learning, and many more. He is the originator of Bhavya Publications and Universal Innovator Lab. Universal Innovator is actively involved in research, innovation, conferences, startup funding events, and workshops. He has served the research field as a Keynote Speaker/Faculty Resource Person/Session Chair/Reviewer/TPC member/postdoctoral supervisor. He is convener and organizer of ICICC conference series. He is currently working at the Department of Computer Science and Engineering, Maharaja Agrasen Institute of Technology, under GGSIPU, Delhi, India. He also serves as series editor in Elsevier and De Gruyter publishing houses.

Purnima Lala Mehta received her bachelor's degree and master's degree in the field of Electronics and Communications Engineering (ECE) from Bharati Vidyapeeth's College of Engineering for Women (BVCOEW), University of Pune, India, and Northcap University (formerly ITM University), Gurgaon, India, respectively. She has received her Doctor of Philosophy (PhD) degree in the field of "Wireless Cellular Communications through Aerial Drones" from Aarhus University, Denmark (ranked 141 by QS World Ranking). She has been working as an Assistant Professor since July 2012 and has around 8 years of teaching and research experience including international exposure at countries like Denmark, Germany, China, etc. Her research interests include the areas of Mobile Computing, Aerial Drone-Based Wireless Communications, Wireless Ad Hoc Networks, Millimeter Wave Communications, Future Generations of Communications, and Business Modeling. She has contributed her papers in multiple peer-reviewed conferences, journals, and books. For her astounding research work in Springer, she has been awarded "Franklin Quarterly Membership" by *London Journals Press*. She has been appointed as reviewer of SCI indexed journal *Wireless Personal Communications*, Springer, and has served as a Resource Person/Panel Speaker at a number of research events like IEEE 5G Summits, IEEE ANTS conference, etc. She has chaired and conducted multiple sessions in peer conferences including IEEE WPMC, Portugal.

Introduction

Deepak Gupta, Victor Hugo C. de Albuquerque, Ashish Khanna, and Purnima Lala Mehta

Abstract Sensors play a crucial role in capturing the measurements from the environment around and on computation produced results for further understanding and analysis of the environment. Sensors are vital for applications in a broad range of industrial operations. The book on *Smart Sensors for Industrial Internet of Things* brings together the latest research in smart sensors technology and exposes the reader to myriad industrial applications that this technology has enabled. The contributions in this book give a broader view of the usage of smart sensor devices covering a wide range of interdisciplinary areas like Intelligent Transport Systems, Healthcare, Agriculture, Drone Communications, and Security. By presenting an insight into smart sensors for industrial IoT, this book directs the reader to explore the utility and advancements in smart sensors and their applications into numerous research fields.

Keywords Smart sensors · Industrial Internet of Things · Transportation · Healthcare · Agriculture · Industry

1 Smart Sensors for Industrial Internet of Things

Sensors play a crucial role in capturing the measurements from the environment around and on computation produced results for further understanding and analysis of the environment. Sensors form a bridge between the production equipment and

D. Gupta (✉) · A. Khanna
Department of Computer Science and Engineering, Maharaja Agrasen Institute of Technology, GGSIP University, Delhi, India
e-mail: deepakgupta@mait.ac.in; ashishkhanna@mait.ac.in

V. H. C. de Albuquerque
University of Fortaleza, UNIFOR, Fortaleza, Brazil
e-mail: victor.albuquerque@ieee.org

P. L. Mehta
IILM Academy of Higher Learning-College of Engineering & Technology, Greater Noida, India

D. Gupta et al. (eds.), *Smart Sensors for Industrial Internet of Things*, Internet of Things, https://doi.org/10.1007/978-3-030-52624-5_1

the surrounding environment [1]. Sensors are vital for applications in a broad range of industrial operations. The Internet of Things (IoT) usually refers to a worldwide network of interconnected heterogeneous objects and recently a new definition of IoT seen as a loosely coupled, decentralized system of cooperating smart objects [2]. Modern industrial wireless sensor networks (IWSNs) integrate the two main networks, i.e., wired and wireless sensor networks using mobile intelligences in smart factories [3].

The book on *Smart Sensors for Industrial Internet of Things* brings together the latest research in smart sensors technology and exposes the reader to myriad industrial applications that this technology has enabled. The book emphasizes several topics in the area of smart sensors in industrial real-world applications. The contributions in this book give a broader view of the usage of smart sensor devices covering a wide range of interdisciplinary areas like Intelligent Transport Systems, Healthcare, Agriculture, Drone Communications, and Security. This book series will demonstrate that this new domain is an outstanding and significant domain that has a brilliant future.

The book reflects specific topics such as smart sensors for industrial IoT, intelligent connected vehicles, green IoT, heart disease prediction, medical image classification, smart fitness diagnosis, Internet of drones, and so on. By presenting an insight into smart sensors for industrial IoT, this book directs the reader to explore the utility and advancements in smart sensors and their applications into numerous research fields.

The book is divided into application-based sections. The first is the Introductory section comprising two chapters, the second session deals with Transportation and Automobile vehicles comprising four chapters, the third section deals with Healthcare covering five chapters, the fourth section focuses on Agriculture and covers four chapters, and lastly, we present some Case Study chapters on topics like the Internet of Drones, Energy Harvesting, and Secure and Smart Homes. The following sections shall walk through a brief introduction to the individual book chapters.

2 Introduction

The second chapter titled "Internet of Things Concept and Its Applications" is an introductory chapter that starts with the basic concept of the Internet of Things (IoT), talking about the network interface and communication of physical objects, devices, and peripherals that can interact and exchange data between one another without depending on human interactions. The chapter covers IoT applications in the domains of Smart Grid, Solid Waste Management, Healthcare, Marine Environment Monitoring, and Protected Agriculture.

The third chapter "Smart Sensors and Industrial IoT (IoT)—A Driver for the Growth of Industry 4.0" discusses smart sensors in the field of industrial Internet of things (IIoT) and presents an analysis of changes that were initiated by applying smart sensors and IIoT in industry to connect technologies in industries, factories, households, and workplaces. The authors further emphasize on the impact of smart sensors and IIoT on the working of the manufacturing industry or business organization and are able to answer about the smart sensor for better understanding with underlying reasons why producers and consumers both are resistant to smart products.

3 Research Solutions—Transportation and Automobile

The first section of chapters dissertates the integration of smart sensors with vehicles supporting the concept of intelligent transport systems. Roadside safety measures and traffic controlling can benefit from multiple sensors deployed intelligent transport systems [4]. The fourth chapter "Smart Sensors for IIoT in Autonomous Vehicles: A Review" focuses on smart sensors for autonomous vehicles and discusses signal conditioning methods such as RF module, BTS, TPMS, capacitive balancing of humidity sensor (in HVAC) used in an autonomous vehicle, and industrial IoT implementation. The fifth chapter "Connected Vehicles—Intelligent Transport Systems" covers the connected vehicle (CV) technology along with an intelligent traffic system to aid in increasing the safety of people on the road. With respect to CVs, the authors have emphasized its benefits, applications, and challenges in this chapter. The sixth chapter "Connected Vehicles—Intelligent Transport Systems" deals with Connected Vehicles in Intelligent Transport Systems (ITS), and the authors present a design of a secure long-range decentralized network of vehicles [5]. In their proposed work, all the vehicles within the range of a kilometer will be able to communicate with each other to solve all the specified problems. In the seventh chapter, "Design of Auto-braking System for Accident Prevention and Accident Detection System Using IoT," the authors focus on decreasing the impact of a collision and establishing communication with the nearby hospital for providing necessary support to the victims during road accidents. Their proposed work is supported by simulations and can be used in any type and size of the vehicle.

4 Research Solutions—Healthcare

Smart devices are a future in healthcare, and smart sensors are essential for smart fitness devices and medical care units. Our next section of the book focuses on smart sensors in healthcare [6]. In the eighth chapter "IoT in Healthcare Perspective and Green IoT," the author presents a new approach to the conjoining of green IoT and healthcare IoT. With the proper usage of green resources, the greenhouse effect can be prevented from occurring. The ninth chapter "IoMT with Cloud-Based Disease Diagnosis Healthcare Framework for Heart Disease" presents the Internet of Medical Things (IoMT) that interlinks a collection of intelligent sensors on the patient's body to observe and interpret multimodal health data, including the patient's physiological and psychological signals [7]. The chapter proposes a new IoMT-based disease diagnosis healthcare framework for heart disease prediction using the BBO-SVM model. In the tenth chapter "Hyperparameter Optimization of Deep Neural Network for Multimodality Fused Medical Image Classification," the authors present a hyperparameter optimization of deep neural networks in multimodality fused medical image classification for medical and industrial IoT. A Multimodality Image Fusion Classification (MMIFC) is proposed in the chapter by the incorporation of image fusion, feature extraction, and classification techniques. The eleventh chapter "Cognitive IoT-Based Smart Fitness Diagnosis and Recommendation System Using 3-Dimensional CNN with Hierarchical Particle Swarm Optimization" proposes a cognitive IoT-based smart fitness diagnosis and recommendation system using a three-dimensional convolutional neural network with a hierarchical particle swarm optimization (PSO) algorithm and has been applied to check the health statuses of exercisers. In the twelfth chapter "IIoT with Cloud Teleophthalmology Based Age-Related Macular Degeneration Disease Prediction Model," a scalable cloud-oriented teleophthalmology structure by an Internet of Medical Things (IoMT) to detect the AMD has projected examination. A projected Optimal Generative Adversarial Network (OGAN) helps to investigate the images to find as well as to compute AMD disease severity. The experimental outcome through their proposal showed the superior performance of the proposed model over the compared methods by attaining a maximum accuracy of 98.03%.

5 Research Solutions—Agriculture

Smart agriculture, precision agriculture [8], and vertical farming [9] are a new age thing in the agricultural market with multiple applications utilizing smart sensors. In the thirteenth chapter of this book titled "Significance of IoT in the Agricultural Sector," the authors emphasize the significance of IoT in the agriculture sector and discuss the evolution, benefit, and uses of IoT in agriculture. The authors have further developed and tested an IoT-based smart irrigation framework dependent on the moisture level of the soil. In the next chapter, i.e., the fourteenth chapter, "Soil Moisture Sensor Nodes in IoT-Based Drip Irrigation System for Water Conservation," the authors present a new sensor network-assisted irrigation system and a rule-based analysis model that has been developed in this research work to enhance the efficiency of water usage. A smart irrigation system can be built with smart sensor networks for collecting field values and can be analyzed using rules for effectively watering the plants. The fifteenth chapter "Precision Agriculture Using Advanced

Technology of IoT, Unmanned Aerial Vehicle" focuses on the illustration and utilization of those advanced technologies for smart farming. Precision agriculture (PA) uses site-specific crop management concept based on measured data using sensors and data analytics to find the root cause of yield reduction. The sixteenth chapter "IoT-Based Brinjal Crop Monitoring System" uses a remote crop monitoring mechanism using LORAWAN in the greenhouse. The motive of this chapter is to enhance the traditional way of agriculture in rural areas with the help of a wireless sensor network using LORAWAN protocol.

6 Research Solutions—Case Studies

The next three chapters are part of specific case studies. The seventeenth chapter "Internet of Drones: An Engaging Platform for IIoT-Oriented Airborne Sensors" introduces the Internet of Drones (IoD) paradigm, and this chapter takes a walk in describing IoD and proposes a state-of-the-art architecture, and its applications, especially that are oriented to industrial IoT. The eighteenth chapter "A Novel Approach on Renewable Energy Harvesting Using the Internet of Things (IoT)" presents renewable energy harvesting using the Internet of Things by proposing an IoT-based system to distribute energy among the solar board. The control methodology is to drive two little DC engines with the goal that the sun picture is kept at the focal point of the four-quadrant photograph indicator detecting the sun position. The final paper of this book titled "Security and Surveillance at Smart Homes in a Smart City Through the Internet of Things" focuses on how security and surveillance are achieved in smart homes. Smart homes provide efficient management with minimum lifetime costs of hardware and facilities. They optimize things such as structures, systems, and services and also manage the interrelationships between these three.

7 Conclusion

To conclude, this book comprises eighteen papers that were selected after a peer review. The main objective of this book is to explore research work involving smart sensors in IoT and industrial IoT applications. The book covers a broad range of applications and scenarios, where IoT and IIoT technologies can be well understood and open more research areas and paradigms for the audience to adapt and follow. Lastly, the book aims to reach through a mass number of amateur researchers, industry experts, researchers, scientists, engineers, and practitioners and help them guide and evolve to advance research practices. We thank our readers and we hope that each chapter in this book gives ideas and deeper understanding into a variety of concepts and research paradigms.

References

1. IEEE: New trends in smart sensors for industrial applications – Part I. IEEE Trans. Ind. Electron. **64**, 7281 (2017)
2. Fortino, G., Trunfio, P.: Internet of Things Based on Smart Objects, Technology, Middleware and Applications. Springer, New York (2014). ISBN: 978-3-319-00490-7
3. Luo, Y., Duan, Y., Li, W., Pace, P., Fortino, G.: Workshop networks integration using mobile intelligence in smart factories. IEEE Commun. Mag. **56**(2), *68–75* (*2018*)
4. Guerrero-Ibáñez, J., Zeadally, S., Contreras-Castillo, J.: Sensor technologies for intelligent transportation systems. Sensors. **18**, 1212 (2018)
5. Mahmood, Z.: Connected Vehicles in the Internet of Things: Concepts, Technologies, and Frameworks for the IoV. Springer Nature Switzerland AG, Switzerland (2020). https://doi.org/10.1007/978-3-030-36167-9
6. Pramanik, P.K.D., Upadhyaya, B.K., Pal, S., Pal, T.: In: Dey, N., Ashour, A.S., Bhatt, C., Fong, S.J. (eds.) Healthcare Data Analytics and Management, pp. 1–58. Academic Press, Cambridge (2019)
7. Alhussein, M., Muhammad, G., Hossain, M.S., et al.: Cognitive IoT-cloud integration for smart healthcare: case study for epileptic seizure detection and monitoring. Mobile Netw. Appl. **23**, 1624–1635 (2018). https://doi.org/10.1007/s11036-018-1113-0
8. Omran, E.-S.E., Negm, A.M.: In: Omran, E.-S.E., Negm, A.M. (eds.) Technological and Modern Irrigation Environment in Egypt: Best Management Practices & Evaluation, pp. 77–105. Springer International Publishing, Cham (2020)
9. Bhowmick, S., Biswas, B., Biswas, M., Dey, A., Roy, S., Sarkar, S.K.: In: Bera, R., Sarkar, S.K., Singh, O.P., Saikia, H. (eds.) Advances in Communication, Devices, and Networking, pp. 521–528. Springer, Singapore (2019)

Internet of Things Concept and Its Applications

Prashant Ahluwalia and Nitin Mittal

Abstract In the current era of digital communication and networking, the term Internet of Things abbreviated as IoT has become very famous. The Internet of Things relates essentially to the network interface and communication of physical objects, devices, and peripherals that can interact and exchange data between one another without depending on human interactions or computer interactions. IoT applications promise to add enormous value to our lives. With newer wireless networks, superior sensors, and revolutionary computing capabilities, for its wallet share, the Internet of Things could be the next frontier in the race.

Keywords Internet of Things · Smart grid · Solid waste management · Healthcare · Marine environment · Protected agriculture · Cloud computing · Transmission Control Protocol (TCP) · Internal Control Protocol (ICP) · Global positioning system (GPS)

1 Introduction

The IoT research is in the early stages, and it has been given separate definitions by researchers. IoT comprises of two sentences: "Internet" and "Thing." "Internet" can be defined as "TCP/IP protocol-based global computer interconnection" and "Thing" is "an unrecognizable object." Thus, "Internet of Things" implies semantically a global network of interconnected objects that are uniquely addressable, based on the Transmission Control Protocol (TCP) and the Internal Control Protocol (ICP) Therefore, it is suitable to define the IoT as "things with identities and virtual personalities operating in smart spaces using smart interfaces to link and interact in social, environmental, and user contexts" [1]. IoT can also be defined at anytime, anywhere, anything, and anyone using any path or service as a link between individuals and things [2, 3]. This means addressing aspects such as convergence,

P. Ahluwalia
Computer Science & Engineering, Chandigarh University, Mohali, India

N. Mittal (✉)
Electronics & Communication Engineering, Chandigarh University, Mohali, India

D. Gupta et al. (eds.), *Smart Sensors for Industrial Internet of Things*, Internet of Things, https://doi.org/10.1007/978-3-030-52624-5_2

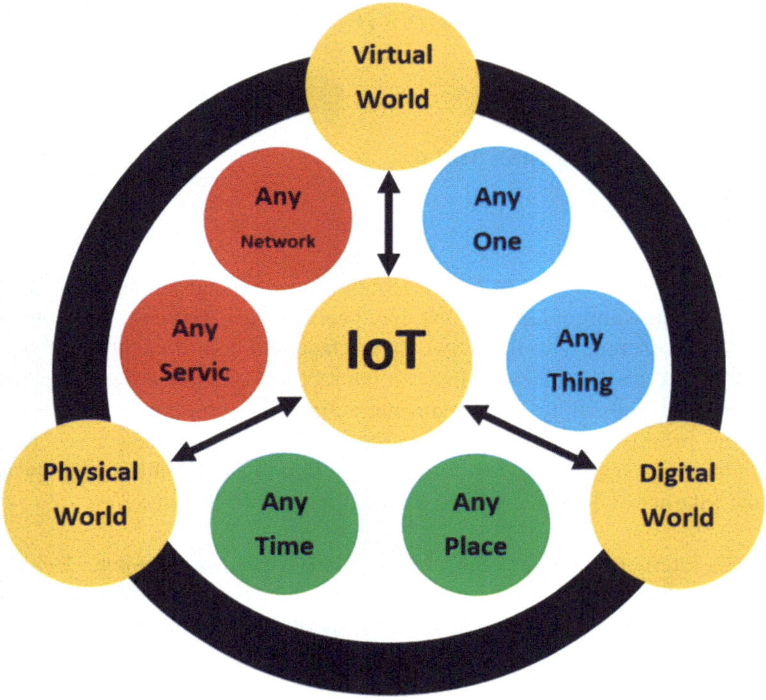

Fig. 1 Internet of Things (IoT) with its connections and related entities

content, collections (repositories), computing, interaction, and connectivity in a context where there is a seamless interconnection between people/humans and things and/or between things (see Fig. 1) [4]. Thus, IoT is a huge dynamic global network infrastructure of internet-enabled physical and virtual objects/entities with internet facilities that involves embedded systems and all sorts of information technologies such as global positioning system (GPS), infrared devices, scanners, RFID tags/devices, sensors, actuators, smartphones and the internet to be sensed, recognized, computing and so on. The International Telecom Union (ITU) published an IoT annual report [5], which extended the concept of IoT in 2005.

1.1 Vision

Today, IoT is used to indicate sophisticated device and service connectivity that extends beyond the traditional machine to include a range of protocols and apps.

There are three visions for IoT [6]:

1. Internet-oriented: it is necessary to create intelligent items in an internet-oriented vision.
2. Semantic-oriented: the amount of sensors accessible will be enormous in semantic-oriented vision and their collected information will be enormous. Thus, for better depictions and comprehension, the raw information must be managed and processed.
3. Things-oriented: in the vision of things, we can monitor any object using sensors and pervasive technology.

2 Characteristics

There are three important characteristics of IoT [7]:

1. Comprehensive sense: Using sensors to gather information from any object whenever and wherever.
2. Intelligent processing: Use cloud computing methods to evaluate enormous quantities of information for object control.
3. Reliable transmission: Accurate and real-time transmission of information via the Internet and communication networks.

3 Applications of IoT

The few applications of IoT are:

1. Smart Grid.
2. Solid Waste Management.
3. Healthcare.
4. Marine Environment Monitoring.
5. Protected Agriculture.

3.1 Smart Grid

The intelligent grid is proposed to tackle electricity grid problems (e.g., bad reliability, high power outages, high greenhouse gas and carbon emissions, economy, security, and energy security) [8]. One of the smart grid ideas is that the smart grid (SG) is a communication network at the top of the energy grid that collects and analyzes information from various energy grid parts to predict the supply and demand that can be used to manage energy [3].

Some of the required functionalities to deploy the smart grid are as follows [9]:

1. *Communication networks*: public, private, wired, and wireless communication networks that can be used as the communication infrastructure for smart grid [10].
2. *Cyber security*: policy-making to guarantee the availability, integrity, and confidentiality of communication and control systems required to manage, operate, and protect intelligent grid infrastructure [11].
3. *Distributed power resources*: use of different kinds of generation (e.g., renewable energy) and/or storage devices (batteries, bi-directional plug-in electric vehicles) connected to distributed systems [12].
4. *Management of distribution grids*: attempt to maximize the performance of parts in distribution systems such as feeders and transformers and integrate them with transmission systems, boost reliability, boost the effectiveness of distribution systems and enhance the management of distributed renewable energy sources [13].
5. *Electrical transport*: integration of large-scale plug-in electrical vehicles [14].
6. *Energy efficiency*: To provide processes for distinct types of clients to change their energy consumption during peak hours and to optimize the equilibrium between power supply and demand [15].
7. *Energy storage*: the use of techniques for direct or indirect energy storage such as pumped hydroelectric storage technology [16].
8. *Wide-range tracking*: monitoring parts of the energy scheme over a big geographic region to optimize their efficiency and avoid issues before they occur [17].
9. *Advanced metering infrastructure (AMI)*: AMI as one of SG's main parts provides a bidirectional communication network between intelligent meters (SMs) and utilities to collect, send, and evaluate information on the usage of energy [18–20].

3.1.1 Requirements for Using IoT in SG

Communication techniques for the use of IoT in SG: Communication techniques can be used to obtain and communicate obtained data on the state of the systems of SG. We have communication technology standards of short-range and long-range. Examples of short-range communication techniques are ZigBee, Bluetooth, and ultra-wideband systems. Power line communications [21], optical fiber, 3G and 4G wireless cellular networks and satellite communications can be used for long-range communications.

3.1.2 Future Research Directions

There are many difficulties that need to be resolved in future study directions in order to attain technical objectives in implementing IoT in SG. Since IoT systems must operate in distinct settings that may have difficult circumstances (e.g., elevated or low temperatures, high voltages, exposure to electromagnetic waves, water work, etc.), they must therefore meet criteria under such circumstances as reliability or compatibility.

3.2 *Solid Waste Management*

Waste management is a name given to a waste collection scheme, which includes transportation, disposal, or recycling. This word is ascribed to waste material generated by a human activity that must be managed to prevent its adverse effect on health and the environment. Most often, it is anticipated that waste will reuse the available resources. Methods of waste management can differ from developed to urban, rural, industrial, and residential. Metropolitan and rural waste management are a municipality's general responsibility, while industrial waste is their responsibility and is managed by themselves.

The different types of considered waste are as follows:

1. *Organic Waste*. It is the trash of organic waste [22].
2. *Recyclable Waste*. It is all the waste that can be used in the technique of conversion to other parts or in raw material production [23].
3. *Industrial Waste*. They are the mainly powerful residues that originate from the industrial production system. It usually consists of residues of raw materials designed for industrial process recycling or reuse [24].
4. *Hospital Waste*. It is the waste that comes from hospitals and medical centers and can contaminate and transmit diseases to people who come into contact with it [25].
5. *Commercial Waste*. Business organizations manufacture it, such as apparel stores, toys, and machinery [26].
6. *Green Waste*. It is the material that mainly results from the pruning of trees, branches, trunks, barks, and leaves in the highways. Because it is organic matter, it can be used to compost and generate organic fertilizer [27].
7. *Electronic Waste*. This is the waste generated by the disposal of consumer electronics products that have stopped operating or become outdated [28].
8. *Nuclear Waste*. It is the one mainly generated by nuclear power plants. It is a very dangerous waste because it is a radioactivity aspect [29].

3.2.1 Available IoT Architecture Reference Models for Waste Management Systems

The Web is powered by the Transmission Control Protocol/Internet Protocol (TCP/IP) architecture to allow network hosts to communicate as they are known. There is also a need for an IoT-based system architecture that always addresses issues such as scalability, interoperability, consistency, quality of service (QoS), etc. According to the author of [30], IoT has access to different models and reference architectures.

Some projects can be identified as one of the predominant reference models, such as RAMI 4.0, the reference architecture for intelligent factories applied to IoT standards. A consortium formed by AT&T, Cisco, General Electric, International Business Machines Corporation (IBM) and Intel, Industrial Internet Reference

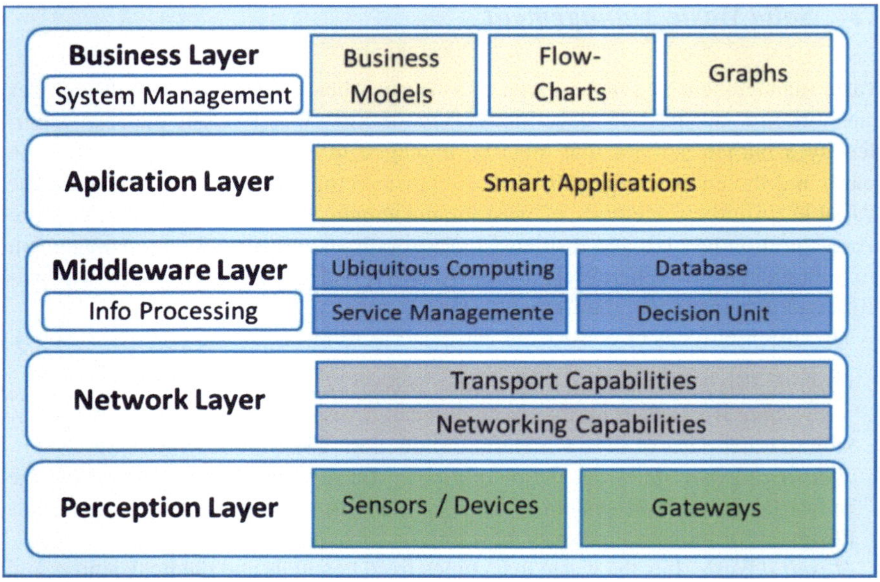

Fig. 2 Layered architecture for waste management systems

Architecture (IIRA), has also launched another venture, while the Internet of Things Architecture (IoT-A) is stimulating an architecture design involving comprehensive system requirements [31].

Most design models rely on a standard architecture based on a requirement assessment or on certain layers that form the basic model of a reference architecture. Figure 2 shows the basic layout architecture. The most basic approach is considered to be a three-layer architecture consisting of implementation, networking, and layers of perception [32]. Recent literature also includes some other models that contribute more to the abstraction of IoT architecture, such as the Service Oriented Architecture (SOA) model, the middleware [30, 33, 34] model, and the five-tier model [35–37].

First, there is a brief discussion of these layers, which, in effect, alternate between the models presented.

1. *Perception Layer*. The perception layer of the IoT architecture is similar to the physical layer of the Open Systems Interconnection (OSI) model as it is based on the hardware level and is responsible for the collection, processing, and transmission of physical information through secure channels to the upper layers. It applies techniques for detecting parameters of physical features through particular sensors such as weight, temperature, humidity, etc., and for collecting object recognition information such as Quick Response codes (QR codes) and RFID.

2. *Network Layer.* The network layer is responsible for moving the measured data in the perception layer to the upper layers where the processing devices are located. It uses ZigBee, Z-wire, GSM, UMTS, Wi-Fi, Infrared, and 6LoWPAN. Besides basic tasks, the network layer also performs data management operations.
3. *Middleware Layer.* The middleware layer is a software layer or even a set of sublayers interconnecting IoT parts otherwise unable to interact, i.e., an interpreter. It plays a significant part in developing fresh techniques and offers competitiveness so that the application layer can communicate with the perception layer and guarantee efficient communication.
4. *Application Layer.* The application layer does not make a significant contribution to the building of an IoT architecture, but it is in this layer that the different equipments are constructed to interact with customers, i.e., where the data is interpreted and available.
5. *Business layer.* This layer is accountable for coordinating the entire IoT scheme, including service-related apps such as supplying the underlying layers with a high-level analysis report and protecting users'privacy. This layer can be responsible for generating graphs and business models.

Since IoT connects everything together for information exchange, traffic and network stores tend to be exponentially improving. The development of IoT apps is thus based on advancing technology and design following a reference model for IoT architecture.

3.3 Healthcare

The Internet of Things (IoT) is certainly one of the study community, government sector, and industry's most interesting subjects. While traditional internet enables communication between a number of restricted devices and humans, IoT connects all kinds of linked "things" into an extensive network of computer-related intelligence without a human being's interference. IoT adoption and the development of wireless communication technologies enable the health conditions of patients to be streamed in real-time to caregivers [38, 39]. Furthermore, many available sensors and portable devices can measure specific human physiological parameters such as heart rate (HR), respiration rate (RR), and blood pressure (BP) through a single touch. Although it is still in the early development stage, businesses and industries have quickly adopted the power of IoT in their existing systems, and they have witnessed improvements in production as well as user experiences [40].

However, the integration of IoT technology in healthcare brings several challenges, including data storage, data management, exchange of data between devices, security and privacy, and unified and ubiquitous access.

One possible solution that can address these challenges is Cloud Computing technology. Figure 3 shows a typical healthcare system that integrates both IoT and

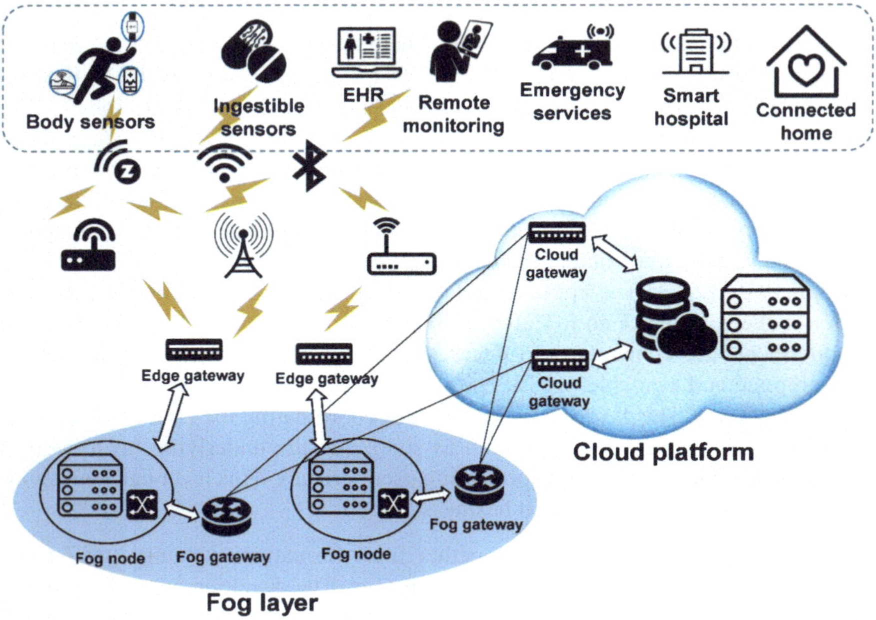

Fig. 3 An overview of a typical IoT and cloud computing-based healthcare system

cloud computing to provide the ability to access shared medical data and common infrastructure ubiquitously and transparently, offering on-demand services, over the network, and performing operations that meet growing needs [41].

IoT delivers proper solutions for various applications that cover all aspects of life such as smart cities [42], smart traffic management, waste management, structural health monitoring, security, emergency services, supply chain, retail, industrial management [43–46], and healthcare.

3.3.1 IoT Framework for Healthcare

The IoT in healthcare framework (IoTHeF) is considered the most fundamental aspect of IoT in healthcare because it helps healthcare applications to completely utilize the IoT and cloud computing. The framework also provides protocols to support the communication and broadcast of raw medical signals from various sensors and smart devices to a network of fog nodes.

As shown in Fig. 4, there are three essential components of IoTHeF, which include topology, structure, and platform. Each component serves a specific function in the IoT healthcare framework, all of which will be discussed in detail in the following sections. The readers are recommended to review proposed IoT architectures in [47, 48] to gain insights into the IoT architectures for healthcare. The systems can collect data about patient health status through multiple sensors. After that,

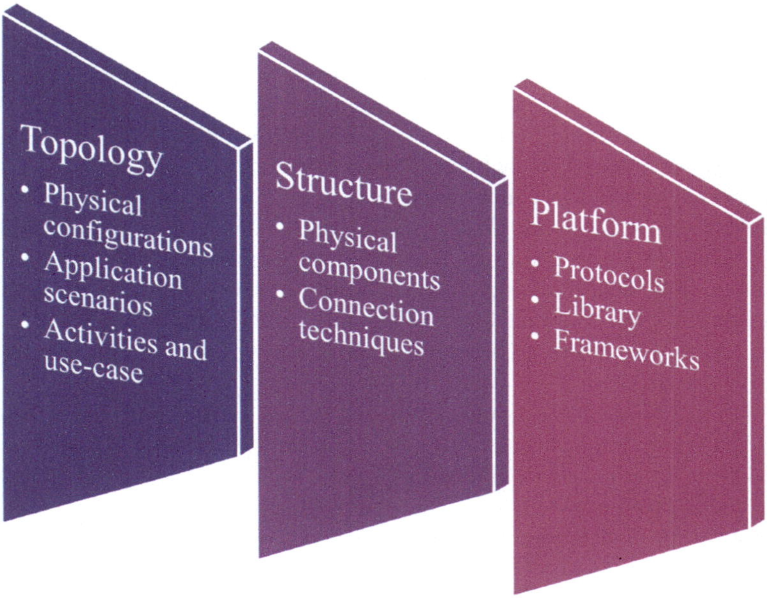

Fig. 4 Three basic components and their main functions in the IoT framework for healthcare

the collected data were transmitted to the remote server for analyzing, and the results were displayed in real time.

The IoTHeF topology handles the arrangement of general IoT components and outlines some standard setups for given application scenarios in the IoTHeF framework. Figure 3 presents a typical IoT and cloud computing in healthcare topology containing three main elements [49]. First of all, a publisher represents a network of connected sensors or hand-held devices in charge of recording patient's vital signs, and continuously sending a considerable amount of raw information such as electrocardiogram (ECG), electromyography (EMG), body temperature, blood glucose (BG), and the volume of air inspired and expired by lungs to a broker. Next, the broker analyzes and stores processed data on the cloud. Finally, a subscriber, who directly monitors patients can access the data from any location and responds immediately when unexpected incidents happen. The IoT HeF framework incorporates individual components into a hybrid computing grid where each component serves a specific purpose on IoT and cloud computing in the healthcare network.

3.4 Marine Environment Monitoring

IoT is a flexible global network architecture with self-configuring capabilities based on ordinary and interoperable communication protocols, where physical and virtual components have identities, physical and virtual features, and are seamlessly integrated into the data network [50]. IoT has recently been widely recognized as a groundbreaking paradigm that can transform our society and industry by seamlessly integrating various devices equipped with sensing, detection, storage, interaction, actuation, and networking capabilities [51]. The WSN (Wireless Sensor Network) plays a key role in IoT. This includes a large number of distributed sensors that are interconnected by wireless connections.

Wireless sensor networks (WSNs) have been widely used over the past few decades as an IoT subset for a variety of smart applications and services, including smart home [52], smart building [53, 54], smart transport [55, 56], smart industrial automation [57, 58], smart healthcare [59], smart grid [60], and smart cities [61]. Similar IoT-based methods of monitoring and protecting marine environments can certainly be used.

While our society and economy grow, the marine environment has received a great deal of attention from scientists and academics. It is very expensive to monitor traditional marine environmental networks such as oceanographic research vessels and hydrological research. Our methods of gathering and reviewing information are time consuming and the data collected are of low resolution. The Internet of Things (IoT) has developed wireless sensor networks (WSNs). IoT has a much higher capacity for data management than WSNs, which allows intelligent object control.

In a typical marine monitoring system based on IoT, various sensors are used to calculate and track precise physical and chemical parameters such as water temperature and pressure, wind direction and velocity, salinity, turbidity, pH, oxygen density, and chlorophyll rates.

Nonetheless, to solve some critical problems, including autonomy, adaptability, scalability, flexibility, and self-healing, it is important to model, build, and deploy an IoT-based marine environment monitoring and protection Scheme [62, 63], following specific requirements for extreme marine environments [64]:

1. High water resistance.
2. Strong robustness in hardware.
3. Low energy consumption and energy harvesting.
4. Stability of radio signal.

In particular, instruments and sensor nodes should be extremely reliable due to hard installation and maintenance; the need for boom and mooring systems; sensor coverage should be closely measured due to large areas [65]; equipment should be installed against potential vandalism behavior.

3.4.1 Overview of IoT in Marine Environment Monitoring

IoT-Based Marine Environment Monitoring Applications

IoT-based marine environment surveillance applications include:

1. Ocean sensing and tracking is a particular marine environment surveillance scheme.
2. Water quality surveillance generally monitors water circumstances and characteristics, including water temperature, pH, turbidity, and conductivity.
3. Monitoring of coral reefs; a system that typically controls coral reef habitats and environment.
4. Monitoring of fish farms by marine (offshore or deep-sea); tracking of water circumstances and characteristics including temperature and pH, measurement of fecal waste and uneaten feed for fish farms, and fishing conditions and operations including the number of dead fish.
5. Wave and current surveillance system for safe and secure waterway navigation measures waves and currents.

Different IoT systems, sensing and control systems, and communication systems are used in separate apps.

Common IoT-Based System Architectures for Marine Environment Monitoring and Protection

The Internet of Things is generally to obtain "knowledgeable, thought able, and controllable" for the surrounding world [66], meaning that the IoT is capable of perceiving, thinking, and controlling the environment by gathering, processing, and analyzing world information. It can create smart decisions that affect the outside world. In the research literature, IoT scientists have suggested various IoT system architectures. Among them, Antao et al. suggested a five-layer system architecture [67].

1. *Perception and Execution Layer*: The layer of perception and execution is the architecture's lower layer. It involves equipment for sensors and actuators with the goal of collecting sensor information and actuating commands.
2. *Data Transmission Layer*: The information transmission layer's primary role is to communicate multiple recorded information via communication networks, mostly mobile or wireless communication networks, to the information processing layer.
3. *Data Pre-Processing Layer*: The pre-processing layer of information is in the center of the IoT system architecture where, using sophisticated data mining techniques, the raw information obtained can be stored and pre-processed. It also finishes aggregation or disaggregation of information, data cleaning and fitting or screening, sharing as necessary, and sometimes triggers alerts or warnings based on predefined guidelines.
4. *Application Layer*: The application layer offers services based on various user-requested apps. This layer's primary aim is to provide intelligent application services to satisfy the requirements of customers. This subsequently includes water quality surveillance, monitoring of coral reefs, marine (offshore or deep-

sea) monitoring of fish farms, wave tracking, and present tracking in IoT-based marine environments [68–70].

5. *Business Layer*: The business layer is the top layer and manages the overall activities and services of the IoT system, including the creation of business models, business logic flowcharts, and graphical representations, based on the data received from the application layer. It also tracks and verifies the outputs of the other four layers in accordance with business models to improve facilities and preserve the privacy of users [68, 70].

Such a layered system architecture provides a good picture of the data/information flow in IoT-based marine environment monitoring and protection systems (Fig. 5).

A General Marine Environment Monitoring Sensor Node

The general architecture of a marine environment monitoring sensor node is shown in Fig. 6. It typically has a buoy device to protect electronic devices against water, and consists of the following four main modules [71]: a sensing module, a microcontroller, a wireless transceiver module, and a power supply module.

Typical Sensors and Sensing Parameters

Sensors are used by generating electrical signals in the form of electrical voltage, current, or frequency to react to modifications in their environments [72]. Typically, there are two types of sensors: physical and chemical sensors. Physical sensors are used to evaluate various physical parameters such as temperature, humidity, pressure, velocity of wind, and direction of wind. Different chemical parameters such as salinity, turbidity, pH, nitrate, chlorophyll, and dissolved oxygen (DO) are measured using chemical sensors. For details, please refer to [73].

Wireless Communication Technologies

A sensor node requires a wireless radio communication module. The access network includes all the devices between the backbone network and the client terminals with a communication range of a few hundred meters to several kilometers.

For the IoT-based marine environment monitoring and protection systems, wireless communication networks have different requirements than other applications, because of the following reasons:

1. Reliability: Radio antenna oscillations and bad ocean weather conditions can cause instability of radio signals.

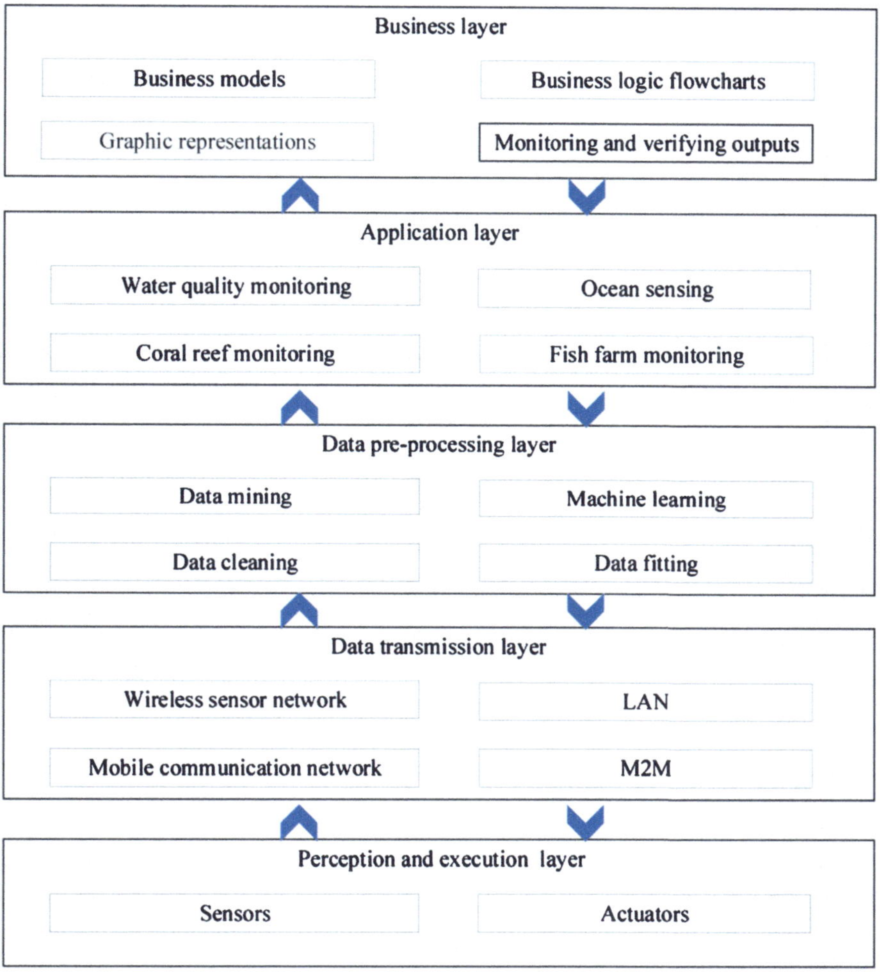

Fig. 5 Common layered architecture for Internet of Things (IoT)-based marine environment monitoring and protection applications

2. Energy efficiency: Low power consumption in stand-alone battery-powered machinery is the key to promoting long-flow and lower maintenance costs.

Multiple wireless communication techniques are typically used in the surveillance and safety scheme of an IoT-based marine environment. Underwater acoustic communication techniques are used in certain particular apps for information collection and communication among marine environmental sensors underwater [74–77].

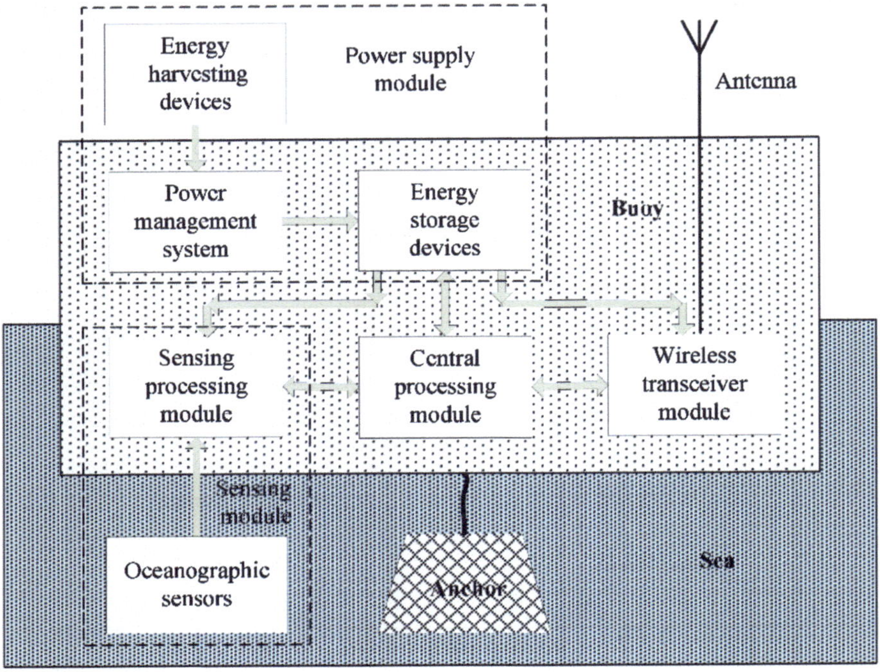

Fig. 6 General architecture of a marine environment monitoring sensor node

3.5 Protected Agriculture

The concept of protected agriculture is relative to open-field agriculture. It uses artificial techniques to change climatic factors such as natural light, temperature, and humidity to create environmental conditions suitable for the growth of animals and plants, enabling them to grow around the clock.

The environment of protected agriculture is completely or largely artificially controlled and which has broken the limits of climate and land conditions for the growth of animals and plants to a certain extent. So, it is also called controllable agriculture. Compared with open-field agriculture, protected agriculture has more potential to apply IoT technology, because it is less affected by climatic and geographical factors. Some mature IoT solutions in other fields can be transferred directly to protected agriculture [78].

With the development of agricultural sensor, wireless communication, cloud computing, machine learning, and Big Data technologies, IoT technology has emerged and is gradually being promoted and applied in the protected agriculture field [79–81]. It is playing an important role in various areas of protected agriculture as it is capable of helping farmers monitor soil condition, climate change, and animal and plant health [82]. If the environmental factor changes beyond the specified

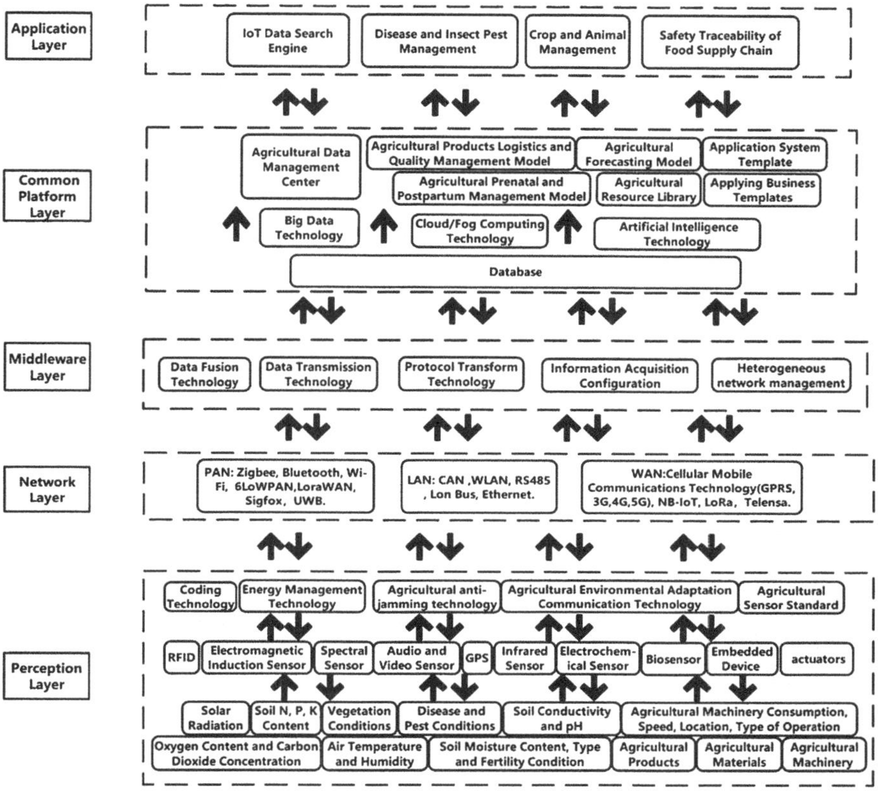

Fig. 7 Structure of IoT in protected agriculture [88]

limit, IoT will automatically send the administrator a warning message to remove the concealed hazard. It can also regulate environmental variables such as temperature, humidity, concentration of carbon dioxide, and lighting depending on the situation of real-time plant growth [83, 84]. Furthermore, cameras in the IoT scheme can capture plant illnesses and insect pests in the greenhouse in real time, help farmers discover issues and take targeted preventive action [85]. Goods such as vegetables can be tracked and visually monitored during transport and storage via GPS, RFID, and other location-based sensors. Supermarket executives use their smartphone or PC to monitor and predict the status of the product and the demand for product on the shelves. Users or consumers can use QR code, barcode, etc. to query the range, origin, processing, and other farm product data. IoT for protected farming can assist build a rural society that is informed, linked, advanced, and adaptable. Low-cost integrated systems can enhance human-physical interaction. Cloud computing, edge computing, and big data are capable of providing precious assessment and decision support. In short, in the coming years, IoT will become a significant instrument for engaging individuals in integrated agriculture that involves providers, farmers, engineers, retailers, merchants, customers, and officials of government [86, 87].

3.5.1 Structure of IoT in Protected Agriculture

Based on the actual situation of protected agriculture and the experience of other scholars, we proposed a five-layer IoT architecture [88]. As shown in Fig. 5, these layers are briefly described below:

1. *Perception layer*: This layer consists of various sensors, terminal devices, agricultural machinery, wireless sensor network (WSN), RFID tags and readers, etc. The common sensors are environmental sensors, animal and plant life information sensors, and other sensors related to agriculture. Through these sensors, information such as temperature, humidity, wind speed, plant diseases, insect pests, and animal vital signs can be obtained. The collected information is simply processed by the embedded device and uploaded to a higher layer through the network layer for further processing and analysis.
2. *Network layer*: The network layer is the infrastructure of IoT, which includes a converged network formed by various communication networks and the internet. The transmission medium can be wired technology such as CAN bus and RS485 bus or wireless technology like Zigbee, Bluetooth, LoRa, and NB-IoT. The network layer not only transmits various kinds of agricultural-related information collected by the perception layer to the higher layer, but also sends the control commands of the application layer to the perception layer, so that the related devices of the sensing layer take corresponding actions.
3. *Middleware layer*: IoT can provide different types of services for different devices. The technical specifications (processor, power supply, communication module) and system of each device are different and different devices cannot be connected and communicated with each other, which leads to heterogeneity problems. The middleware layer aggregates, filters, and processes received data from IoT devices, which greatly reduces the processing time and cost of the above issues and provides developers with a more versatile tool to build their applications. It also simplifies the steps of new service development and new device deployment which enables them to integrate more quickly into older architectures, improving the interoperability of IoT.
4. *Common platform layer*: The common platform layer is responsible for the storage, decision-making, summary, and statistics of agricultural information and the establishment of various algorithms and models for agricultural production process such as intelligent control, intelligent decision making, diagnostic reasoning, early warning, and prediction. This layer is composed of cloud computing, fog computing, edge computing, Big Data, machine learning algorithm, other common core processing technologies as well as its establishment model.
5. *Application layer*: The application layer is the highest level of the architecture and the place where IoT's value and utility are most apparent. There are lots of

intelligent platforms or systems in this layer for the environmental monitoring and control of plants and animals, the early warning and management of diseases and insect pests, and agricultural product safety traceability, which can improve production efficiency as well as save time and cost.

3.5.2 IoT Applications in Protected Agriculture

Plant Management

Compared with open-field agriculture, protected agriculture provides a relatively suitable and controllable environment for crop growth by greenhouse technology, which to some extent is free from the constraints of the natural environment and promoted the intensive and efficient use of agricultural resources. However, spatial and temporal variability of crops' growth environmental parameters in the protected agriculture are strong and affect each other [89, 90], which made it difficult to adapt to the growth of different types of plants in different growth stages by traditional cultivation and environmental regulation. Therefore, it needs higher accuracy in the aspect of monitoring and control. Many works have attempted to design and test types of monitoring and control systems to adjust greenhouse environmental parameters such as air temperature and humidity, light intensity and CO_2 concentration based on IoT and their results have proven it is technically and economically feasible [91]. At the low level of IoT development, the environmental data were simply processed and usually presented in sheet and plot form [92, 93]. Later, some studies collected huge amounts of data to set up various models based on plant growth or greenhouse climate [94, 95], which contributed to predicting the crop yield and environment parameter changes to help farmers better manage greenhouse. CAI et al. created an IoT-based remote monitor system based on low- cost intelligent greenhouse. The system used a single chip machine and sensors from STM32F103 to gather environmental information and LPWAN transmitted data to the cloud. Security connection with TLS was created, which lowered system expenses while maintaining data transmission security.

At the same moment, they introduced a database of time series to store information in order to save huge storage space [96]. He et al. indicated a smart NB-IoT-based temperature control system for the greenhouse. The relative error of collecting greenhouse environmental information was less than 1%, the average accuracy of control was 3.57% (±1.0 ° C), the transmission range was not restricted and the automatic temperature control of plant development was performed [97].

Nowadays, with the development of cloud computing, ML, etc., IoT solution can easily achieve smart data processing and analyzing at low cost and in a convenient way [98, 99]. At the same time, greenhouse technology has also undergone several generations of upgrades and has now evolved into highly mechanized and automated plant factories. Deng et al. have implemented a closed-loop control system in a salad-cultivating plant factory based on the kinetic model [100]. Both digital simulation and real-time results demonstrated the system can work stably under the internal variations and external disturbances [101] Zamora-Izquierdo et al. developed a smart farming IoT platform based on edge and cloud computing, which was designed for soilless culture greenhouses at a low cost. The platform was composed of local, edge, and cloud parts: the local part dealt with data gathering and automatic control through Cyber-Physical Systems, the edge part took main management tasks and could improve the stability of these systems, and the cloud part was in charge of data analytics. Compared with a regular open control, the platform saved more than 30% water [102]. Katsoulas et al. have designed an online irrigation system for hydroponic greenhouse crops and their results indicated it increased water and fertilizer use efficiency by 100% [103]. In an orchid greenhouse, Liao et al. introduced an IoT-based scheme to monitor environmental variables and phalaenopsis growth status [101]. They created an algorithm for image processing to estimate the phalaenopsis's leaf region and recognized the connection between plant leaf development and environmental variables. The suggested scheme could provide elevated spatiotemporal resolution quantitative data for floral farmers and help to update future phalaenopsis farming strategies.

Diseases and insect pests bring a great threat to the growth of crops as and traditional technology and chemical prevention and control has a certain lag and negative impact [104, 105]. IoT's enhanced use has given crop disease and pest control more effective and smarter solutions. Many kinds of IoT sensors can obtain real-time information on place, greenhouse environmental status, plant development, and plague condition anywhere, helping farmers keep track of crop pests and illnesses. Then, all raw image and data are sent to cloud centers and later processed and analyzed by various models and algorithms based on different diseases and pests [106, 107]. Finally, these cloud centers generally provide farmers with the following services: disease or pest identification, disaster prediction and warning and recommended governance measures from expert systems. Tirelli et al. proposed a pest insect trap automatic monitoring system using ZigBee technology, which can estimate the insect density by collected data from different sites and send a warning message to farmers when it exceeded the set value [108]. Ahouandjinou et al. proposed a pest monitoring system which detected the presence of pests by the acquisition of ultrasound and assisted others in building up protocols for early exterminate of the pests [109]. Foughali et al. presented a potato late blight prevention and decision support system using cloud IoT and helped the agriculturists take effective action to treat this disease [110]. Both Semios and Spensa Company launched their own integrated pest management system, which was able to count the number of pests by images, as well as characterize and capture the insects. At present, pest and disease warning research mainly provides medium and long-term warning based on historical data, which can provide macro guidance for crop production but low timeliness. Therefore, future research should focus on online monitoring as well as diagnosis and early warning of agricultural diseases.

Animal Farming

Livestock and aquatic product farming are an important part of protected agriculture and an area where IoT applications have achieved good results. To achieve good control effects in animal breeding, IoT should not only overcome harsh environmental factors, but also pay attention to the effects of animal behaviors [111, 112]. IoT has been applied in monitoring and management of environment, animal, feed, and farming process [113, 114]. The livestock monitoring items include information such as body temperature, weight, behavior, exercise volume, food intake, disease information, and environmental factors, which can help people understand animal's own physiological and nutritional status and adaptability to external environmental conditions. In aquaculture, management projects focus on water quality such as dissolved oxygen content, water temperature, and pH value [115, 116] because water quality greatly affects the growth of aquatic animals. With animal growth and nutrient optimization model and intelligent IoT equipment, it is possible to realize automatic feeding and optimal control of feeding time and intake according to their growth cycle, individual quality, feeding cycle, and eating situation [117, 118]. The Osborne Industrial Company has produced TEAM automated electronic sow feeding (ESF) stations including pregnancy stations and estrus detection stations. The ESF stations identify the sow through the RFID tags worn by each sow and deliver the corresponding quantity and type of feed based on information such as parity, lyrical condition, and gestational age. The estrus detection stations can detect rutting sows in the sow population and their detection accuracy is 7% higher than the manual ones. Encinas et al. presented a distributed monitoring IoT system for water quality monitoring. Their model was able to help fisheries acquire water quality parameters such as pH and temperature data in real time to optimize pond resources and prevent unwanted conditions [119]. Soto-Zarazúa et al. proposed an automated recirculation tilapia farming system based on the fuzzy algorithm. The results of this work showed that it saved water by 97.42% and the water quality environmental parameters were controlled within the acceptable range of tilapia culture [120]. Many researches have focused on analysis of animal behavior, health care and diagnosis, and warning of diseases based on IoT [121–123]. Yazdanbakhsh et al. proposed an intelligent livestock surveillance system. They attempted many machine learning algorithms to process raw data of healthy and ill cows and finally obtained good results used a wavelet-domain ensemble classifier with 80.8% sensitivity and 80% specificity [124]. Wens Group, the largest livestock breeding enterprise in China, took the lead in carrying out research on the animal husbandry based on IoT and built the corresponding system for monitoring livestock vital signs, behavior, and breeding environment information [125]. Liu et al. collected a variety of penaeus vannamei information such as real-time data on the culture environment, disease image data, and expert disease diagnosis and treatment and finally, constructed a remote intelligent diagnosis model based on IoT [126]. Liu et al. suggested a web-based mixed nutrition model to accurately predict gibel carp development, feed quantity, and nitrogen phosphorus output [127]. Due to the absence of appropriate sensors and models, the present internet diagnostics and early warning of animal diseases are at an early point. Future research should focus on improving the Big Data depth algorithm and establishing comprehensive animal behavior and disease models with upgraded hardware and artificial intelligence algorithms. A few studies have made in-depth research in monitoring animal odor and hazardous gas produced during the breeding process such as ammonia gas and sulfur dioxide [128]. Pan developed an electronic nose network system for monitoring and real-time analysis of odors from livestock farms. They placed e-noses in and around the farm to build up a wireless network to measure odor compounds as well as environmental parameters [129].

Agri-Food Supply Chain Traceability

Nowadays, agricultural products/food safety issues are receiving worldwide attention and their safety traceability is one solution accepted by all parties of agri-food domains. Governments in many countries and regions have promulgated laws and regulations to promote the establishment of food traceability system and strengthen the supervision of agricultural products/food safety. The agri-food supply chain traceability IoT-based system can ensure food safety and quality at each link of the production, from the cropland to the consumer (Fig. 8), which could help consumers establish confidence in food safety and contribute to sustainable development of the whole food industry [130].

In past years, a lot of countries have established the traceability platform for meat, milk, fish, and agri-food products based on IoT [131–133]. However, the fresh food cold chain logistics traceability has also drawn widespread attention [134, 135]. RFID technology still played an important role in agri-food supply chain traceability due to its small size and low cost [136]. As a novel technology, near field communication (NFC) has been progressively developed and used because of its safe and simple operation [137, 138].

A common problem in the development of IoT is its asynchronous heterogeneous data flow and distributed features. This requires the traceability system to establish uniform and accurate identification naming rules to facilitate quick and unique retrieval of information on a farm product [139]. As the deployment of IoT infrastructure is completed, the supply chain will move toward virtualization which is no longer required physical contact. The virtualization of an agri-food supply chain helps administrators better monitor, control, plan, and optimize food supply chain processes [140]. Wang et al. proposed a pre-warning system for food safety based on mining and IoT association rules. It included four modules: source of data, analysis of warnings, response, and feedback from emergencies. First, through data processing and analysis of food traceability data for safety evaluation, they accessed food critical information. On this basis, the scheme used the related data mining technique to detect the link between them and then acquired the outcomes of the assessment of the danger of food safety. The system subsequently produced a corresponding early warning reaction based on this outcome [141].

Fig. 8 Schematic representation of the agri-food supply chain from the grower to consumer

Recently, the increasing incidents of food falsification have not only brought about great economic losses, but also undermined consumers' confidence in food safety. The blockchain technology with decentralization, non-tamperability, development transparency, and machine autonomy features brings new solutions to the above problems [142, 143]. Feng reported a new decentralized traceability system based on IoT and blockchain technology, which can provide an open, transparent, neutral, safe, and reliable information platform for producers, governments, consumers, and other stakeholders [144]. Leng et al. combined the supply chain and blockchain technology to propose a double chain architecture agricultural traceability system by studying the dual chain structure and its storage mode, resource rent-seeking and matching mechanism and consensus algorithm [145].

Now, with the innovation and maturity of related technologies, it is not difficult for researchers to develop a complete set of traceability systems. The rise of artificial intelligence technology enables existing traceability systems to provide automation, intelligence, and human services to businesses and consumers. Chen adopted fuzzy cognitive maps and fuzzy rule methods to come up with an autonomous tracing system for backward design in food supply chain based on IoT, which better-traced food product problems [146]. However, there are still some problems in food safety traceability. It is worth noting that the current agricultural product safety traceability system only focuses on a certain level of supply chain or a certain type of goods. With the participation of multiple projects or multiple sessions, we believe that the focus of future research should be on more complex and systematic supply chains.

3.5.3 Future Prospects

For universities and research institutes, cloud-based agricultural intelligent decision-making models, multi-source data-based information fusion algorithms, agricultural Big Data mining technologies, distributed intelligent processing systems and lightweight IoT authentication, encryption, and authorization mechanisms will be the key research directions for the future. With an increased presence of IoT technology in protected agriculture, its potential for the refined management of crop, livestock, and aquatic animals will be recognized. Besides, as a technical means to monitor the production, processing, circulation, and consumption of agricultural products, IoT will play an increasingly important role in food safety.

4 Conclusion

In this survey, we have discussed the concept of IoT, its vision and characteristics. After that we present a comprehensive review of the state-of-the-art in IoT deployment for smart grid, solid waste management, healthcare, marine environment management, and protected agriculture applications. Further IoT promises to help us lead comfortable lives and monitor different things from a remote location. Data from refrigerators, cameras, lights, cars, social media sites, cell phones, TVs, cameras, and everything are stored, monitored, recorded, and re-used to better the future. Granted that IoT connected devices do a lot of good, but still there is plenty of data at play here and this a malicious person's heaven. Despite having so many applications, various challenges that are halting the growth of IoT technology in real life world are scalability, high power consumption, non-reliable network connectivity, lack of data security, and last but not the least cost is also an important issue. In the end we can say that the 'Internet of Things' is a reality and will continue to be a bigger reality. The key is to balance it well, stay connected without losing your privacy.

References

1. Towards a Definition of the Internet of Things (IoT). Available online: https://iot.ieee.org/images/files/pdf/IEEE_IoT_Towards_Definition_Internet_of_Things_Revision1_27MAY15.pdf
2. Ghasempour, A., Moon, T.K.: Optimizing the number of collectors in machine-to-machine advanced metering infrastructure architecture for internet of things-based smart grid. In: In

IEEE Green Technologies Conference (IEEE GreenTech 2016), Kansas City, MO, USA, pp. 51–55 (2016)
3. Ghasempour, A.: Optimum number of aggregators based on power consumption, cost, and network lifetime in advanced metering infrastructure architecture for smart grid internet of things. In: IEEE Consumer Communications and Networking Conference (IEEE CCNC 2016), Las Vegas, NV, USA (2016)
4. Internet of Things Strategic Research Roadmap. Available online: http://www.internet-of-things-research.eu/pdf/IoT_Cluster_Strategic_Research_Agenda_2009.pdf
5. The Internet of Things. Available online: https://www.itu.int/net/wsis/tunis/newsroom/stats/The- Internet-of-Things-2005.pdf
6. Singh, D., Tripathi, G., Jara, A.J.: A survey of Internet-of-Things: future vision, architecture, challenges and services. In: IEEE World Forum on Internet of Things (WF-IoT), Seoul, Korea, pp. 287–292 (2014)
7. Miao, Y., Bu, Y.: Research on the architecture and key technology of Internet of Things (IoT) applied on smart grid. In: International Conference on Advances in Energy Engineering (ICAEE), Beijing, China, pp. 69–72 (2010)
8. Ghasempour, A.: Optimized advanced metering infrastructure architecture of smart grid based on total cost, energy, and delay. In: IEEE Conference on Innovative Smart Grid Technologies (IEEE ISGT 2016), Minneapolis, MN, USA, pp. 1–6 (2016)
9. NIST Releases Final Version of Smart Grid Framework. Available online: https://www.nist.gov/smartgrid/ upload/NIST-SP-1108r3.pdf
10. Mouftah, H.T., Erol-Kantarci, M., Rehmani, M.H.: Communication architectures and technologies for advanced smart grid services. In: Mouftah, H.T., Erol-Kantarci, M., Rehmani, M.H. (eds.) Transportation and Power Grid in Smart Cities: Communication Networks and Services, 1st edn, pp. 217–245. Wiley, Hoboken, NJ, USA (2019)
11. Zhao, Z., Chen, G.: An overview of cyber security for smart grid. In: IEEE 27th International Symposium on Industrial Electronics, Cairns, Australia, pp. 1127–1131 (2018)
12. Refaat, S.S., Abu-Rub, H., Trabelsi, M., Mohamed, A.: Reliability evaluation of smart grid system with large penetration of distributed energy resources. In: IEEE International Conference on Industrial Technology (ICIT), Lyon, France, pp. 1279–1284 (2018)
13. Refaat, S.S., Mohamed, A., Kakosimos, P.: Self-healing control strategy, challenges and opportunities for distribution systems in smart grid. In: IEEE 12th International Conference on Compatibility, Power Electronics and Power Engineering (CPE-POWERENG 2018), Doha, Qatar, pp. 1–6 (2018)
14. Jarvis, R., Moses, P.: Smart grid congestion caused by plug-in electric vehicle charging. In: IEEE Texas Power and Energy Conference (TPEC), College Station, TX, USA, pp. 1–5 (2019)
15. Rafiei, S., Bakhshai, A.: A review on energy efficiency optimization in smart grid. In: Conference on IEEE Industrial Electronics Society, Montreal, QC, Canada, 25–28 Oct 2012, pp. 5916–5919 (2012)
16. Akaber, P., Moussa, B., Debbabi, M., Assi, C.: Automated post-failure service restoration in smart grid through network reconfiguration in the presence of energy storage systems. IEEE Syst. J. **13**(3), 3358–3367 (2019)
17. Qi, F., Yu, P., Chen, B., Li, W., Zhang, Q., Jin, D., Zhang, G., Wang, Y.: Optimal planning of smart grid communication network for interregional wide-area monitoring protection and control system. In: IEEE International Conference on Energy Internet (ICEI), Beijing, China, 21–25 May 2018, pp. 190–195 (2018)
18. Ghasempour, A.: Optimum packet service and arrival rates in advanced metering infrastructure architecture of smart grid. In: IEEE Green Technologies Conference (IEEE GreenTech 2016), Kansas City, MO, USA, 6–8 Apr 2016, pp. 1–5 (2016)
19. Ghasempour, A., Gunther, J.H.: Finding the optimal number of aggregators in machine-to-machine advanced metering infrastructure architecture of smart grid based on cost, delay, and energy consumption. In: IEEE Annual Consumer Communications & Networking Conference (IEEE CCNC 2016), Las Vegas, NV, USA, 9–12 Jan 2016, pp. 960–963 (2016)

20. Ghasempour, A.: Optimized scalable decentralized hybrid advanced metering infrastructure for smart grid. In: IEEE International Conference on Smart Grid Communications (IEEE Smart Grid Comm 2015), Miami, FL, USA, 2–5 Nov 2015, pp. 223–228 (2015)
21. Yang, Q., Wang, H., Wang, T., You, L., Lu, L., Liew, S.C.: Powerline-PNC: boosting throughput of powerline networks with physical-layer network coding. In: IEEE International Conference on Smart Grid Communications, Miami, FL, USA, 2–5 Nov 2015, pp. 103–108 (2015)
22. Kawai, K., Mai Huong, L.T.: Key parameters for behaviour related to source separation of household organic waste: a case study in Hanoi, Vietnam. Waste Manag. Res. **35**, 246–252 (2017)
23. Seyring, N., Dollhofer, M., Weißenbacher, J., Bakas, L., McKinnon, D.: Assessment of collection schemes for packaging and other recyclable waste in European Union-28 Member States and capital cities. Waste Manag. Res. **34**, 947–956 (2016)
24. Zobel, T.: ISO 14001 adoption and industrial waste generation: the case of Swedish manufacturing firms. Waste Manag. Res. **33**, 107–113 (2015)
25. Ali, M., Wang, W., Chaudhry, N., Geng, Y.: Hospital waste management in developing countries: a mini review. Waste Manag. Res. **35**, 581–592 (2017)
26. Bacot, H., McCoy, B., Plagman-Galvin, J.: Municipal commercial recycling barriers to success. Am. Rev. Public Adm. **32**, 145–165 (2002)
27. Krzywoszynska, A.: Waste? You mean by-products! From bio-waste management to agroecology in Italian winemaking and beyond. Sociol. Rev. **60**, 47–65 (2012)
28. Babu, B.R., Parande, A.K., Basha, C.A.: Electrical and electronic waste: a global environmental problem. Waste Manag. Res. **25**, 307–318 (2007)
29. Gan, L., Yang, S.: Legal context of high level radioactive waste disposal in China and its further improvement. Energy Environ. **28**, 484–498 (2017)
30. Khan, R., Khan, S.U., Zaheer, R., Khan, S.: Future Internet: the Internet of Things architecture, possible applications and key challenges. In: International Conference on Frontiers of Information Technology, Islamabad, India, 17–19 Dec 2012, pp. 257–260 (2012)
31. Weyrich, M., Ebert, C.: Reference architecture for the Internet of Things. IEEE J. Mag. **33**, 112–116 (2016)
32. Chuah, J.W.: The Internet of Things: an overview and new perspectives in systems design. In: IEEE Conferences, International Symposium on Integrated Circuits (ISIC), Singapore, pp. 216–219 (2014)
33. Yang, Z.: Study and application on the architecture and key technologies for IOT. In: International Conference on Multimedia Technology, Hangzhou, China, 26–28 July 2011, pp. 747–751 (2011)
34. Wu, M., Lu, T.J., Ling, F.Y., Sun, J., Du, H.Y.: Research on the architecture of Internet of Things. In: International Conference on Advanced Computer Theory and Engineering (ICACTE), Chengdu, China, 20–22 Aug 2010, pp. V5-484–V5-487 (2010)
35. Atzori, L., Iera, A., Morabito, G.: The Internet of Things: a survey. Comput. Netw. **54**, 2787–2805 (2010)
36. Tan, L., Wang, N.: Future Internet: the Internet of Things. In: International Conference on Advanced Computer Theory and Engineering (ICACTE), Chengdu, China, 20–22 Aug 2010, pp. V5-376–V5-380 (2010)
37. Chaqfeh, M.A., Mohamed, N.: Challenges in middleware solutions for the Internet of Things. In: International Conference on Collaboration Technologies and Systems (CTS), Denver, CO, USA, 21–25 May 2012, pp. 21–26 (2012)
38. Abidi, B., Jilbab, A., Haziti, M.E.: Wireless sensor networks in biomedical: wireless body area networks. In: Europe and MENA Cooperation Advances in Information and Communication Technologies, pp. 321–329. Springer, Berlin/Heidelberg, Germany (2017)
39. Xu, Q., Ren, P., Song, H., Du, Q.: Security enhancement for IoT communications exposed to eavesdroppers with uncertain locations. IEEE Access. **4**, 2840–2853 (2016)

40. Scuotto, V., Ferraris, A., Bresciani, S.: Internet of Things: applications and challenges in smart cities: a case study of IBM smart city projects. Bus. Process. Manag. J. **22**, 357–367 (2016)
41. Stergiou, C., Psannis, K.E., Kim, B.G., Gupta, B.: Secure integration of IoT and cloud computing. Future Gener. Comput. Syst. **78**, 964–975 (2018)
42. Dang, L.M., Hassan, S.I., Im, S., Moon, H.: Face image manipulation detection based on a convolutional neural network. Expert Syst. Appl. **129**, 156–168 (2019)
43. Nguyen, T.N., Thai, C.H., Luu, A.T., Nguyen-Xuan, H., Lee, J.: NURBS-based postbuckling analysis of functionally graded carbon nanotube-reinforced composite shells. Comput. Methods Appl. Mech. Eng. **347**, 983–1003 (2019)
44. Nguyen, T.N., Thai, C.H., Nguyen-Xuan, H., Lee, J.: NURBS-based analyses of functionally graded carbon nanotube-reinforced composite shells. Compos. Struct. **203**, 349–360 (2018)
45. Nguyen, T.N., Lee, S., Nguyen-Xuan, H., Lee, J.: A novel analysis-prediction approach for geometrically nonlinear problems using group method of data handling. Comput. Methods Appl. Mech. Eng. **354**, 506–526 (2019)
46. Dang, L.M., Hassan, S.I., Im, S., Mehmood, I., Moon, H.: Utilizing text recognition for the defects extraction in sewers CCTV inspection videos. Comput. Ind. **99**, 96–109 (2018)
47. Yang, Z., Zhou, Q., Lei, L., Zheng, K., Xiang, W.: An IoT-cloud based wearable ECG monitoring system for smart healthcare. J. Med. Syst. **40**, 286 (2016)
48. Almotiri, S.H., Khan, M.A., Alghamdi, M.A.: obile health (m-health) system in the context of IoT. In: IEEE International Conference on Future Internet of Things and Cloud Workshops (FiCloudW), Vienna, Austria, 22–24 Aug 2016, pp. 39–42 (2016)
49. Oryema, B., Kim, H.S., Li, W., Park, J.T.: Design and implementation of an interoperable messaging system for IoT healthcare services. In: IEEE Annual Consumer Communications & Networking Conference (CCNC), Las Vegas, NV, USA, 8–11 Jan 2017, pp. 45–52 (2017)
50. Van Kranenburg, R.: The Internet of Things: a critique of ambient technology and the all-seeing network of Rfid. Institute of Network Cultures, Amsterdam, The Netherlands (2008)
51. Yang, C., Shen, W., Wang, X.: Internet of Things in manufacturing: an overview. IEEE SMC Mag. **4**, 6–15 (2018)
52. Cook, D.J., Crandall, A.S., Thomas, B.L., Krishnan, N.C.: Casas: a smart home in a box. Computer. **46**, 62–69 (2013)
53. Finch, E.: Is ip everywhere the way ahead for building automation? Facilities. **19**, 396–403 (2001)
54. Komninos, N., Philippou, E., Pitsillides, A.: Survey in smart grid and smart home security: issues, challenges and countermeasures. IEEE Commun. Surv. Tutor. **16**, 1933–1954 (2014)
55. Talcott, C.: Cyber-physical systems and events. In: Software-Intensive Systems and New Computing Paradigms, pp. 101–115. Springer, Berlin/Heidelberg, Germany (2008)
56. Yongfu, L., Dihua, S., Weining, L., Xuebo, Z.: A service-oriented architecture for the transportation cyber-physical systems. In: 31st Control Conference (CCC), Hefei, China, 25–27 July 2012, pp. 7674–7678 (2012)
57. Ungurean, I., Gaitan, N.-C., Gaitan, V.G.: An IoT architecture for things from industrial environment. In: International Conference on Communications (COMM), Bucharest, Romania, 29–31 May 2014, pp. 1–4 (2014)
58. Wang, C., Bi, Z., Da Xu, L.: IoT and cloud computing in automation of assembly modeling systems. IEEE Trans. Ind. Inform. **10**, 1426–1434 (2014)
59. Catarinucci, L., De Donno, D., Mainetti, L., Palano, L., Patrono, L., Stefanizzi, M.L., Tarricone, L.: An IoT-aware architecture for smart healthcare systems. IEEE Internet Things J. **2**, 515–526 (2015)
60. Yan, Y., Qian, Y., Sharif, H., Tipper, D.: A survey on smart grid communication infrastructures: motivations, requirements and challenges. IEEE Commun. Surv. Tutor. **15**, 5–20 (2013)
61. Dameri, R.P., Rosenthal-Sabroux, C.: Smart City: How to Create Public and Economic Value with High Technology in Urban Space. Springer, Cham, Switzerland (2014)

62. Boonma, P., Suzuki, J.: An adaptive, scalable and self-healing sensor network architecture for autonomous coastal environmental monitoring. In: IEEE Conference on Technologies for Homeland Security, Woburn, MA, USA, 16–17 May 2007, p. 18 (2007)
63. Hadim, S., Mohamed, N.: Middleware: middleware challenges and approaches for wireless sensor networks. IEEE Distrib. Syst. Online. **7**, 1 (2006)
64. Albaladejo, C., Sánchez, P., Iborra, A., Soto, F., López, J.A., Torres, R.: Wireless sensor networks for oceanographic monitoring: a systematic review. Sensors. **10**, 6948–6968 (2010)
65. Cardei, M., Wu, J.: Energy-efficient coverage problems in wireless ad-hoc sensor networks. Comput. Commun. **29**, 413–420 (2006)
66. Zhang, X.: Application of Internet of Things technology in agricultural production. In: International Symposium on Big Data and Artificial Intelligence, Hong Kong, China, 29–30 Dec 2018, pp. 269–274 (2018)
67. Antão, L., Pinto, R., Reis, J., Gonçalves, G.: Requirements for testing and validating the industrial Internet of Things. In: IEEE International Conference on Software Testing, Verification and Validation Workshops (ICSTW), Vasteras, Sweden, 9–13 April 2018, pp. 110–115 (2018)
68. Khan, R., Khan, S.U., Zaheer, R., Khan, S.: Future Internet: The Internet of Things architecture, possible applications and key challenges. In: International Conference on Frontiers of Information Technology (FIT), Islamabad, India, 17–19 Dec 2012, pp. 257–260 (2012)
69. Yang, Z., Yue, Y., Yang, Y., Peng, Y., Wang, X., Liu, W.: Study and application on the architecture and key technologies for IoT. In: International Conference on Multimedia Technology (ICMT), Hangzhou, China, 26–28 July 2011, pp. 747–751 (2011)
70. Wu, M., Lu, T.-J., Ling, F.-Y., Sun, J., Du, H.-Y. Research on the architecture of Internet of Things. In: International Conference on Advanced Computer Theory and Engineering (ICACTE), Chengdu, China, 20–22 Aug 2010, pp. V5:484–V485:487 (2010)
71. Anastasi, G., Conti, M., Di Francesco, M., Passarella, A.: Energy conservation in wireless sensor networks: a survey. Ad Hoc Netw. **7**, 537–568 (2009)
72. Porter, J., Arzberger, P., Braun, H.-W., Bryant, P., Gage, S., Hansen, T., Hanson, P., Lin, C.-C., Lin, F.-P., Kratz, T.: Wireless sensor networks for ecology. AIBS Bull. **55**, 561–572 (2005)
73. Xu, G., Shen, W., Wang, X.: Applications of wireless sensor networks in marine environment monitoring: a survey. Sensors. **14**, 16932–16954 (2014)
74. Jiang, Z.: User density in a cluster underwater acoustic network. Int. J. Intell. Control Syst. **17**, 31–40 (2012)
75. Jiang, Z.: Underwater acoustic networks–issues and solutions. Int. J. Intell. Control Syst. **13**, 152–161 (2008)
76. Zhu, Y., Peng, Z., Cui, J.-H., Chen, H.: Toward practical mac design for underwater acoustic networks. IEEE Trans. Mobile Comput. **14**, 872–886 (2015)
77. Zixuan, Y., Zhifang, W., Chang, L.: Research on marine environmental monitoring system based on the Internet of Things technology. In: IEEE International Conference on Electronic Information and Communication Technology (ICEICT), Harbin, China, 20–22 Aug 2016, pp. 121–125 (2016)
78. Jensen, M.H., Malter, A.J.: Protected Agriculture: A Global Review. World Bank Publications, Washington, DC, USA (1995)
79. Mekala, M.S., Viswanathan, P.: A survey: smart agriculture IoT with cloud computing. In: International Conference on Microelectronic Devices, Circuits and Systems (ICMDCS), Vellore, India, 10–12 Aug 2017, pp. 1–7 (2017)
80. Kamilaris, A., Prenafeta-Boldú, F.X.: Deep learning in agriculture: a survey. Comput. Electron. Agric. **147**, 70–90 (2018)
81. Wolfert, S., Ge, L., Verdouw, C., Bogaardt, M.-J.: Big data in smart farming—a review. Agric. Syst. **153**, 69–80 (2017)
82. Dlodlo, N., Kalezhi, J.: The Internet of Things in agriculture for sustainable rural development. In: International Conference on Emerging Trends in Networks and Computer Communications (ETNCC), Windhoek, Namibia, 17–20 May 2015, pp. 13–18 (2015)

83. Pang, Z., Chen, Q., Han, W., Zheng, L.: Value-centric design of the internet-of-things solution for food supply chain: value creation, sensor portfolio and information fusion. Inf. Syst. Front. **17**, 289–319 (2015)
84. Liu, P., Li, B., Yu, X.: Intelligent control system of cucumber production in the greenhouse based on Internet of Things. In: International Conference on Cloud Computing and Security, Haikou, China, 8–10 June 2018, pp. 395–406 (2018)
85. Ma, J., Li, X., Wen, H., Fu, Z., Zhang, L.: A key frame extraction method for processing greenhouse vegetables production monitoring video. Comput. Electron. Agric. **111**, 92–102 (2015)
86. Talavera, J.M., Tobón, L.E., Gómez, J.A., Culman, M.A., Aranda, J.M., Parra, D.T., Quiroz, L.A., Hoyos, A., Garreta, L.E.: Review of IoT applications in agro-industrial and environmental fields. Comput. Electron. Agric. **142**, 283–297 (2017)
87. Verdouw, C.N., Robbemond, R.M., Verwaart, T., Wolfert, J., Beulens, A.J.M.: A reference architecture for IoT-based logistic information systems in Agri-food supply chains. Enterp. Inf. Syst. UK. **12**, 755–779 (2018)
88. Ge, W., Zhao, C.: State-of-the-art and developing strategies of agricultural internet of things. Trans. Chin. Soc. Agric. Mach. **45**, 222–230 (2014) (in Chinese)
89. Teitel, M., Atias, M., Barak, M.: Gradients of temperature, humidity and CO_2 along a fan-ventilated greenhouse. Biosyst. Eng. **106**, 166–174 (2010)
90. López, A., Valera, D.L., Molina-Aiz, F.D., Peña, A.: Sonic anemometry to evaluate airflow characteristics and temperature distribution in empty Mediterranean greenhouses equipped with pad–fan and fog systems. Biosyst. Eng. **113**, 334–350 (2012)
91. Sreekantha, D.K., Kavya, A.M.: Agricultural crop monitoring using IOT—a study. In: International Conference on Intelligent Systems and Control (ISCO), Coimbatore, India, 5–6 Jan 2017, pp. 134–139 (2017)
92. Kodali, R.K., Jain, V., Karagwal, S.: IoT based smart greenhouse. In: IEEE Region 10 Humanitarian Technology Conference (R10-HTC), Agra, India, 21–23 Dec 2016, pp. 1–6 (2016)
93. Mat, I., Kassim, M.R.M., Harun, A.N., Yusoff, I.M.: IoT in precision agriculture applications using wireless moisture sensor network. In: IEEE Conference on Open Systems (ICOS), Langkawi, Malaysia, 10–12 Oct 2016, pp. 24–29 (2016)
94. Carmassi, G., Bacci, L., Bronzini, M., Incrocci, L., Maggini, R., Bellocchi, G., Massa, D., Pardossi, A.: Modelling transpiration of greenhouse gerbera (*Gerbera jamesonii* H. Bolus) grown in substrate with saline water in a Mediterranean climate. Sci. Hortic. **156**, 9–18 (2013)
95. Singh, M.C., Singh, J.P., Singh, K.G.: Development of a microclimate model for prediction of temperatures inside a naturally ventilated greenhouse under cucumber crop in soilless media. Comput. Electron. Agric. **154**, 227–238 (2018)
96. Changqing, C., Hui, L., Wenjun, H.: Internet of agriculture-based low cost smart greenhouse remote monitor system. In: Chinese Automation Congress (CAC), Zhejiang, China, 30 Nov 2018, pp. 3940–3945 (2018)
97. He, C., Shen, M., Liu, L., Okinda, C., Yang, J., Shi, H.: Design and realization of a greenhouse temperature intelligent control system based on NB-IoT. J. South China Agric. Univ. **39**, 117–124 (2018)
98. Dan, L.I.U., Xin, C., Chongwei, H., Liangliang, J.: Intelligent agriculture greenhouse environment monitoring system based on IOT technology. In: IEEE International Conference on Intelligent Transportation, Big Data and Smart City, Halong Bay, Vietnam, 19–20 Dec 2015, pp. 487–490 (2015)
99. Elijah, O., Rahman, T.A., Orikumhi, I., Leow, C.Y., Hindia, M.N.: An overview of Internet of Things (IoT) and data analytics in agriculture: benefits and challenges. IEEE Internet Things J. **5**, 3758–3773 (2018)
100. Deng, X., Dou, Y., Hu, D.: Robust closed-loop control of vegetable production in plant factory. Comput. Electron. Agric. **155**, 244–250 (2018)

101. Liao, M.S., Chen, S.F., Chou, C.Y., Chen, H.Y., Yeh, S.H., Chang, Y.C., Jiang, J.A.: On precisely relating the growth of Phalaenopsis leaves to greenhouse environmental factors by using an IoT-based monitoring system. Comput. Electron. Agric. **136**, 125–139 (2017)
102. Zamora-Izquierdo, M.A., Santa, J., Martínez, J.A., Martínez, V., Skarmeta, A.F.: Smart farming IoT platform based on edge and cloud computing. Biosyst. Eng. **177**, 4–17 (2019)
103. Katsoulas, N., Bartzanas, T., Kittas, C.: Online professional irrigation scheduling system for greenhouse crops. Acta Hortic. **1154**, 221–228 (2017)
104. Larsen, A.E., Patton, M., Martin, E.A.: High highs and low lows: elucidating striking seasonal variability in pesticide use and its environmental implications. Sci. Total Environ. **651**, 828–837 (2019)
105. Silva, V., Mol, H.G.J., Zomer, P., Tienstra, M., Ritsema, C.J., Geissen, V.: Pesticide residues in European agricultural soils—a hidden reality unfolded. Sci. Total Environ. **653**, 1532–1545 (2019)
106. Pixia, D., Xiangdong, W.: Recognition of greenhouse cucumber disease based on image processing technology. Open J. Appl. Sci. **3**, 27 (2013)
107. Ahmed, R.Z., Biradar, R.C., Chaudhari, S.S.: Cluster-based data aggregation for pest identification in coffee plantations using wireless sensor networks. Comput. Electr. Eng. **56**, 591–607 (2016)
108. Tirelli, P., Borghese, N.A., Pedersini, F., Galassi, G., Oberti, R.: Automatic monitoring of pest insects traps by Zigbee-based wireless networking of image sensors. In: 2011 IEEE International Instrumentation and Measurement Technology Conference, Binjiang, China, 10–12 May 2011, pp. 1–5 (2011)
109. Ahouandjinou, A.S.R.M., Kiki, P.M.A.F., Assogba, K.: Smart environment monitoring system by using sensors ultrasonic detection of farm pests. In: 2017 2nd International Conference on Bio-engineering for Smart Technologies (BioSMART), Paris, France, 30 Aug–1 Sept 2017, pp. 1–5 (2017)
110. Foughali, K., Fathallah, K., Frihida, A.: Using cloud IOT for disease prevention in precision agriculture. Procedia Comput. Sci. **130**, 575–582 (2018)
111. Kim, Y.-B., Choi, D.-O.: Design of business management system for livestock pens based of IoT. J. Korean Entertain. Ind. Assoc. **8**, 207–216 (2014)
112. Ilapakurti, A., Vuppalapati, C.: Building an IoT framework for connected dairy. In: 2015 IEEE first international conference on big data computing service and applications, Redwood City, CA, USA, 30 Mar–2 Apr 2015, pp. 275–285 (2015)
113. Oriwoh, E., Sant, P., Epiphaniou, G.: Guidelines for internet of things deployment approaches—the thing commandments. Procedia Comput. Sci. **21**, 122–131 (2013)
114. Chen, J., Sung, W., Lin, G. Automated monitoring system for the fish farm aquaculture environment. In: 2015 IEEE International Conference on Systems, Man and Cybernetics, Kowloon, China, 9–12 Oct 2015, pp. 1161–1166 (2015)
115. Vijayakumar, N., Ramya, R.: The real time monitoring of water quality in IoT environment. In: 2015 International Conference on Innovations in Information, Embedded and Communication Systems (ICIIECS), Coimbatore, India, 19–20 Mar 2015, pp. 1–5 (2015)
116. Wiranto, G., Mambu, G.A., Hiskia, Hermida, I.D.P., Widodo, S.: Design of online data measurement and automatic sampling system for continuous water quality monitoring. In: 2015 IEEE International Conference on Mechatronics and Automation (ICMA), Beijing, China, 2–5 Aug 2015, pp. 2331–2335 (2015)
117. Scalera, A., Brizzi, P., Tomasi, R., Gregersen, T., Mertens, K., Maselyne, J., Van den Weghe, H.: The PigWise project: A novel approach in livestock farming through synergistic performances monitoring at individual level. In: The Conference on Sustainable Agriculture Through ICT Innovation, Turin, Italy, 24–27 June 2013, pp. 1–8 (2013)
118. Memon, M.H., Kumar, W., Memon, A., Chowdhry, B.S., Aamir, M., Kumar, P.: Internet of Things (IoT) enabled smart animal farm. In: IEEE International Conference on Computing for Sustainable Global Development, New Delhi, India, 16–18 Mar 2016, pp. 2067–2072 (2016)

119. Encinas, C., Ruiz, E., Cortez, J., Espinoza, A.: Design and implementation of a distributed IoT system for the monitoring of water quality in aquaculture. In: 2017 Wireless Telecommunications Symposium (WTS), Chicago, IL, USA, 26–28 Apr 2017, pp. 1–7 (2017)
120. Soto-Zarazúa, G.M., Peniche-Vera, R., Rico-García, E., Toledano-Ayala, M., Ocampo-Velázquez, R., Herrera-Ruiz, G.: An automated recirculation aquaculture system based on fuzzy logic control for aquaculture production of tilapia (*Oreochromis niloticus*). Aquacult. Int. **19**, 797–808 (2011)
121. Stutz, P.V., Golani, L.K., Witkin, J.M.: Animal models of fatigue in major depressive disorder. Physiol. Behav. **199**, 300–305 (2019)
122. Shinde, T.A., Prasad, J.R.: IoT based animal health monitoring with naive Bayes classification. Int. J. Emerg. Trends Technol. **1**, 8104–8107 (2017)
123. Jingqiu, G., Zhihai, W., Ronghua, G., Huarui, W.: Cow behavior recognition based on image analysis and activities. Int. J. Agric. Biol. Eng. **10**, 165–174 (2017)
124. Yazdanbakhsh, O., Zhou, Y., Dick, S.: An intelligent system for livestock disease surveillance. Inform. Sci. **378**, 26–47 (2017)
125. Benhai, X., Liang, Y., Xiaohua, P., Zhengang, Y.: Review on application of Internet of Things technology in animal husbandry in China. Trans. Chin. Soc. Agric. Eng. **31**, 237–246 (2015)
126. Liu, S., Xu, L., Li, D., Duan, Q., Wei, X.: Research on remote system for disease diagnosis of *Penaeus vannamei* based on internet of things. J. China Agric. Univ. **19**, 189–195 (2014) (in Chinese)
127. Liu, X., Sha, Z., Wang, C., Li, D., Bureau, D.P.: A web-based combined nutritional model to precisely predict growth, feed requirement and waste output of gibel carp (*Carassius auratus* gibelio) in aquaculture operations. Aquaculture. **492**, 335–348 (2018)
128. Keeffe, S.O., Manap, H., Dooly, G., Lewis, E.: Real-time monitoring of agricultural ammonia emissions based on optical fibre sensing technology. Sensors, 2010 IEEE (2010)
129. Pan, L., Yang, S.X.: An electronic nose network system for online monitoring of livestock farm odors. IEEE-ASME Trans. Mech. **14**, 371–376 (2009)
130. Costa, C., Antonucci, F., Pallottino, F., Aguzzi, J., Sarriá, D., Menesatti, P.: A review on Agri-food supply chain traceability by means of RFID technology. Food Bioprocess Technol. **6**, 353–366 (2013)
131. Clemens, R.L.: Meat traceability in Japan. Iowa Ag Rev. **9**, 2 (2015)
132. Li, H., Zhang, B., Zhang, L., Xue, Y., He, M., Ren, C.: A food traceability framework for dairy and other low-margin products. IBM J. Res. Dev. **60**, 1–8 (2016)
133. Chen, C., Xu, X.: Design and application of traceability and supervision platform for broiler based on Internet of Things. Trans. Chin. Soc. Agric. Eng. **33**, 224–231 (2017)
134. Ndraha, N., Hsiao, H.-I., Vlajic, J., Yang, M.-F., Lin, H.-T.V.: Time-temperature abuse in the food cold chain: review of issues, challenges, and recommendations. Food Control. **89**, 12–21 (2018)
135. Sichao, L., Xifu, W.: Toward an intelligent solution for perishable food cold chain management. In: 2016 7th IEEE International Conference on Software Engineering and Service Science (ICSESS), Beijing, China, 26–28 Aug 2016, pp. 852–856 (2016)
136. Zhao, G., Yu, H., Wang, G., Sui, Y., Zhang, L.: Applied Research of IOT and RFID Technology in Agricultural Product Traceability System, pp. 506–514. Springer International Publishing, Cham, Switzerland (2015)
137. Pigini, D., Conti, M.: NFC-based traceability in the food chain. Sustainability. **9**, 1910 (2017)
138. Badia-Melis, R., Mishra, P., Ruiz-García, L.: Food traceability: new trends and recent advances. A review. Food Control. **57**, 393–401 (2015)
139. Liu, Y., Wang, H., Wang, J., Qian, K., Kong, N., Wang, K., Zheng, L., Shi, Y., Engels, D.W.: Enterprise-oriented IoT name service for agricultural product supply chain management. Int. J. Distrib. Sens. Netw. **11**, 308165 (2015)
140. Verdouw, C.N., Wolfert, J., Beulens, A.J.M., Rialland, A.: Virtualization of food supply chains with the internet of things. J. Food Eng. **176**, 128–136 (2016)

141. Wang, J., Yue, H.: Food safety pre-warning system based on data mining for a sustainable food supply chain. Food Control. **73**, 223–229 (2017)
142. Galvez, J.F., Mejuto, J.C., Simal-Gandara, J.: Future challenges on the use of blockchain for food traceability analysis. Trac-Trend. Anal. Chem. **107**, 222–232 (2018)
143. Kshetri, N.: 1 Blockchain's roles in meeting key supply chain management objectives. Int. J. Inf. Manag. **39**, 80–89 (2018)
144. Feng, T.: A supply chain traceability system for food safety based on HACCP, blockchain & Internet of things. In: 2017 International Conference on Service Systems and Service Management, Dalian, China, 16–18 June 2017, pp. 1–6 (2017)
145. Leng, K., Bi, Y., Jing, L., Fu, H.-C., Van Nieuwenhuyse, I.: Research on agricultural supply chain system with double chain architecture based on blockchain technology. Future Gener. Comput. Syst. **86**, 641–649 (2018)
146. Chen, R.-Y.: Autonomous tracing system for backward design in food supply chain. Food Control. **51**, 70–84 (2015)

Smart Sensors and Industrial IoT (IIoT): A Driver of the Growth of Industry 4.0

Vijay Prakash Gupta

Abstract Smart sensors and Industrial Internet of Things (IIoT) are innovative tools in the current business environment and are considered drivers of Industry 4.0, as well as factories, households, and workplaces. Using smart sensors in a variety of ways, companies can improve and grow their business (Bahrin, M.A.K., Othman, M.F., Azli, N.N., Talib, M.F.: Industry 4.0: a review on industrial automation and robotic. J. Teknol. **78**(6–13), 137–143 (2016)). In the current business environment, it has been seen that smart sensors act as catalysts and drivers for taking a company to the top of its industry. The use of smart sensors and IIoT in business practices in various sectors of the economy has a very positive impact on delivering value and quality products and services, and it also helps in reducing costs, increasing production output, and enhancing efficiency.

The combination of smart sensors and IIoT represents the ordering and arrangement of a sensor, microprocessor, and communication technology for the purpose of converting environmental inputs (e.g., humidity, weight, liquid detection, temperature) into readable data and transmitting the data to a centralized repository.

Once implemented at scale, the combination of sophisticated sensors and increased computational power will enable new ways to analyze data and gain actionable insights to improve many areas of operations.

In this chapter, the author will present an analysis of changes initiated by applying smart sensors and IIoT in industry to connect technologies in industries, factories, households, and workplaces. The objective of this chapter is to seek answers to the question of how smart sensors and IIoT act as innovative tools driving growth in the current business environment and to explore the role and significance of smart sensors and IIoT in daily life and, most importantly, in business and manufacturing. The author will also try to analyze the impact of smart sensors and IIoT on operations in manufacturing or business organizations and discuss smart sensors so as to arrive at a better understanding of the underlying reasons why both producers and consumers are resistant to smart products.

Keywords Smart sensors · Business · Production process · Smart manufacturing · ICT · IIoT · Industry 4.0

V. P. Gupta (✉)
Institute of Technology & Science, Ghaziabad, Uttar Pradesh, India

D. Gupta et al. (eds.), *Smart Sensors for Industrial Internet of Things*, Internet of Things, https://doi.org/10.1007/978-3-030-52624-5_3

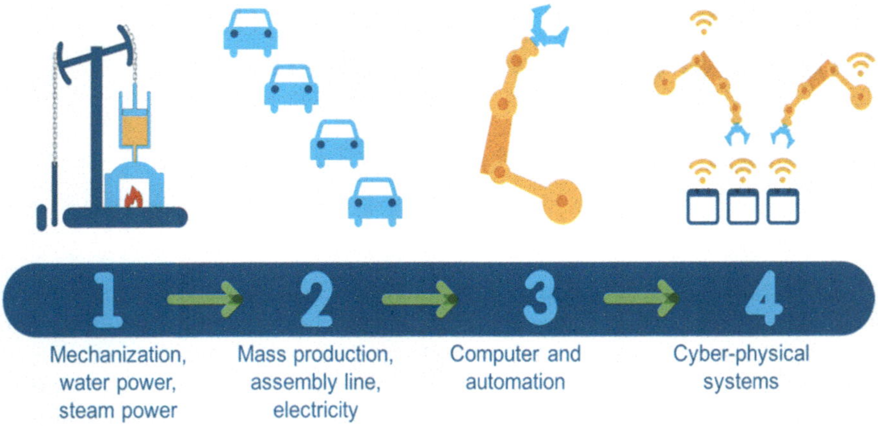

Fig. 1 Phases of Revolution of Industry. (Source: Cisco Canada Blog-Manufacturing-Industry 4.0-Jennifer Rideout)

1 Introduction (Fig. 1)

Modern industry in its current manifestation has been shaped by four major transformations in the Industrial Revolution. The first Industrial Revolution was characterized by a shift from agriculture, i.e., an agrarian economy, to mechanization, where producers used handmade tools, water power, and steam power to produce goods. In the second phase of the Industrial Revolution, producers started mass production and assembly line production processes with the help of electrical power driven machinery. This second stage (in the early twentieth century) gave birth to the real sense of industrialization. Following the mechanization of industry, electronics and communication systems were adopted in various industries for the production of mass quantities of goods, and this stage was the era of optimized and automated production processes.

Today we are experiencing what some are calling Industry 4.0, i.e., the fourth stage of the Industrial Revolution in which production is done in smarter ways using innovation, digitalization, and smart sensors, along with the Industrial Internet of Things (IIoT), which makes Industry 4.0 a reality.

Thus, Industry 4.0 symbolizes the beginning of the Fourth Industrial Revolution ([1]; Qin et al. 2016). It represents an era of digitalization, innovation, and the automation of production processes and includes IIoT to enable real-time machine-level monitoring to make the production process proactive.

Uses of IIoT in manufacturing is considered a core element of industrial transformation because it represents a set of semantic machine-to-machine communication, IoT, and blockchain technologies that integrate virtual space with the physical world ([2]).

1.1 Fundamental Aspects of Industry 4.0 and IIoT

Industry 4.0 and IIoT cover the following fundamental aspects:

1. ***Digitalization and integration of vertical and horizontal value chains***: This emphasizes the development of customized products, business-to-business (B2B) relationships, and intersectional operation of supply chains.
2. ***Digitalization of products and services***: This supports industry or business through intelligent networks to support the effective and rapid launch of new products and services and make business more modern and profitable.
3. ***Introduction of innovative digital business models***: The integration of digitalization and innovation, along with IIoT, will lead to the development of solutions for new and integrated real-time availability and control of production processes across industries.

Thus, the aforementioned fundamental aspects have brought about a radical transformation of industry, i.e., transformation from traditional industries to new production technologies, i.e., Industry 4.0.

2 What Is Industry 4.0?

It has been observed that in the last decade, too many updates and innovative technologies have been in development, for example, automation, digitalization, IoT, blockchain, big data, cloud storage, and mobile computing. All of these innovative technologies are considered pillars of Industry 4.0, which turns traditional methods of production into modern methods of production and promotes a vision for making industry smart.

Industry 4.0 can be treated as an umbrella of technological trends like advanced robotics, artificial intelligence (AI), smart sensors, cloud computing, and IoT, which can be used collectively to make industry modern and smart.

Industry 4.0 is more agile in managing risk and improving the efficiency, distribution, and integrity of data in the current business environment.

2.1 How Does Industry 4.0 Help Business?

In today's competitive environment, businesses must seize on the benefits and embrace the changes in tools and technology of production and business operations. Industry 4.0 may be a solution to enhance standardization, efficiency, quality, and utilization of factory operations at minimum cost. Given the current business environment, most industries are now adopting IIoT solutions and seeing improvements in production efficiency and output at minimum cost and improved quality.

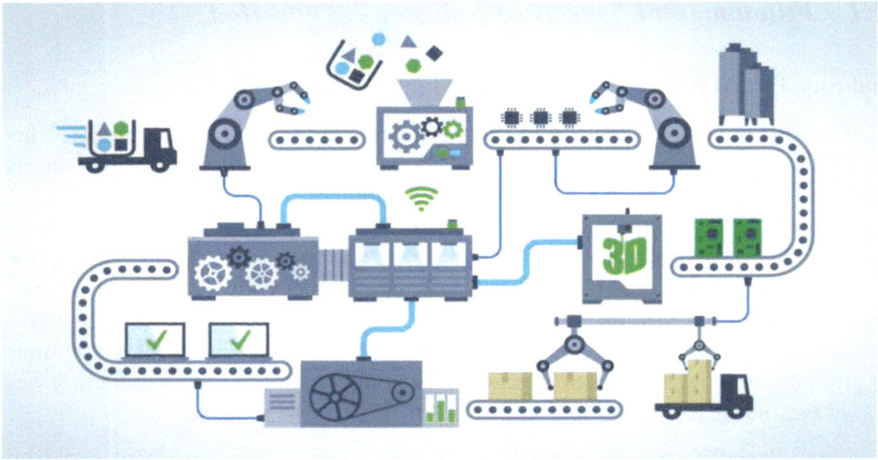

Fig. 2 Industry 4.0: Solutions for Proper Factory Operations. (Source: Cisco Canada Blog-Manufacturing-Industry4.0-Jennifer Rideout)

Industry 4.0 better connects production, processes, people, and physical systems with machines and producers, allowing industry and businesses to become more efficient, agile, and collaborative than ever. The combination of technologies like AI, IoT, and robotics helps businesses and manufacturers to craft new, more productive, and less wasteful business models. Thus, they benefit industry through faster business cycle times and lower costs of production (Fig. 2).

3 What Are Smart Sensors?

"A smart sensor is driven by the microprocessor and includes various features such as communication capability and on-board diagnostics that provide information to a monitoring system and/or operator to increase operational efficiency and reduce maintenance costs."

- A smart sensor is a device built as a component of IoT that converts input from the physical world to perform predefined functions and then processes data before passing the data on to a digital data stream for transmission to a gateway.
- Smart sensors are able to perform work more accurately and help in the automated collection of various data with zero error among the accurately recorded information. Normally, such devices are used for monitoring and control mechanisms in a wide variety of contexts.

3.1 Characteristics of Smart Sensors

- The main feature of smart sensors is their ability to perform self-diagnosis by monitoring and observing signals.
- Smart sensors have various inbuilt features, for example, self-detection, smart calibration, digital sensor data, sensing ability, and communication for remote monitoring and remote configuration.
- They reduce human intervention and management control in various systems.
- The main benefit of smart sensors is that they make tasks easier to perform through integration and ensure isolation in harsh industrial environments.
- Smart sensors use computers to interpret data in a logical manner.
- They increase work productivity.

3.2 Uses of Smart Sensors

1. **Monitor equipment and system performance**: Smart sensors are already being used in various industries and are having a huge impact on manufacturing. They will lead to more economic growth be used to monitor equipment and system performance, which will reduce waste and mismanagement of production processes.
2. **Generate and collect data**: Smart sensors are also used to generate and collect data by connecting different kinds of devices and systems. This creates seamless connectivity throughout plants, allowing manufacturers to aggregate all generated data.
3. **Predict machinery failure and trigger maintenance protocols**: They help predict failure in machinery and production processes and maintain production protocol.
4. **Speed the flow of information**: Smart sensors help to speed up the flow of information related to making real-time changes that increase production output. The data generated by sensors increase the transparency and flow of information, boosting profitability.

3.3 Advantages/Benefits of Smart Sensors in Industry

- They help to expand performance possibilities through total vertical and horizontal control.
- They help to boost productivity and efficiency and better manage resources.
- They optimize and integrate production processes for automation within the production process.

- They help to identify and address technical issues before they become bigger problems.
- They help to provide up-to-date information relevant to aspects of production and business processes.
- They increase efficiency to achieve mass production and decrease assembly costs.

4 Industrial Internet of Things

IIoT refers to interconnected components and sets of connections of network devices, the Internet, big data, blockchain, and sensors, including production management, that allow for data collection, exchange, and analysis to facilitate increases in industrial efficiency and productivity without waste and deficiencies for the economic benefit of industry.

4.1 Characteristics of IIoT

- Fast and flexible: near zero lag in response time.
- Allows for accurate analysis of data, leading to effective decision-making.
- Allows for regular monitoring and gives complete control over production processes.
- Able to detect defects in products and production process.
- Allows for safe data transfer from one production unit to another.
- Powered by software, sensors, and electronic devices, along with Internet connectivity.

4.2 Uses of IIoT

1. **Digital/connected factory:** IoT facilitates the flow and transfer of operational information to other connected production lines in manufacturing processes. This enables production managers to assess and manage factory units from remote areas and take advantage of process automation and optimization. Along with this, it also establishes a better chain of commands and helps to identify the tasks and accountability of managers and workers [3].
2. **Facility management**: The use of IoT in manufacturing equipment enables condition-based maintenance alerts. There are so many machine tools which are very and that are especially designed to function within certain heat and pressure boundaries. IoT sensors can actively monitor machines and send an alert to a production manager. By ensuring the prescribed working environment for machinery, manufacturers can conserve energy and eliminate machine downtime.

3. **Production flow monitoring**: IoT in the manufacturing process is used to monitor production processes from start to the packaging of finished goods. This complete monitoring of the process in real time helps to make adjustments in production and operation management to better manage the production process. Moreover, the constant monitoring and focus on the production process helps to identify deficiencies and waste and helps production managers eliminate waste and unnecessary work in progress.
4. **Logistics and supply chain optimization**: IIoT is used to access real-time supply chain information by tracking materials, equipment, and products as they move through the supply chain, which helps manufacturers collect and feed delivery information to enterprise resource planning (ERP) systems.
5. **Plant safety and security**: IoT helps to identify breakdowns or other defects in machinery and tools in plants, which can lead to improvements in worker safety and security in plants and avoid accidents or injuries in the course of daily operations [4].

4.3 Advantages/Benefits of IIoT

There are various advantages of IIoT that help us understand how IIoT is useful for business and industrial production.

- It helps to monitor the automatic and complex process of manufacturing.
- It helps in maintaining 360° quality control.
- It improves customer satisfaction and customer experience by providing value-based products.
- It helps in the development of an integrated ERP system that spans not only inventory and planning but also supply chains and manufacturing processes.
- It gives real-time insight that helps businesses to make faster decisions about their operations.

5 Types of Smart Sensors Used in Industry

Smart sensors and technologies are changing the face of the planet. All industries have begun to adapt to the new, smart world. Manufacturers are also making many gains by incorporating smart technology into their processes. Factories can become more streamlined, make products more efficiently, and even become more eco-friendly just by installing smart technology. Below are four smart sensor options that may bring a factory from the old era into the smart world.

5.1 Smart Tracking Sensors

Smart tracking sensors track the production system and processes and monitor factory production. They collect data and information at every step of the manufacturing process and report on the interactions among all machines on the factory floor. These sensors also send data to factory operators about the operation of machines. As a result, any problem within the production line or production process will be automatically detected and communicated to the responsible person. Before the advent of smart tracking sensors, factories relied on workers to keep a vigilant eye on the assembly process. That left plenty of room for the kind of human errors that smart tracking sensors eliminate.

5.2 Energy Management System

Energy management refers to the process of obtaining important and useful data about where, when, how, and why energy is used within a production process, allowing producers to increase efficiency, reduce costs, and improve sustainability. The energy problem remains one of the most important issues in production facilities, so smart sensors are used to forestall the loss of energy during the production process. These smart energy management sensors are used to track and analyze a factory's energy usage.

5.3 Machinery Health Sensors

Improving the reliability of machinery requires technology for analyzing all kinds of data. Machinery health sensors track the health of machinery itself, instead of the health of the production process overall. The goal of these sensors is to boost the work and lifetime of machinery installed in a production process [5]. When machinery health sensors detect defects in a machine, they send an alert that an urgent repair is needed. They can also predict which problems are likely to occur in the near future, allowing them to be addressed before they become serious issues. This enables the factory to avoid being forced to slow or shut down production, saving money and time [6].

5.4 Radio-Frequency Identification Sensors

Radio-frequency identification is a type of technology whereby digital data encoded in radio-frequency identification tags or smart labels are captured by a reader via radio waves. RFID captures data and stores them in a database.

In production processes RFID helps in the identification of data, which are often attached to individual pieces or collections of inventory; radio-frequency identification sensors scan these tags as they go through the production line or production process. Used together, RFID tags and sensors identify parts, prevent inventory loss, and ultimately ensure the integrity of end products. They also eliminate the necessity for manual data entry, yet another way to bring factories into the smart era.

6 Catalyst for the Growth of Smart Sensors and IIoT in Industry

- Cities are constantly growing, and rural areas are transforming into urban areas, and urban areas are transforming into smart cities, and population density is continuously increasing the necessity for smart sensors, which have become one of the important components of economic development.
- In this dynamic environment, due to increasing demand for technologies and machines, smart sensors assist to solve all the problems and promote resource efficiency through numerous solutions, such as smart energy, smart building, smart transport, and others.
- Sensors have become one of the most useful technologies and are extensively used in various business and industrial applications. Nowadays, due to the usefulness & utility of smart sensors, people are more demanding of it [7].
- Smart sensors are upgraded versions of traditional sensors, and their efficient and energy-saving features act as a driver of growing market demand in end-user industries.
- The growing market demand and application of IIoT act as a catalyst for the growth of the smart sensor.
- Moreover, regulatory bodies and government also play a crucial role in the growth of smart sensors across the world by adopting liberalized trading policies and promoting a favorable business environment, which incentivizes multinational corporation (MNC) to invest, which is predicted to drive the market and industrial growth.
- Global smart sensors market is very dynamic & sensitive and therefore the major players have adopted various strategies for market expansion and new product launches by mergers and acquisitions and strategic alliances to extend the use of smart sensors in the domestic market.
- Amazon's Alexa and Echo represent novel developments that use smart sensors and IIoT technology, like voice, sound, light, temperature, and pressure-based sensors. They have resulted in revolutionary changes to living standards.

With the assistance of smart sensors and IIoT in Alexa, your home is transformed into a smart home. Like its voice recognition capability, playing music on a clap, streaming music and videos, dimming lights and turning on and off fans, turning on water, automatic locking of gates and doors, and getting the latest news and weather updates without lifting a finger are now possible.

- By observing the success results of Amazon Alexa and Amazon Echo the demand for smart sensors & IIoT are increasing day by day and industries are leading to the development of the latest technologies to capture more profit and market share.
- Due to an aversion to common sensors by users, companies are now introducing new, small smart sensors enabled with IoT at low cost that are used in various industries, such as automobiles, airlines, hospitals, malls, railway station, market, hotels, and restaurants, highways, residential areas, medicine, agriculture, food processing, and chemical industries.

7 Smart Sensors and IIoT Markets: Growth, Trends, and Forecast

The technological development of Industry 4.0 and the application of smart sensors in business and manufacturing sectors are anticipated to be incremental and are growing day by day. It has been observed that adopting new technologies and international collaboration are some of the important key factors for the growth of Industry 4.0. What follows are a forecast of market and business growth and trends and a presentation of a market overview.

7.1 Market Overview

From the various data and sources, it has been observed that the compounded annual growth rate (CAGR) of the smart sensors market was valued at USD 36.62 billion in the year 2019 and it is forecasted to reach USD 102.10 billion by the end of 2025. It is expected that the growth may be of 18.82% during the forecast period of (2020-2025) [8].

The increasing use and application of information and communication technology (ICT) and IoT act as key drivers for the growth of the smart sensor and IIoT markets.

7.2 Sectorwise Analysis and Global Market Trends of IoT Market

7.2.1 Global IoT Market Trends and Forecasting (Fig. 3)

According to the preceding figure and market research on IoT that focuses on growth trends in the IoT market across the globe, the CAGR up to 2026 is projected to be 25.68%.The projection data show how dramatically the IoT market is

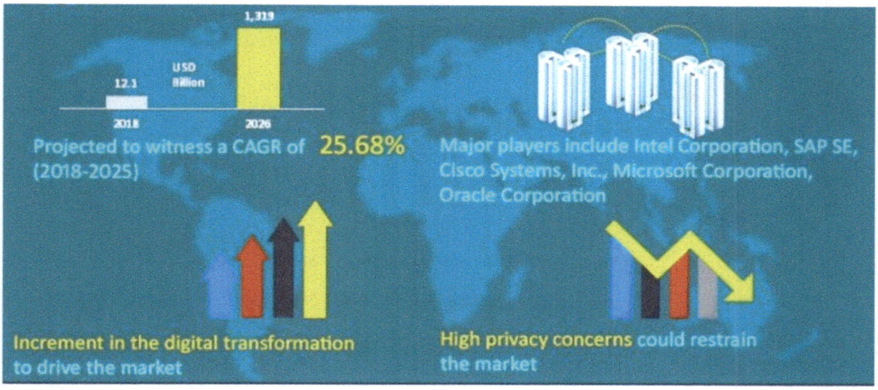

Fig. 3 Global IoT Market Trends and Forecasts. (Source: www.verifiedmarketresearch.com)

increasing across the globe. The major players in the industry are Intel Corporation, SAPSE, Cisco Systems, Microsoft Corporation, Oracle Corporation, and others. These companies will be the leaders in the IoT market.

7.2.2 Sectorwise Analysis of IoT Market

IoT connected machines are collectively adopted by all nations across the globe in various industries of different sizes and vertical configurations to reduce the workload by automating processes and increasing efficiencies. This is also adopted with the objective to reduce the cost & time as well as to provide a superior and digitized experience.

As per the report published by Maximize Market Research and shown in Fig. 4, the connected machines market was valued at around US$201.33 billion in 2019 and is expected to reach around US$1.2 trillion by 2027, at a CAGR of 25% during the forecast period 2020–2027 among the major sectors that are connected through IoT across the globe is the automobile sector which leads to the growth of the aviation industry"

8 Smart Sensors and IIOT as a Driver of Industrial and Economic Growth

Due to successive technological advancements, developments, and innovations, the global industrial landscape has drastically changed the face of the industrial sector. Industry 4.0 has played a major role in the transformation of traditional industries into intelligent ones by integrating innovative technologies.

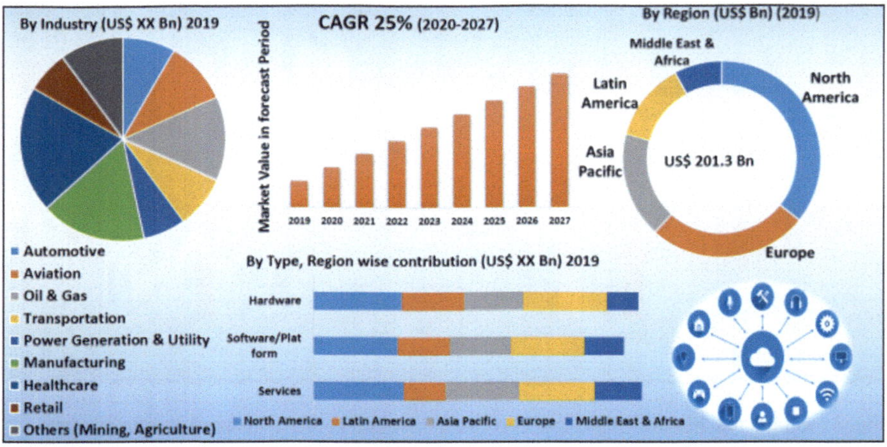

Fig. 4 Global Industrial IoT Connected Machine Market Size across the Globe. (Source: www.maximizemarketresearch.com)

With the integration of physical machinery by IoT, the industrial environment has become one of smart factories and intelligent manufacturing units. IoT is one of the fastest growing technologies that has significantly contributed to the emergence of Industry 4.0. Industry 4.0 and IoT provide multiple solutions, applications, and services. Hence, they can improve the quality of life and yield significant personal, professional, and economic opportunities and benefits in the near future.

9 Summary

Apart from industrial revolution, smart sensors and IIoT have also contributed to a large number of other technology revolutions. Innovation and digitalization in industrial units have given rise to automation, which has resulted in the growth of industrial output and the transformation of societies worldwide.

Smart sensors and IoT have become some of the most useful technologies in industry and are extensively used components in various applications of business and industry.

Smart sensors and IoT are upgraded versions of traditional sensors and their efficient and energy-saving features drive the growing market demand in end-user industries.

Thus, smart sensors and IoT help to improve productivity and boost revenues and profits for organizations and ultimately contribute to economic growth and development, which is why they are considered drivers of economic growth and industrial production.

References

1. Alexopoulos, K., Makris, S., Xanthakis, E., Sipsas, K., & Chryssolouris, G. (2016). A concept for context-aware computing in manufacturing: the case of the white goods. International Journal of Computer Integrated Manufacturing. 1–11. https://doi.org/10.1080/0951192X.2015.1130257.
2. Hozdić, E.: Smart factory for industry 4.0: a review. Int. J. Mod. Manuf. Technol. **7**(1), 28–35 (2015)
3. https://www.automationworld.com/factory/iiot/blog/13318945/industry-40what-does-it-mean-to-your-operations
4. https://www.newgenapps.com/blog/8-uses-applications-and-benefits-of-industrial-iot-in-manufacturing
5. Zhang, Y., Gu, Y., Vlatkovic, V., Wang, X.: Progress of smart sensor and smart sensor networks. In: Fifth World Congress on Intelligent Control and Automation (IEEE Cat. No. 04EX788), vol. 4, pp. 3600–3606. IEEE, Washington, DC (2004)
6. Gupta, V.P., Arora, A.K.: Automation in healthcare services. In: Applications of Deep Learning and Big IoT on Personalized Healthcare Services, pp. 1–19. IGI Global, USA (2020)
7. Prosser, S., Schmidt, E.: Smart sensors for industrial applications. Sens. Rev. **17**(3), 217–222 (1997)
8. https://www.maximizemarketresearch.com/market-report/iot-connected-machines-market/53219/

Smart Sensors for IIoT in Autonomous Vehicles: Review

Suresh Chavhan, Ravi Arun Kulkarni, and Atul Ramesh Zilpe

Abstract Autonomous vehicle is designed to perform all various operations synchronously to work under necessary regulations, liability, acceptance, and safety with Industrial IoT. The autonomous vehicle mainly works with three sensors, such as Camera, RADAR, and LIDAR. These sensors gather the information data like image, distance, waves and provide it to the electronic controllers, where it processes to take the appropriate actions. During the processing of the information, various parameters, such as latency, packet loss, and bandwidth, are evaluated using the historical and available information. The information data generated by the sensors are processed using various signal conditioners to meet the required parameters. In this chapter, some of the signal conditioning methods, such as RF module, BTS, TPMS, and capacitive balancing of humidity sensor (in HVAC), are used in autonomous vehicle and Industrial IoT implementation. The communication modes and their aspects from Industrial Internet of Things are discussed and existing methods for the systems are analyzed from the signal conditioning perspective.

Keywords RADAR · LIDAR · RF · BTS · TPMS · HVAC

1 Introduction

In an autonomous vehicle, the electronic control unit works on a feedback control system. It takes input from sensors and output to the actuators. The electronic controller matchups the input data with the predefined set of data from the lookup table [1]. To process the information received, the input must be in the specified format; to make the output of sensors in required format, various operations like analyzing, processing, identifying, removing, etc., are needed to perform on the information (i.e., signal conditioning). Many of the times the data received from

S. Chavhan (✉)
Autonomous Research Center, Vellore Institute of Technology, Vellore, Tamil Nadu, India
e-mail: suresh.chavhan@vit.ac.in

R. A. Kulkarni · A. R. Zilpe
Department of Automotive Electronics, School of Electronics Engineering, Vellore Institute of Technology, Vellore, Tamil Nadu, India

D. Gupta et al. (eds.), *Smart Sensors for Industrial Internet of Things*, Internet of Things, https://doi.org/10.1007/978-3-030-52624-5_4

sensors are not from identical sources. The output of the sensors is in low voltage value and the processor needs the input in digital format. The controller operates with multiple data at a time thus it sets different parameters for different outputs. To get the desired output, the controller operates by various signal conditioning methods based on sensors. These conditioning methods are conditioned on the acquired signals depending upon the applications. Currently, there are various conditioning methods [2–6], such as linearization, amplification, excitation, filtering, analog to digital conversion, etc., which are used for conditioning the signals. Data acquired using sensors, such as TPMS, and ultrasonic sensor, are processed using signal conditioning methods to obtain usable data set in predefined parameters (Table 1).

1.1 TPMS

Tire pressure measurement is an add-on feature for ABS; it measures the exact pressure of tire and aids in determining the tractive force at the individual tire. By knowing this tractive force, ABS applies exact braking force to each tire individually and achieves effective breaking [3]. To measure the exact pressure of tires and transmit real-time data, the pressure sensor (Table 1) and transmitter are located inside the tire. This pressure measurement information is carried to the driver and displayed in the cabin of the car.

1.2 Ultrasonic Sensor

Autonomous vehicle uses this sensor in the parking system or as a short-distance measurement sensor at low speeds. Ultrasonic sensor (Table 1) corresponds to the size and shape of the vibrating surface (i.e., emitting the ultrasound) frequency at which it vibrates [4]. It uses sonic waves, in the range of 20–40 kHz, to measure the distance of an object. Its principle of operation is based on the measurement of the time of flight (ToF) of the sonic wave from when it is emitted until the echo is received. ToF sensor is capable of measuring distance and reflectivity of full resolution. ToF pixel has a high dynamic range.

Table 1 Signal conditioning methods and sensors

System	Signal conditioning	Sensor
TPMS	Encoding and decoding pressure value	Pressure sensor
RADAR	Transmitting and receiving of pulse	Ultrasonic sensor
LIDAR	Scattering of light in 2D plane	Leaser light, IR sensor
Combine digital signal	Base transceiver (BTS)	RF module, multiplexing
HVAC	Capacitive balancing	Humidity sensor

1.3 Combined Digital Signal for Wireless Service

Combined Digital Signal for Wireless Service Over Sheared Transport Medium: Wireless carriers provide wireless communication service to wireless communication customers. To improve the wireless communication service provided to such customers, some wireless carriers have begun implementing or are considering implementing a distributed antenna system (DAS) [6, 7] to extend the range of their existing wireless network infrastructure such as BTS (Table 1).

1.4 Capacitive Balancing of Relative Humidity Sensors

Effective heating and ventilation are achieved with the capacitive balancing of relative humidity and integrated signal conditioning in the air control unit. The humidity sensor (Table 1) inside the cabin of the vehicle continuously monitors the amount of water vapors present and measures relative humidity by putting a thin strip of metal oxide between two electrodes [8, 9]. Electrical capacity of metal oxide changes with a relative humidity of the atmosphere. This allows the HVAC unit to regulate the blow out air temperature to adjust blower speeds and open recirculation vents to control and prevent windshield fogging.

2 Tools Used in the System

The development of a modern computing systems where digital objects can be uniquely identified and can be able to communicate real-time data with controllers and process the data on the basis of which automated actions are taken requires the need for a combination of new and effective technologies which is only possible through an integration of different technologies which can make the objects to be identified and communicate within the system.

2.1 Pressures Sensor in TPMS

In TPMS the remote sensing module is comprised of a pressure sensor, a signal processor, and an RF transmitter [10, 11]. The TPMS is a system comprising of a pressure sensor and a controller connected to the vehicle's electronic control unit. The pressure sensor is capable of transmitting a signal through wireless transmission antennas on the vehicle side respectively provided in the vicinity of the specified tire (Fig. 1).

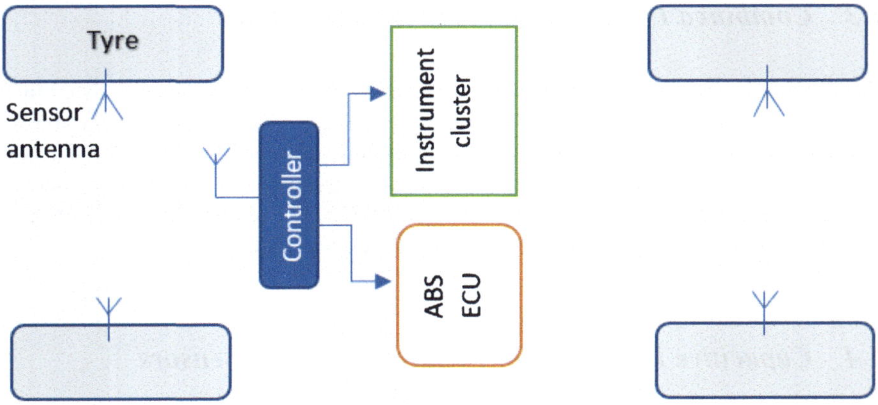

Fig. 1 Analytical model of TPMS system in vehicle. This shows a figure consisting of blocks representing tire, sensor antenna, receiver antenna, controller, and instrument cluster connected via RF module and CAN networking

The pressure sensor unit in each tire is provided with a unique serial number to identify the readings of an individual tire. The controller reads out the pressure values according to the sensor identification number and forwards the same to the ABS control unit and instrument cluster.

TPMS is designed to measure the air pressure of each tire and transmitting the measurement result to the controller on the vehicle in the form of wireless signal transmits requested by controller to sensor unit of a tire corresponding to the transmission antenna on the vehicle side at a predetermined position at a predetermined timing, and upon receipt of this, the sensor unit transmits an answer signal including the measurement result to the controller [12]. The controller on the vehicle side reads out the measurement result and controls the output of tire pressure and indication mark (!) in the case of abnormal air pressure. In this system, when a transmission antenna on the vehicle side at a predetermined position transmits a request signal so as to communicate with the sensor unit of a corresponding tire, depending on the case, the sensor unit of another tire may receive the request signal and return an answer signal, so that the controller on the vehicle side cannot receive an answer signal normally and it may receive an answer signal from another tire by mistake, hence to make it impossible to judge which tire this signal of air pressure corresponds to. To solve this problem of getting communication intricate, an inherent identification code is set for every tire, and the identification code included in a request signal is transmitted from sensor unit of a tire. The identification code request signal is collated with the identification code previously stored, only when the results match the code, an answer signal is returned. In this case, however, it takes much time to have communication because of the time required to match the identification code and response to the stimule will be deteriorated resulting delay to monitor the tire pressure.

2.2 Ultrasonic Sensor

The ultrasonic sensor module comprises one transmitter and receiver. The transmitter can deliver 40 kHz ultrasonic sound while the maximum receiver is designed to accept only 40 kHz sound waves. The receiver ultrasonic sensor that is kept next to the transmitter shall thus be able to receive reflected 40 kHz, once the module faces any obstacle in front.

An ultrasonic sensor measures the distance of an object by transmitting and receiving an ultrasonic sound wave. Transducer emits the signal and receives ultrasonic pulses that relay back information about an object's proximity. High-frequency sound waves reflect from boundaries to produce distinct echo patterns. The working principle of this module is simple. It sends an ultrasonic pulse out at 40 kHz which travels through the air and if there is an obstacle or object, it will bounce back to the sensor. By calculating the travel time and the speed of sound, the distance can be calculated (Fig. 2).

Transducer transmits the signal and receives ultrasonic pulses that relay information about the proximity of an object. In order to produce distinct echo patterns, high-frequency sound waves reflect from boundaries. This module's functioning principle is simple. It sends out an ultrasonic pulse at 40 kHz that travels through the air and will bounce back to the sensor if there is an obstacle or object. The distance can be determined by measuring the travel time and the sound frequency. The calculated distance of the object is utilized to determine the time remaining to hit the object with respect to the current vehicle speed. This inputs guide to feed the acceleration and braking inputs in order to prevent the accident. This data is communicated to the control unit within vehicle networking, requires the fast and accurate wireless communication media as 5G over the fast internet facilities (IIoT).

When applied to the ultrasonic transducer, an electrical pulse of high voltage vibrates across a specific frequency spectrum and generates a burst of sound waves. Whenever any obstacle comes before the ultrasonic sensor, the sound waves reflect in echo form and generate an electrical pulse. It calculates the time taken to receive the echo from sending sound waves. The echo patterns will be compared to the sound wave patterns to determine the condition of the detected signal (Fig. 3).

Whenever any obstacles come ahead of the ultrasonic module it calculates the time taken from sending the signals to receiving them since time and distance are related for sound waves passing through air medium at 343.2 m/s. Upon receiving the signal MC program while executed displays the data, i.e., the distance measured on LCD interfaced with the microcontroller in cm.

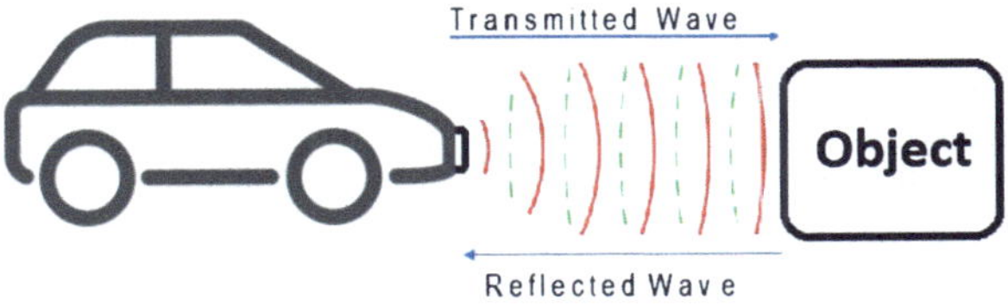

Fig. 2 Locating object using ultrasonic sensor

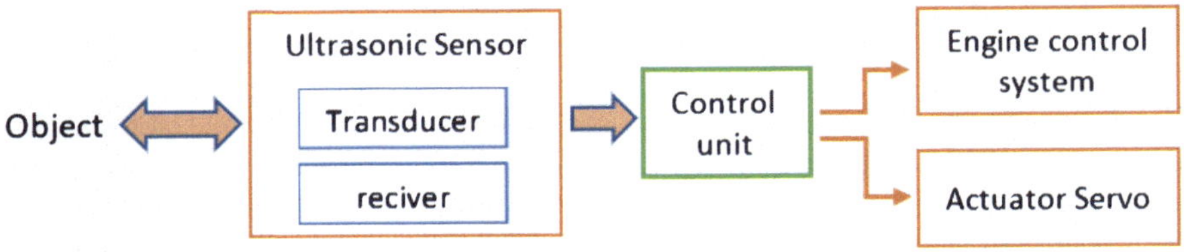

Fig. 3 Block diagram of Ultrasonic sensor

2.3 *Infrared for Distance Measurement*

To measure the distance in 2D plane, infrared sensor is used in the LIDAR system of autonomous vehicle to avoid the obstacle coming in the desired path. The main advantage of using this sensor is quick response and cost-effectiveness compared to ultrasonic sensors, also it has non-linear characteristics and it shows reflectance of the object layers. Using reflected light intensity IR sensor estimates the distance of an object in the vicinity and it is useful in the low lighting conditions because these sensors are not vision-based and also can detect transparent obstacles.

2.4 *CDS for Wireless Service Over Sheared Transport Medium*

CDS system over shared transport media consists of (BTS) multiple base transceiver stations (BTSs) that radiate radio frequency (RF) signals from an antenna to form a cell and/or cell sector.

To make sensors to response faster there is a need to handle the numerous data generated by sensors in the vehicle. Analyzing and processing such huge data requires a faster network than the existing technology. To deal with multiple data at an instance demands for high bandwidth signal. These requirements are met by wireless communication via a sheared transport medium between a base transceiver station (BTS) and a remote antenna entity (RAE). The present infrastructure includes multiple base transceiver (BTS) which radiates the bands of radio frequency signals for the receiver [13]. A wireless carrier is extending the range of wireless network strength by communicating with the distributed antenna systems to many base-shared transport media. Every DAS includes multiple remote antenna entities (RAEs), a dedicated transport medium that connects a BTS to multiple remote antenna entities of the DAS, and one or more antennas connected to each remote entity. A BTS can provide RF signals to one or more remote antenna entities, in turn, the multiple antennas connected to the remote antenna entities, which radiates the received radio frequency signal by extending the cell or cell sector coverage to the receiver, are in the vicinity of each remote entity from the BTS (Fig. 4).

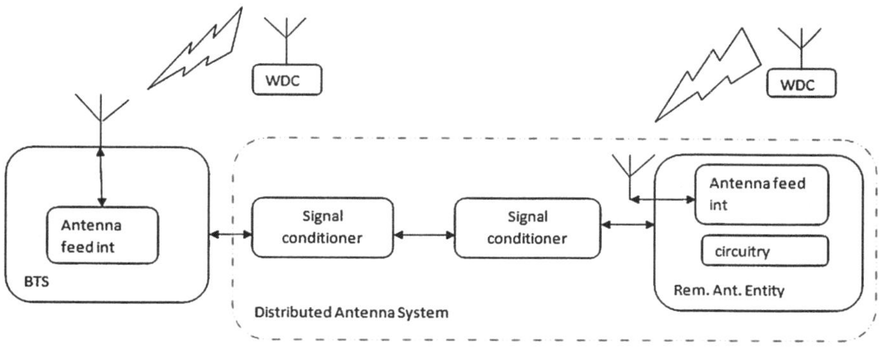

Fig. 4 An exemplary system for carrying out the signal

In current usage, for safety improvement in autonomous cars system requirements are communicated directly with each other by shearing location, velocity, and direction to other vehicles. This will enable to avoid accidents and also provide information about roadside conditions. To achieve this, the industry should be able to agree a technological standard for vehicle-to-vehicle communication. With implementing IoT for data transmitting and information sharing between base transceiver and road side, unit vehicle safety and data processing can be improved. For example, a wireless communication is utilized to give a correspondence way between a BTS and the remote reception apparatus elements of the DAS. Transmission channels and IoT are important in light of the limit of the in-vehicle data transmission medium.

Using an active 4G internet, data signals and the control data balance an RF bearer which is then sent through a radio wire feed from the BTS to a receiving wire for transmission over the air. In the event that rather it is wanted to send the data flag and control data through a devoted vehicle medium to a remote substance of a DAS, at that point the simple RF sign is sent through the radio wire feed from the BTS to another element which at that point sends the simple RF signal over the committed vehicle medium. On the other hand, a substance may test the simple RF sign sent through the reception apparatus feed, at a rate at any rate twice that of the transmission capacity, as indicated by the Nyquist hypothesis, to make a digitized portrayal of the simple RF signal, at that point send the digitized RF sign to a remote element of the DAS over the devoted vehicle medium. In either case, the subsequent transmission capacity required to send the sign from the BTS to the remote substance is a lot more prominent than the data flag and control data created within the BTS. A devoted vehicle medium is essential between the BTS and the remote substance of the DAS to take into account correspondence of the high-transfer speed simple RF signal or the digitized portrayal of the simple RF signal [3–7, 14]. Conveying signals between a BTS and a remote element of a DAS utilizing such high data transfer capacity of a devoted vehicle medium, be that as it may, is unfortunate, as it very well may be very expensive to give and keep up such a committed vehicle medium.

2.5 Capacitive Balancing of Humidity Sensor

Capacitive balancing of humidity sensor consists of humidity sensor, hygrometer, analog to digital convertor, low pass filter, oscillator sensor, microcontroller, and control unit attached to the air conditioning system (Fig. 5).

Humidity sensors are capacitance sensors that measure the amount of moisture in the air. The information from the sensor both regulates the volume of air projected onto the windows to reduce misting and manages the humidity levels inside the car to enhance climate comfort. These sensors are typically mounted at the base of the rear-view mirror. From the data delivered by the humidity and temperature sensor, the HVAC system calculates the dew point temperature of the air. Some systems use an infrared sensor that remotely measures the windshield and side window temperatures as well. The performance of the sensor can degrade over time and cause the sensor to malfunction and give false readings. In this situation the system shows the code stored in the HVAC module (Fig. 6).

The oscillator is to relate the change in capacitance to a voltage. A low pass filter removes the oscillation frequency leaving the voltage response and gain amplifies the signal to the desired range. This system provides the analog signal measured by the humidity sensor which is further converted into a digital signal by using the analog-to-digital converter. This signal provides voltage response values as input to the microcontroller. Microcontroller compares the input with predefined set of values and feeds output to air conditioning devices in automotive.

Principle of operation: The humidity sensor is a small capacitor consisting of a hygroscopic dielectric material placed between a pair of electrodes. Most capacitive sensors use a plastic or polymer as the dielectric material, with a typical dielectric constant ranging from 2 to 15. When no moisture is present in the sensor, both this constant and the sensor geometry determine the value of the capacitance.

At normal room temperature, the dielectric constant of water vapor has a value of about 80, a value much larger than the constant of the sensor dielectric material. Therefore, the absorption of moisture by the sensor results in an increase in sensor capacitance. At equilibrium conditions, the amount of moisture present in a

Fig. 5 Capacitive balancing plates

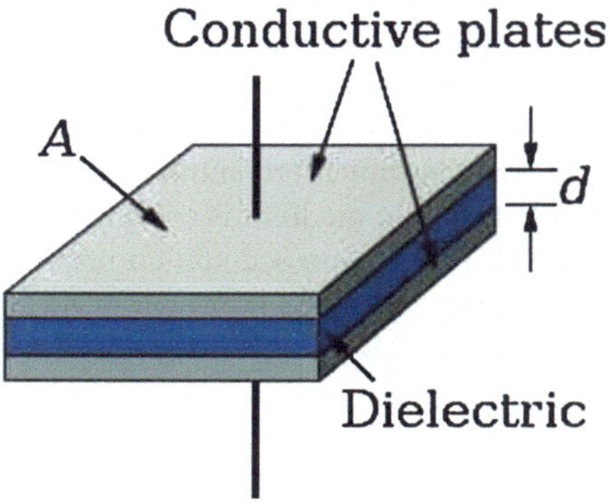

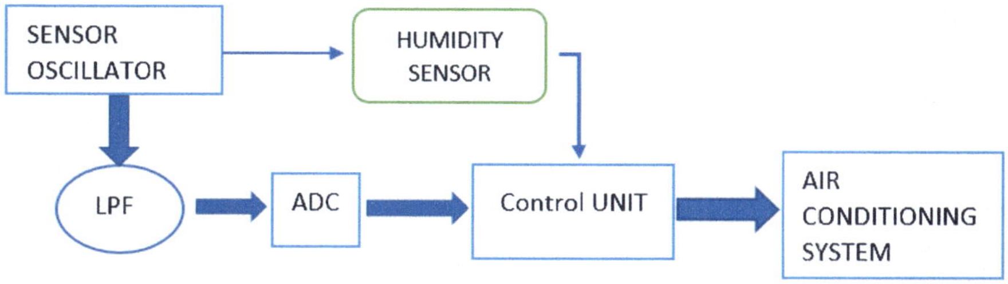

Fig. 6 System architecture for HVAC capacitive balancing system

hygroscopic material depends on both ambient temperature and ambient water vapor pressure. This is true also for the hygroscopic dielectric material used in the sensor. By definition, relative humidity is a function of both the ambient temperature and water vapor pressure. There is a direct relationship between relative humidity, the amount of moisture present in the sensor, and sensor capacitance. This relationship is at the base of the operation of a capacitive humidity instrument. As we recall our relative humidity basics, we remember that relative humidity is the ratio of the actual water vapor pressure present compared to the maximum water vapor pressure (saturation vapor pressure) possible at a given temperature. The dielectric material varies at a rate that is related to the change in relative humidity.

3 Complexity

In TPMS the signal transmitted by each sensor is needed to update and keep real-time data to calculate the exact braking force required to apply at each tire. In ultrasonic sensors, multiple objects in different locations may not be detected. Multiple objects at different height make it difficult to determine the obstacle within the vicinity of the sensor. In a combined digital signal system, the base of the major problem is the bandwidth of a transmitted signal and high-speed data transmission is required to receive the required data decoding and sending it to the particular receiver with a high-speed data transmission technique. In the humidity sensor, the change in atmospheric temperature and pressure results in a change in comfort conditions and needs to update the humidity value in real-time according to the surrounding condition.

Complexity in some of the existing IoT applications is huge amount of data transmitting to the controller. Mechanism such as poling and interrupts results in delay. For safety critical systems, in-vehicle data communication demands high-speed data transmission, with less latency. Technique such as Dedicated Short-range Communications (DSRC) and 5G is applicable for connecting autonomous vehicles between the cars and infrastructure, most appropriate in the critical, real-time and safety-related applications. The communication in autonomous vehicle industry with IoT particularly 5G technology is looked forward as the main feature for autonomous cars.

4 Conclusion

In an autonomous vehicle there are various systems which require to deal with multiple data generated from different sensors at instance. Real-time data processing and continuous updating it with sensors and controllers is achieved with implementing Internet of thing applications. Manipulation of data is required to trim it down for system specific parameters and signals are conditioned in such a way to make data usable in less time with quick response. There is multiple way to condition the signal but the selection of the particular signal conditioning method is depending upon the system integration with physical condition and other operating parameter requirements. While integrating the actuator output with the feedback (response of control loop), signal processing is done with considering system requirements. System design engineers of autonomous vehicle designers have a wide scope in selecting proper signal conditioning methodology to meet the design requirements and it varies with the priority of the data to be processed and cost-effectiveness of the system.

References

1. Yalung, C.N., Adolfo, M.S.L.: Analysis of obstacle detection using ultrasonic sensor. Int. Res. J. Eng. Technol.. e-ISSN: 2395-0056. **4**(1), (2017)
2. Dawood, Y.S., Almaged, M., Mahmood, A.: Autonomous model vehicles: Signal conditioning and digital control design. International Journal of Engineering and Innovative Technology (IJEIT), **8**(3), 18–24 (2018)
3. Isa, K.B., Jantan, A.B.: An autonomous vehicle driving control system. Int. J. Eng. Educ. **21**(5), 855 (2005)
4. Nabipoor, M., Majlis, B.Y.: A new passive telemetry LC pressure and temperature sensor optimized for TPMS. J. Phys. Conf. Ser. **34**(1), 770 (2006)
5. Song, H.J., Hsu, H.P., Wiese, R., Talty, T.: Modeling signal strength range of TPMS in automobiles. In: IEEE Antennas and Propagation Society Symposium, June, vol. 3, pp. 3167–3170. IEEE, Washington, DC (2004)
6. Davis, R.A., Richard, A.K.I., Foote, S.R., Foster, R., Honeywell International Inc.: Methods and systems for capacitive balancing of relative humidity sensors having integrated signal conditioning. US Patent 6,867,602, 2005
7. Becker, J.C., Simon, A.: Sensor and navigation data fusion for an autonomous vehicle. In: Proceedings of the IEEE Intelligent Vehicles Symposium 2000 (Cat. No. 00TH8511), October, pp. 156–161. IEEE, Washington, DC (2000)
8. Koehler, B., Hentges, G., Mueller, W.: Improvement of ultrasonic testing of concrete by combining signal conditioning methods, scanning laser vibrometer and space averaging techniques. NDT&E Int. **31**(4), 281–287 (1998)
9. Prasad, S., Dianda, J.R., Sprint Spectrum, L.P.: Signal conditioner and method for communicating over a shared transport medium a combined digital signal for wireless service. US Patent 7,634,250, 2009
10. Toyoda, I., Corp, D.: Capacitive humidity sensor. US Patent 6,647,782, 2003
11. Krasniqi, X., Hajrizi, E.: Use of IoT technology to drive the automotive industry from connected to full autonomous vehicles. IFAC-PapersOnLine. **49**(29), 269–274 (2016)

12. Hotta, S., Yoshimura, T.: Vehicular remote-control system and tire pressure monitoring system. US Patent 7181189B2
13. Andria, G., Attivissimo, F., Giaquinto, N.: Digital signal processing techniques for accurate ultrasonic sensor measurement. Measurement. **30**(2), 105–114 (2001)
14. Koehler, B., Hentges, G., Mueller, W.: A novel technique for advanced ultrasonic testing of concrete by using signal conditioning methods and a scanning laser vibrometer. In: NDT in Civil Engineering 1997 Conference Proceedings, pp. 123–134

Vehicular Intelligence: A Study on Future of Mobility

Anish Kumar Sarangi and Ambarish Gajendra Mohapatra

Abstract IoT is redefining the way we look at things today. "Things" today are able to communicate and with AI being deployed at the edge, a whole new window of possibilities has opened up. Things with communication capabilities which we called Intelligent Internet of Things are able to bring transformations to healthcare, manufacturing, transportation, automobiles, and many other industries in many ways. Connected vehicles is one outcome of the IoT and intelligence of edge devices which is bringing infotainment, remote diagnosis, entertainment, automatic braking, lane detection, collision avoidance and lot of other systems that are being helpful in increasing the passenger safety, pedestrian safety and at the same time decreasing congestion, traffic delays as a result of which reducing the effect environmental pollution due to over burning of fuel during congestion. Road accidents today are one of the top ten reasons of human death according to various reports. In this kind of scenario, connected vehicle technology along with intelligent traffic system can help a lot in increasing the safety of people on road.

Keywords Connected vehicles · Intelligent transportation system · Industrial IoT · Internet of Things

1 Introduction

Today traffic congestion is one of the most challenging problems faced by many cities in and around the world. More than 75 million units of cars are sold every year. However, road infrastructure doesn't expand at the same rate. Be it developing or a developed country, issue of traffic congestion and safety of commuters are the problems on which every local government body has been trying to solve over the years. According to various reports, road accidents are among top ten causes of

A. K. Sarangi (✉)
Fusion Practices Pvt. Ltd., Pune, Maharashtra, India

A. G. Mohapatra
Department of Electronics and Instrumentation Engineering, Silicon Institute of Technology, Bhubaneswar, Odisha, India

D. Gupta et al. (eds.), *Smart Sensors for Industrial Internet of Things*, Internet of Things, https://doi.org/10.1007/978-3-030-52624-5_5

death in the world and eighth biggest cause in India. Hence, in this scenario, engineers and scientists have come up with connected vehicles, intelligent transport systems, and autonomous cars to bring advanced level of features to solve various problems pertaining to traffic congestion and driver safety.

Today in the age of IoT, "things" around us has the capability to communicate with each other which collects data and performs actions based on that data without requiring any human intervention. Due to this, it has opened a huge door of opportunities for innovators around the world to bring automation, safety, failure detection, and many more possibilities in almost all the industries, be it healthcare, manufacturing, logistics, energy, etc. Transportation industry is also one such industry where IoT has brought revolutions and writing the future of it by bringing in concepts like Intelligent Transportations Systems and Connected Vehicles which would make transportation safer, congestion-free, pollution-free, and decrease fuel usage. A connected vehicles consists of embedded sensors, transducers, and communication devices which helps in communicating with devices within the car as well outside, be it services, networks, servers, cloud, etc. This opens up a wide range of applications like car infotainment, traffic safety, in-car entertainment, remote diagnostic and telematics, global positioning system (GPS), autonomous vehicles, and many more. Cars including advanced driving assistance technology and intelligent transport systems with the capability of ad-hoc network connections can be considered as connected cars. Today connected cars market is skyrocketing with the market size reaching to around USD 131.9 billion by 2019 [1] and many researchers around the world are also contributing to the field and bringing new innovations every day by exploring and creating different means of connectivity.

One of other reasons of bringing connected vehicles is air pollution and carbon emission which inflicts severe damage on the environment. In the USA, due to congestion itself, the total cost of travel time due to congestion was USD 121 billion in the year 2011, and CO_2 emission due to congestion was 56 billion pounds [2]. As mentioned earlier connected vehicles has the potential to remove traffic congestion and thus, can also remove air pollution and cost incurred for the same using onboard electronics [3] and smart systems along with intelligent traffic management system [4].

Again with huge innovations that have come up in the field of mobile communication, people are able to get high-speed mobile Internet and the same connectivity that they used to get using cables and leased lanes. This type of connectivity has also been made available in cars which has opened many aspects of connected cars. Today enormous cars on road are Internet-enabled; however, it is estimated to grow to 90% by 2020 [5]. This Internet connectivity will enhance the safety of cars like theft detection [6] and tracking and online diagnosis via cloud online servers. In order to help reader understand abbreviations, a table containing those has been listed in Table 1.

Table 1 List of abbreviations

Abbreviation	Meaning
V2S	Vehicle to Sensor Onboard
V2X	Vehicle to Environment
V2V	Vehicle to Vehicle
V2P	Vehicle to Pedestrian
V2C	Vehicle to Cloud
ITS	Intelligent Transport System
ECU	Electronic Control Unit
ISM	Industrial Scientific Medical
RFID	Radio Frequency Identification
UWB	Ultra-Wideband
CAN	Control Area Network
LOS	Line of Sight
VANET	Vehicular Ad-hoc Network
DSA	Dynamic Spectrum Access
DSRC	Dynamic Short Range Communication

2 Types of Connectivity

Connected vehicle refers to the vehicle enabled with communication capabilities which can communicate with external as well as internal devices. According to the type of connectivity the connections have been classified into V2S (Vehicle to Sensor onboard), V2V (Vehicle to Vehicle), V2I (Vehicle to Infrastructure), V2X (Vehicle to External Environment), V2C (Vehicle to Cloud), and V2P (Vehicle to Pedestrian). These kinds of interactions need various data pipelines in the information system so to make connected vehicle transportation a more informed, intelligent, and safer means of transportation. Connected vehicles also forms basic building blocks of Internet vehicles which enables dynamic communication for sharing, computing, gathering, generating, and secure release of information among enabled devices within or outside the vehicle thereby paving the way for next-generation intelligent transport system [7].

2.1 Intra-Vehicle Communication

With increase in intelligence being incorporated in the vehicle, many sensors and intelligent transducers are added to the vehicle in order to monitor various parameters of different parts of the vehicle like tires, engine, cooling system, braking system, etc., for autonomous operation of vehicles. It is forecasted that the number of sensors that would be incorporated per vehicle would be 200 by 2020 [8]. Reading from sensors is sent to the Electronic Control Unit (ECU) for computation and sometimes feedback is also sent for further actions. The transmission of information from sensors to ECU requires an intra-vehicle network which needs to be designed very carefully. Some solutions include controller area network (CAN), FlexRay, TTEthernet, etc., which require a cable connection between sensors and ECU [9, 10].

2.1.1 CAN Bus

Control Area Network (CAN) bus was introduced to the Society of Automotive Engineer Congress in 1986 in Detroit, developed by BOSCH, whose engineers found that the existing communication protocols with serial bus didn't have the capabilities to meet the evolving requirement of the automotive industry. It is a serial data communication use for real-time data communication among different sensors and actuators and ECU within the vehicle. CAN bus provides ease of addition of nodes as it follows bus-based network topology? Each CAN node has a CAN transceiver which establishes a communication link between the nodes and CAN bus. CAN bus is not an address-based protocol rather it is message-based protocol, i.e., message generated by any node in the network is received by all other nodes present in the network. Moreover, it allows a particular node to request a message from any node present in the network. Message-based protocol eliminates the need of reprogramming the entire node which would have been required in case of address-based protocol. Architecture of CAN bus has been shown in Fig. 1.

However, wired communication increases complexity while providing after-sale services and also the weight of wires is significant which adds up to the vehicle's overall mass [11] making it less efficient in terms of fuel usage. Moreover, there are certain places like tire, steering wheels, etc., where sensors are not placed just

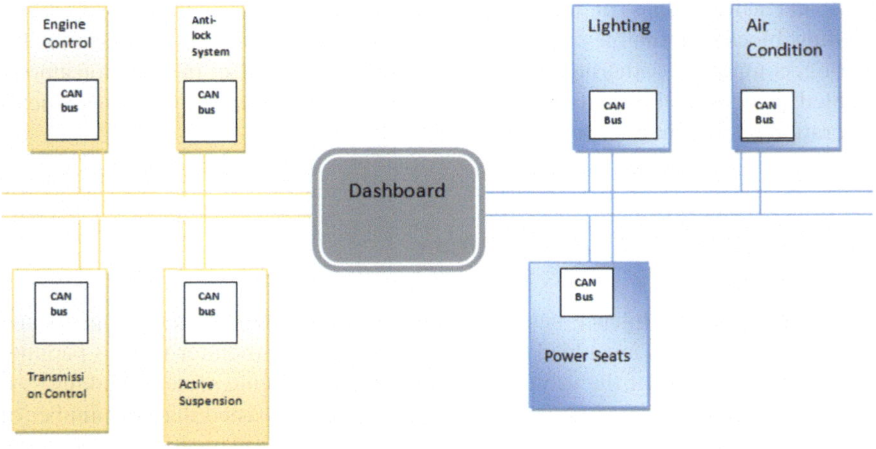

Fig. 1 Architecture of CAN bus

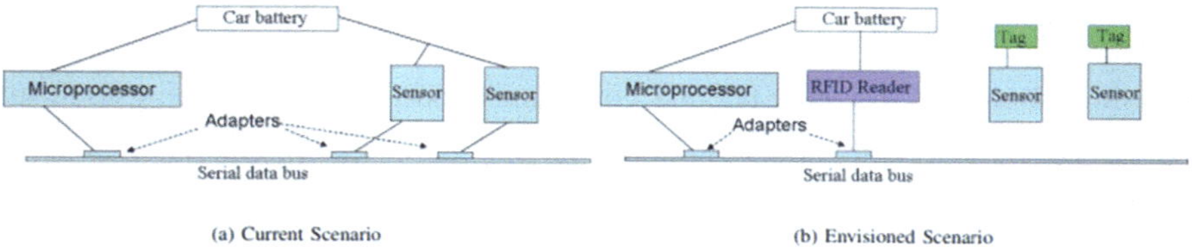

Fig. 2 Wired architecture in vehicles for communication

because a wired connection is not possible for those spots. Wired connection architecture used in the vehicle is shown in Fig. 2.

However, with the rise in Wireless Sensor Network technology, there are various communication devices and protocols which are attached to sensors and ECU for intra-vehicle communication today. This has drastically reduced the complexity as well as cost which was encountered while using wired communication.

Wireless sensor used for intra-vehicle communication is different in terms of characteristics as compared to one used in typical wireless sensor network which has constrains of power and generally follow mesh topology with multi-hop communication between source and destination. In contrast to that wireless communication involved in intra-vehicle communication follow star topology and its generally one-hop transmission between onboard sensors and ECU. Network topology is also stationary which doesn't change with time. Again power constrains are also not present as devices are directly connected to the power system of vehicles.

Even with many advantages there are certain challenges to it as well.

- Communication devices are placed in a very limited space which makes communication environment very harsh and requires proper characterization of channels. Again communication between sensor and ECU is non-line of sight which makes it even difficult [12, 13].
- Latency requirement in V2S is very less and should be reliable. Decisions taken by ECU are done in milliseconds and in order to do that data should be available to it in some milliseconds as well.
- In traffic if there are a number of cars densely present with intra-communications happening in each of them, there is a possibility of interference in the signal which would make communication and data transmission difficult in each of them.
- Another aspect of wireless communication in V2S is security. Unprotected and unencrypted data transmission may lead to potential data theft and malicious attack [14].

To enable this communication various state-of-the-art communication technologies are used today. Some of the technologies enabling wireless communication are discussed below.

2.1.2 Bluetooth

Bluetooth is one of the most popular communication technologies being used in today's world. Be it smart phones, headsets, speakers, smart appliances, smart light, etc., almost every other consumer device is using Bluetooth technology for near field communications. Bluetooth devices work in 2.4 GHz ISM (Industrial Scientific Medical) band with a raw speed transmission speed of 1 Mb/s. Gaussian frequency shift-keying is used for modulation and its baud rate is 1 Msymbol/s [15]. For intra-vehicular communications Bluetooth Piconet is formed which is constituted of a collection of Bluetooth devices which has intra-communication capabilities among them. A Piconet has one device marked as master and others as slave. The maximum number of slaves can be 7 which sums up the number of Bluetooth devices in a Piconet to be 8 in number which is again a constraint as far as V2S is concerned which on-boards large number of sensors and required its data to be communicated to ECU. Although there is a provision for additional devices which can be added to the network, which is called parked, which only work when master asks it to transmit. Whatsoever, in intra-vehicle communication, all the communication devices in the network must be active and needs to communicate continuously. Again Bluetooth runs on ISM band of 2.4 GHz, which is used by other communication devices as well like WiFi, Zigbee, etc. This may result in the interference of signals due to the presence of other devices. However, Bluetooth 3.0 has the ability to run in 6–9 GHz which was created in 2010. Today Bluetooth has evolved to Bluetooth 5.0 which is now called BLE (Bluetooth Low Energy) which primarily uses 2.4 GHz frequency band.

2.1.3 Zigbee

Using Zigbee for V2S is one of the most promising communication technologies which can be used in connected vehicles. It follows IEEE 802.15.4 standard protocol and uses ISM band (868 MHz, 915 MHz, and 2.4 GHz) for data transmission [16]. These are low powered radio devices which are cheaper in terms of cost and better in terms of energy efficiency. Data rate of Zigbee is low and transmits data up to distance varying from 10 to 100 m and beyond in single-hop depending on transmission power and whether communication is happening in line of sight or non-line of sight. The data rate defined for Zigbee is 250 Kbps. Zigbee follows mesh topology, star topology, and tree topology for communication in the network. Security is provided using 128-bit key encryption for secure data communication. Transmission of data in Zigbee happens in two modes: unicast mode and broadcast mode. Broadcast is one when data is transmitted to each and every nod present in the network. But, in unicast mode, data is transmitted to a unique destination using a unique address in the network. Basically this is a one to one data transmission. When the destination node is far away from the source, transmission takes multiple hops across the node before reaching the destination. In V2S context, unicast mode of communication is required as data transmission is done from sensors to ECU. Real implementation of Zigbee for intra-vehicle wireless sensor has been shown in [17] and interference of Bluetooth during communication is also presented. Some of the advantages of using Zigbee for V2S is that it can be readily used for data communication and with a very little change in the scheme, strong link can be established between source and destination for communication purposes. Moreover, it requires a very limited power which doesn't put a lot of strain on power system of vehicles. However, latency is one of the prime characters which is considered while implementing wireless sensor network anywhere in the world. The detailed study of the latency of Zigbee can be found in [18]. Again there are certain engineering issues present which would come up while connecting the existing CAN bus.

2.1.4 RFID (Radio Frequency Identification)

As the name suggest, RFID uses electromagnetic waves for data transfer which is contained in the tags or transponder. This was initially used for the identification of a particular item which was attached with an RFID tag. RFID has two components: RFID tags and RFID reader. Again tags are of two types: passive tags and active tags. Passive tag doesn't have its own power source whereas active tag has its own power source. Passive tag transfers data only when RFID reader is brought near to the tag which energized the tag and enabled it to transfer data. On the other hand, active tags with own power source can be used to transfer data to some hundred meters. It uses different frequency bands depending on the application it is used for. For experimental purposes it is operated in ISM band. Ref. [19] has shown experimental verification and feasibility of using RFID technology in wireless intra-vehicle communication. The basic proposition is that sensors present at various parts of the vehicle will be connected to the RFID tags and the RFID reader will be connected to ECU where all the data would be made available for processing and decision making. RFID reader energizes RFID tags with a pulse of energy signal which it sends to each tag in order to obtain sensor data. The use of passive RFID tags suggests that no extra power source would be required for communication purpose. However, there are certain challenges of using this technology as well. The study tells that it might happen that tags placed at different parts of the cars may not get energized and consequently fails to transmit data. Again, huge power loss during transmission is difficult to solve. Moreover, the critical transmission of data is not guaranteed during simultaneous transmission of data due to possible interference and collision. However, using active RFID tags with sensor, which already require power, can solve maximum problems which are faced by usage of passive tags.

2.1.5 Ultra-Wideband

This communication technology uses very low power and has the capability of transmitting information to short ranges and uses a frequency band from 3.1 to 10.6 GHz and at a high data rate, as high as 480 Mb/s [11]. The most significant difference between other conventional communication technique and ultra-wideband (UWB) technology is that other technologies use pulse modulation, phase modulation, frequency modulation or amplitude modulation for data transmission, whereas ultra-wideband uses pulse radio energy generated at a specific interval of time thereby occupying a large bandwidth, enabling pulse-position or time modulation. One of the biggest advantages is that it uses a different bandwidth than that of conventional communication technologies which makes communication less susceptible to inference. Other advantages of using UWB are:

- No multipath fading
- Localization with a high accuracy can be obtained
- Low cost
- Low processing requirement
- Resistance to channel fading
- No interference due to conventional communication signals around

These all reasons make UWB very suitable for intra-vehicle wireless communication which has been thoroughly studied in [20]. Study has shown a high reliability of data transmission along with low energy requirement which is essential for sending sensor values from different parts of the vehicle to the ECU. Implementation of UWB for V2S in [20] involved sending of speed sensor data of four wheels to the ECU. A high reliability of data transmission has been observed in the implementation. Other implementations depicted UWB performance of data transmission from different parts and regions of the car like from engine, passenger area, beneath the chassis, etc., as characteristic of channels vary as accordance to the location from where data is transmitted due to the presence of hindrances between sensors and ECU. Finding the most suitable physical layer transmission is a critical thing, given an underlying wireless model.

2.1.6 mmWave

mmWave has got its name primarily because of its short wavelength in the order of millimeter. It uses a frequency spectrum of 7 GHz spanning from 57 to 64 GHz for commercial purposes and has the ability of high data transmission rate at around Gbps for multimedia applications like uncompressed video streaming, high-speed downloading, etc. [21]. This kind of high-speed network is very important when it comes to transmitting sensors data which may generate huge amount of data at any instant of time. Sensors like Lidar and high-resolution cameras used in autonomous vehicles generate around 40–100 megabytes per second. Hence, the transmission rate should also be matching to the generation speed so that ECU can get appropriate data for taking decisions in the minimum time possible. mmWave has been standardized by IEEE 802.15.3c and IEEE 802.11ad. The main setback of mmWave is that it suffers a higher propagation loss as compared to that of other lower frequency band technologies which primarily work in 2.4 and 5 GHz of the frequency band. Hence, a powerful directional antenna has to be used in order to compensate those propagation losses, especially in non-line of sight transmissions, which is the case in intra-vehicular communications. Some of the implementations have been done for small-scale parameters, large-scale parameters and studies have been done to show impacts of passenger positions and antenna placement on transmission [22]. The study shows a promising application of mmWave for V2S, primarily because of high data transmission rate. However, in many countries, like India and China, 57–64 GHz frequency band is not available for public use as yet. Hence, licensing those spectrums and using it for V2S purpose would be a very costly affair.

2.2 Inter-Vehicle Connectivity

Inter-vehicle connectivity is one of the most important pillars of connected vehicles. By means of communication between vehicles, or what is called V2V, each vehicle would not be isolated on road as far as information sharing is concerned. Onboard sensors inside the vehicles generate a tremendous amount of data which remains only to the vehicles; however, this information can be shared with the neighboring vehicles via VANET (Vehicular ad-hoc Network) which forms dynamic connections between vehicles in a particular area. This enables a vehicle to share information with another vehicle which is multi-hops away. Without any external network infrastructure requirement, on road safety techniques can be deployed like collision detection, warning for a lane change, hazardous location notification, traffic congestion warning, slow vehicle movement warning, etc.

Vehicular Ad-hoc Network (VANET) is today one of the topics which has grabbed the attention of various researchers around the world to work on it and to find the best possible wireless communication link for data sharing among the vehicles. One of the most important challenges that come into the picture is the environment where these communications happen. Typically, communication devices work best when the source of message and destination, to which message is directed to, are in the line of sight (LOS). However, in inter-vehicle communication scenario, source and destination are at non-line of sight (NLOS) due to the presence of various obstacles like building, walls, vehicles on the road, etc. Buildings on the corner of the road may prove to be a hindrance to the line of sight (LOS) connection between vehicles, whereas a truck on the highway may also prove to be the same. Due to these hindrances, a lot of packet loss and high signal attenuation takes place [23]. However, there is a lack of universal communication channel which can be used in each and every scenario. The channels that exist today are very scenario-specific and have its own advantages and disadvantages. Some of the characteristics of VANET which possesses a significant impact on the inter-vehicle connectivity can be listed below:

- The topology of the network is highly dynamic in nature and is fast changing due to the high mobility of vehicle and difference in the trajectory of movement of each vehicle.

- As a result of the highly dynamic topology of the network and a limited range of VANET, frequent partitioning of the network may take place which may result in disconnection in communication thereby hampering data transmission.
- In a real scenario, a lot of obstacles like vehicles, buildings, etc., which lead to irregular and unreliable links to a mobile vehicle.
- Due to the presence of a high volume of connected vehicles, interference due to the presence of a large amount of networks may account to loss of wireless connections.

Even though there are technical limitations, there are many advantages of V2V communication which can be listed as below.

- Vehicle mobility can be map restricted and can be predicted to a certain level.
- Providing route optimization and directions to the drivers so as to get the fleet to reach the destination as fast as possible.
- Increased safety by providing driver with lane change assistance, road condition notification, collision avoidance assistance, etc.
- Power constrains are not present for communication purpose and each vehicle has enough processing capabilities.
- With the help of the Global Positioning System, it is easy to track vehicles with high precision.

Some of the technologies enabling V2V connectivity or inter-vehicle connectivity are described in the next sub-section of this section.

2.2.1 DSRC/Wave

DSRC stands for Dedicated Short Range Communication which is the prime technology that enables inter-vehicular connectivity. It is a 5.9 GHz spectrum band with bandwidth of 75 MHz which was primarily designed for low latency information exchange directly between the vehicles and the infrastructure that falls along. It may constitute nearby cars or traffic infrastructure guiding the vehicles for the smooth passage of vehicles. This protocol has been described in Wireless Access in Vehicular Environment (WAVE) as well as in IEEE 802.11p for MAC layers and family of IEEE 1609 for upper layers. Efforts are underway by various industries, governments, and research organizations to bring in this technology and implement it in a large scale in various regions around the world. There are primarily three layers which form WAVE, i.e., physical layer, data link layer, and application layer. The architecture of DSRC has been shown in Fig. 3.

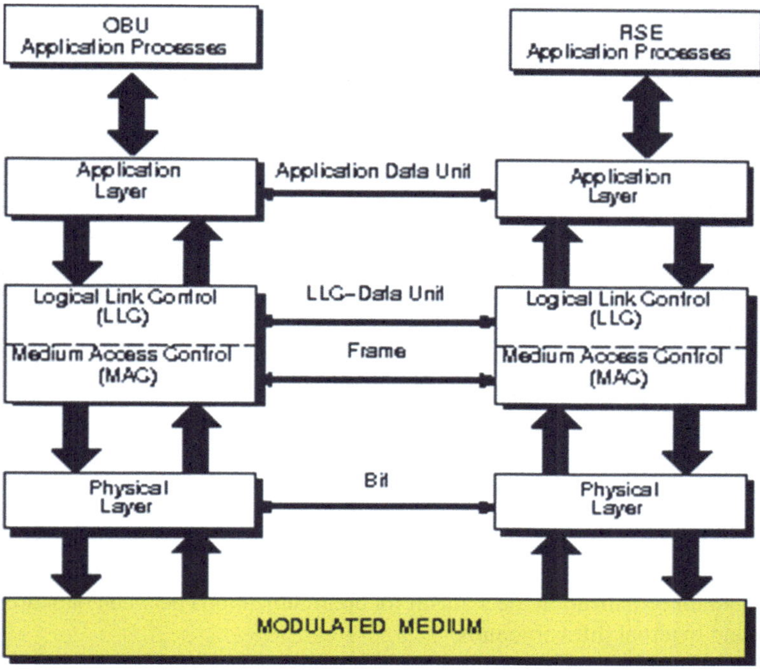

Fig. 3 Architecture of DSRC (Source: http://www.wirelesscommunication.nl/reference/chaptr01/dtmmsyst/dsrc/dssr4.htm)

Physical Layer

This layer has been adopted from IEEE 802.11a/g and has a default uplink and downlink of about 250 and 500 Kbits/s and can be increased to up to 3–27 Mbps in a 10 MHz channel. Enormous studies and tests have been done regarding the performance of DSRC. However, there are certain challenges to it. Some of the names which have been there are: (1) Unreliable connectivity during non-line of sight communications and obstructions in the communication channel. (2) Interference due to the presence of multiple communication channels between infrastructures and vehicles. (3) Intermittent loss rate behavior during data transmission which is also known as the gray-zone phenomenon. Studies are being conducted for continuous evolution and modification in DSRC physical layer for better communication reliability and connectivity.

Data Link Layer

This layer is further divided into Logical Link Control (LLC) layer and Medium Access Control (MAC) layer. LLC layer follows the IEEE 8088.2 specification which is responsible for connectionless and connection-oriented and connectionless communication services. MAC layer is primarily responsible for the efficient contention system to avoid data collision, disturbance and to resolve data collision in multi-lane environments. However, there is a reliability issue in communication especially in safety-related applications. Hence, certain MAC designs suggest handshaking mechanisms wherein a dedicated connection is established before any data is communicated. However, this makes it very slow in nature. This mechanism is not applied during broadcast mode where the receiver only listens for data and transmits data only in a given time slot.

Application Layer

A service interface is provided by application layer for applications and application multiplexing (to be able to send multiple data with a single channel), fragmentation, and a standard initialization mechanism for communications between onboard electronics and external infrastructures.

2.2.2 Dynamic Spectrum Access

As discussed earlier, demand for infotainment is ever increasing and hence higher bandwidth of network demand is increasing as well. With a dedicated bandwidth, it is very difficult to satisfy the quality of service (QoS). Again, in urban areas, due to the high density of vehicles, scarcity of spectrum is very common. Studies in cognitive radio have made way for Dynamic Spectrum Access (DSA) to be a complimentary solution to DSRC. Vehicles are allowed to communicate with one another temporarily and spatially over a licensed communication spectrum that are vacant at a given time [24]. Certain researchers also showcased the possibility of using vacant TV communication space [25] whose spectrum ranged between 54 and 698 MHz. The results suggest that penetration and propagation power of these signals are far superior in nature. The use of TV whitespace by DSA has been standardized under IEEE 802.11af and IEEE 802.22 for Wireless Local Area Network.

DSA communication system is based on cognitive radio technology as mentioned earlier and along with it used intelligent radio and adaptive radio which allows DSA to adapt to external factors and environments so that the best possible communication can be served. It uses these technologies to dynamically switch between frequency spectrum in a given location and for a given period of time. It primarily consists of three subsystems: (1) Two-layer control channel subsystem. (2) Multi-hop data communication subsystem. (3) Spectrum sensing and channel switching subsystem. This is worth noting here that shortage of spectrum in urban areas has widely been mitigated by the use of DSA, especially in highways where vehicles are densely populated. However, some challenges like efficient MAC protocol design and QoS for safety, entertainment, and infotainment application is yet to be handled in a better way.

2.3 *Vehicle to Internet*

Due to the advancement in Internet connectivity and technologies enabling it, the industry is widely using it in various applications to bring a connected eco-system to solve a particular problem. Various manufacturers and researchers are using the Internet today to bring connectivity to vehicles as well, enabling vehicles to get information from infrastructure surrounding it. Researchers and academia have been working on bringing Internet connectivity to the vehicles and so far have been successful to a certain extent. Companies like Tesla have been giving entertainment, telematics, infotainment, etc., services to their car users via the Internet. Mobile cellular network and WiFi are the most promising technologies which are seen today as far as V2I connectivity is concerned. 3G and 4G technologies already being a very stable technology, it is proving out to bring connectivity to vehicles pretty well. Some works even saw the usage of low cost outdoor WiFi access point on the road side to provide Internet connectivity to be pretty useful [26]. They have also mentioned some of the shortcomings like inconsistent connectivity and overhead processing time by the programs inside the processors. However, some of the state-of-the-art technologies being used in the industries are discussed in the next section.

2.3.1 Created Connectivity

The connectivity brought in due to the presence of tethered Internet connection from a mobile phone via WiFi, Bluetooth, USB, etc. doesn't come built-in with the vehicle itself. One of the most prevalent technologies enabling tethering is Mirror Link which is supported by Car Connectivity Consortium (CCC) and ICT companies like Nokia and Sony. This enables users to bring in Internet connectivity via USB, WiFi, or Bluetooth to the vehicles which brings infotainment services to the vehicle. Some examples are like Toyota Touch 2 by Toyota, Apple's car play for Iphone tethering, etc. Some of the aftermarket devices also bring in Internet connectivity. For example, General Motors OnStar brings in connectivity by using onboard built in mobile phone and provides services based on subscriptions.

2.3.2 Connectivity Built-in

Many car manufacturers have come up with their built-in infotainment and communication devices which use state-of-the-art 4G LTE communication services to bring Internet connectivity to vehicles without the need of having any kind of tethering mobile phones. However, this requires a collaborative effort between car manufacturers as well as cellular network companies to make this possible. Some of the examples include Audi Connect and BMW-connected Drive which provides services like driver assistance, call center service, vehicle telematics, etc., which is possible through Internet connectivity. With research, development, and maturity that has come to 4G LTE technology, LTE connected cars have evolved in many folds. Verizon and Alcatel-Lucent are leading from the front in research and development of LTE-connected cars. Still debates on the best solution in terms of connectivity in cars are on. Built-in solutions prove to be a very good option as far as connectivity stability, configurations, and customized services are concerned as compared with brought-in connectivity. However, embedded devices face limitation in-terms of hardware updates and evolutions.

2.3.3 Drive-Thru Connectivity

Thousands of WiFi hotspots have already been deployed at various places across the world which provides last-mile Internet connectivity to the millions of Internet users. This particular architecture can prove to be a very cost-effective solution to bring in the Internet to the vehicles on the go. Basic idea is that, when vehicles come in the area covered by a WiFi hotspot, it can connect to it and access Internet to provide Internet-oriented services. Hence, as this happens only when vehicles come in the range covered by WiFi hotspot, it is called drive-thru connectivity. It becomes effective for the fact that it utilizes existing infrastructure in order to bring Internet connectivity. Again, with developments in Hotspot 2.0 which has the capability of providing seamless and secure connections to the end users, drive-thru connectivity has a huge potential of growth and adoption possibility. However, it poses certain limitations. Generally, in drive-thru connections, unlike stationary connection indoor environments, end users are mobile and moving at a considerable amount of speed. Apart from that, the time for which connectivity is established is very low and reduces further with the reduction in the coverage area of hotspot. Establishing a robust connectivity at such a short interval is a challenge. Again, before actual data is transferred, processes like authentication, connection establishment, and IP configurations cannot be walked over with. Apart from that wireless connectivity has its own disadvantages like channel attenuation, shadowing, and channel fading which contribute to communication failure.

Some of real world studies in [27, 28] where drive-thru communication (based on IEEE 802.11b and 802.11g, respectively), was tested, in a highway environment where access points, two in number, was placed closed to each other were conducted. The driving speed of the vehicle was varied and performance was measured.

Some of the important observations which have come out of this study are that drive-thru communication happens to have primarily three-phased characteristics depending on its distance from the access point. It can be named as entry phase, productive production phase, and exit phase. Throughput obtained during entry and exit phase is low and is highest in the production phase. The reason to that is because of delay due to connection establishment, data loss, weaker signal, and overestimation of rate. In [29], the study of backhaul capability of drive-thru system has been done in free highway where interference due to the presence of other vehicles and non-line of sight scenarios has been mitigated. A wireless channel having a backhaul capability of 1 Mb/s reduces the amount of data delivered from 92 to 25 MB. Again, if a delay of 100 ms is present in one way, web services are significantly affected and degraded due to additional delay of request-response communication model of HTTP. Some improvement strategies have been put forward in certain studies which include better MAC design, cooperation between vehicles on road, multi-hop data transmission in V2V, improving transport protocols for improving on transmission disturbances and influence of non-line of sight scenarios due to presence of other vehicles, etc.

3 Applications and Benefits of Connected Vehicles

Application of connected vehicles has been proficiently been proposed and many research and development is underway to bring out the best possibilities of connected vehicles which can bring in additional life-saving safety features, telematics, infotainment, entertainment, etc., to the vehicles. Some applications are listed below.

- It brings in additional safety for the people sitting in the vehicle which results in several life-threatening situations by sharing information about the environment in which it's been rode. It gives information about nearby vehicles so that alerts can be generated or brakes can be applied by the vehicle in-case of any mishap that occurs.
- It gives alert to drivers in sensitive locations where drivers should drive slow like school area, hospitals, blind turns, etc., via vehicle to infrastructure connectivity.
- Real-time traffic alerts and suggestions can be given to drivers so that one can take alternate route with less congestion.
- Real-time vehicle metrics can be sent to the manufacturers so that they can perform better servicing on vehicle and also can send reminders to vehicle owners regarding servicing needs based on vehicle condition rather than time between servicing.
- The metrics sent can further be used in increasing the efficiency of the engine so that it consumes less fuel and give better performance in various parameters.

- Less fuel consumption leads to less generation of carbon dioxide and harmful gases to the environment thereby reducing pollution. Apart from that, less traffic congestion will also lead to reduction in vehicular air pollution.
- Understanding traffic patterns in areas can help in planning road expansion.
- Real-time traffic data can also help in guiding emergency vehicles like fire extinguishing vehicle, ambulance, etc., to reach a particular destination faster.

4 Challenges of Connected Vehicles

Connected vehicles have their own challenges when it comes to commercialization in this first evolving and advancing world. Some of the challenges are described below:

- Challenges exist in terms of bringing lasting connectivity to the vehicles and for this car makers need to collaborate with mobile and cellular network provider.
- Car manufacturers have to train their sales people on connected cars and vehicles who can in turn make customers understand about various technologies which has been deployed in the car and how those benefits help them.
- There is a preference difference as far as built-in or brought-in connectivity is concerned. Even though built-in devices provide better connectivity, yet, people like to have full access to their contacts and music and hence, prefer to provide Internet connectivity to vehicles by tethering via WiFi, Bluetooth, USB, etc.
- The Internet connectivity to the car comes with an additional bill of Internet services for vehicles. Now the question arises, who would be paying off this bill? Tesla started up the roadster model with inbuilt Internet connectivity free. However, it has been recently announced that some of its features which require higher Internet speed would require customers to pay a different bill. With customers who have been used to pay only one time for new vehicles, will it be acceptable for them to pay additional service bill?

5 Conclusion

Connected vehicle has the ability to revolutionize the way drive vehicles by opening windows of possibilities wherein applications are infinite, just by bringing the phenomenon of connectivity of data among and within vehicles. It also has a lot of contribution toward autonomous cars. Today, autonomous cars used sophisticated hardware and sensors onboard the vehicle to sense the environment and take actions accordingly. First problem to this is that the cost of vehicle having autonomous capabilities rises drastically. Again, power consumed in order to power these sensors onboard is also very high which drains the batteries, that otherwise would have increased the distance covered by vehicles per charge. However, if connectivity comes to a highly reliable state without interruption, V2X can be used by autonomous cars to take decisions by getting information about the environment. However, this is not happening in the near future. Enormous studies have to be conducted to get more reliability and robustness as far as connectivity of vehicles is concerned everything that concerns it.

References

1. Transparency Market Research: Connected Car Market—Global Industry Analysis, Size, Share, Growth, Trends and Forecast, 2013–2019. Transparency Market Research, Albany, NY (2013)
2. Schrank, D., Eisele, B., Lomax, T.: TTI's 2012 Urban Mobility Report. Texas A&M Transport. Inst. and Texas A&M Univ. Syst., College Station, TX (2012)
3. Olaverri-Monreal, C., Gomes, P., Fernandes, R., Vieira, F., Ferreira, M.: The see-through system: a VANET-enabled assistant for overtaking maneuvers. In: Proc. IEEE Intell. Veh. Symp. (IV), San Diego, CA, USA, June, pp. 123–128 (2010)
4. Anda, J., LeBrun, J., Ghosal, D., Chuah, C.-N., Zhang, M.: VGrid: vehicular adhoc networking and computing grid for intelligent traffic control. In: Proc. IEEE 61st Veh. Technol. Conf. (VTC'05 Spring), Stockholm, Sweden, June, pp. 2905–2909 (2005)
5. Telefnica: Connected car industry report (2013)
6. Ramadan, M., Al-Khedher, M., Al-Kheder, S.: Intelligent anti-theft and tracking system for automobiles. Int. J. Mach. Learn. Comput. **2**(1), 88–92 (2012)
7. Liu, N.: Internet of vehicles: your next connection. Huawei WinWin. **11**, 23–28 (2011)
8. Pinelis, M.: Automotive sensors and electronics: trends and developments in 2013. In: Automot. Sensors Electron. Expo, Detroit, MI, USA, October (2013)
9. D'Orazio, L., Visintainer, F., Darin, M.: Sensor networks on the car: state of the art and future challenges. In: IEEE Des. Autom. Test Eur. Conf. Exhib. (DATE), pp. 1–6 (2011)
10. Tuohy, S., Glavin, M., Jones, E., Trivedi, M., Kilmartin, L.: Next generation wired intra-vehicle networks, a review. In: Proc. IEEE Intell. Veh. Symp. (IV), Gold Coast, Australia, June, pp. 777–782 (2013)
11. Qu, F., Wang, F.-Y., Yang, L.: Intelligent transportation spaces: vehicles, traffic, communications, and beyond. IEEE Commun. Mag. **48**(11), 136–142 (2010)
12. Tsai, H.-M., et al.: Feasibility of in-car wireless sensor networks: a statistical evaluation. In: Proc. IEEE SECON, San Diego, CA, USA, June, pp. 101–111 (2007)
13. Moghimi, A.R., Tsai, H.-M., Saraydar, C.U., Tonguz, O.K.: Characterizing intra-car wireless channels. IEEE Trans. Veh. Technol. **58**(9), 5299–5305 (2009)
14. Zhang, T., Antunes, H., Aggarwal, S.: Defending connected vehicles against malware: challenges and a solution framework. IEEE Internet Things J. **1**(1), 10–21 (2014)
15. Bisdikian, C.: An overview of the Bluetooth wireless technology. IEEE Commun. Mag. **39**(12), 86–94 (2001)
16. Zigbee specification. ZigBee Alliance (2004 December) [online]. Available from: http://www.zigbee.org
17. Tsai, H.-M., et al.: Zigbee-based intra-car wireless sensor networks: a case study. IEEE Wirel. Commun. **14**(6), 67–77 (2007)
18. Ahmed, M., et al.: Intra-vehicular wireless networks. In: Proc. IEEE Globecom Workshops, Washington, DC, USA, November, pp. 1–9 (2007)
19. Tonguz, O.K., Tsai, H.-M., Talty, T., Macdonald, A., Saraydar, C.: RFID technology for intra-car communications: a new paradigm. In: Proc. IEEE Veh. Technol. Conf. (VTC'06 Fall), Montreal, QC, Canada, September, pp. 1–6 (2006)
20. Niu, W., Li, J., Liu, S., Talty, T.: Intra-vehicle ultra-wideband communication testbed. In: Proc. IEEE Mil. Commun. Conf. (MILCOM'07), Orlando, FL, USA, October, pp. 1–6 (2007)

21. Qiao, J., Cai, L.X., Shen, X.S., Mark, J.W.: Enabling multi-hop concurrent transmissions in 60GHzwireless personal area networks. IEEE Trans. Wirel. Commun. **10**(11), 3824–3833 (2011)
22. Sawada, H., et al.: A sixty GHz intra-car multi-media communications system. In: Proc. IEEE 69th Veh. Technol. Conf. (VTC Spring'09), Hilton Diagonal Mar, Spain, April, pp. 1–5 (2009)
23. Boban, M., Vinhoza, T.T., Ferreira, M., Barros, J., Tonguz, O.K.: Impact of vehicles as obstacles in vehicular ad hoc networks. IEEE J. Sel. Areas Commun. **29**(1), 15–28 (2011)
24. Wang, T., Song, L., Han, Z.: Coalitional graph games for popular content distribution in cognitive radio VANETs. IEEE Trans. Veh. Technol. **62**(8), 4010–4019 (2013)
25. Altintas, O., et al.: Demonstration of vehicle to vehicle communications over TV white space. In: Proc. IEEE Veh. Technol. Conf. (VTC Fall), San Francisco, CA, USA, September, pp. 1–3 (2011)
26. Bychkovsky, V., Hull, B., Miu, A., Balakrishnan, H., Madden, S.: A measurement study of vehicular internet access using in situ wi-fi networks. In: Proc. ACM MobiCom, Los Angeles, CA, USA, September, pp. 50–61 (2006)
27. Ott, J., Kutscher, D.: Drive-thru Internet: IEEE 802.11 b for automobile. In: Proc. IEEE INFOCOM, Hong Kong, China, March (2004)
28. Ott, J., Kutscher, D.: The drive-thru architecture: WLAN-based Internet access on the road. In: Proc. IEEE 59th Veh. Technol. Conf. (VTC'04 Spring), May, pp. 2615–2622 (2004)
29. Gass, R., Scott, J., Diot, C.: Measurements of in-motion 802.11 networking. In: Proc. Seventh IEEE Workshop Mobile Comput. Syst. Appl., Semiahmoo Resort, WA, USA, August, pp. 69–74 (2005)

Connected Vehicles: Intelligent Transport Systems

Navneet Yadav and Rama Kanta Choudhury

Abstract The world so far has witnessed an exponential growth of vehicle on road. With the increase in the number of vehicles, there are evidently a large number of problems encountered on the road. Some of these problems could be named as populated and polluted roads, violation of traffic rules, road accidents, excessive wastage of fuel, delay in emergency services, etc. The proposed work intends to alleviate these problems by designing a long-range decentralized network of vehicles. All the vehicles within the range of a kilometer will be able to communicate with each other to solve all the specified problems. The major focus of the work will be on designing a secure network within the range of a few hundred meters. A specified code word will be broadcasted from each vehicle in such a way that there occurs no data collision between transmissions from different vehicles. First and foremost the transmitted code will specify if the vehicle is facing an emergency situation or not. The additional information in the code will include the address of the vehicle in network, checksum, and parity to ensure accurate transmission of data.

Keywords Network free · Transceiver · Vehicle to vehicle communication

1 Introduction

"Driving is like Baseball, it's the one who gets home safely that counts."
—Tommy Lasorda

According to a theory by Frigyes Karinthy, known as the Six Degree of Separation, all living things and everything else in the world is six or fewer steps away from each other. Considering the theory correct for the events and mishappenings as well, with the exponential rise in the population, there has been a significant rise in the workforces (especially in densely populated countries such as India or China), leading to more and more number of vehicles coming on the road, and hence causing various issues such as traffic jams, accidents, road rages, violation of traffic rules, illegal racing, and many more.

N. Yadav (✉) · R. K. Choudhury
Maharaja Agrasen Institute of Technology, Delhi, India

D. Gupta et al. (eds.), *Smart Sensors for Industrial Internet of Things*, Internet of Things, https://doi.org/10.1007/978-3-030-52624-5_6

In many parts of the world drivers have only one rule in their mind, that there isn't any rule. From roads transforming into an F1Formula race track to pedestrians taking a stroll on busy roads, one can watch the madness unfold.

On the other end of the horizon are the traffic jams which go on to prove that the roads are a great leveler. You may own a first-class SUV or an Auto-Rickshaw, it doesn't matter as long as you reach your destination. Every vehicle since the invention of wheel can easily be seen on Indian roads leading to a situation of traffic congestion.

Apart from the cases of jams and congestions, one of the biggest risks a vehicle owner takes while parking it outside his/her house or office, is will it stay safe? According to the data received from the National Crime Records Bureau, in a city like Delhi, there has been an average record of more than 15,000 automobile theft registered every year, out of which hardly a few hundreds are recovered. Rest remain missing forever or are crushed in the trash.

While driving on the road, there have been situations, when there is an emergency on road, and there is no one to ask for help. This mayday situation has been experienced by every experienced driver at least once. The most difficult task in such a situation is to ask for help from fellow drivers on road, especially on highways where vehicles are running at a very high pace.

To solve such and many other problems, authors propose a solution of a decentralized vehicular network that can help drivers within a range of a kilometer to connect with each other.

As suggested by Dr. A.P.J. Abdul Kalam, "*If it is not working out, try to find a solution by discussing with the people around*", the decentralized vehicular network allows the vehicles to communicate with each other. Each vehicle shares the traffic situation around it, the GPS coordinates of its movement, need for SOS services (if any), request to keep roads free (in case of emergency vehicles), and many other similar data that doesn't affect the privacy of the driver and can help him stay safe while driving at the same time.

In this chapter ahead, different sections explain the basic features of the device, how it works, the fundamentals of communication taking place between vehicles, different modules, controllers and sensors used, and other relevant information regarding the presented work.

2 Components and Modules

This section describes the different modules and components being used in the hardware, along with their specifications and other necessary information used in this work.

2.1 nRF24L01+

A single-chip transceiver module nRF24L01 works on 2.4–2.4835 GHz ISM frequency band and is an open-source transceiver module. Figure 1 shows its exterior and Fig. 2 shows its internal block diagram. It comes with an embedded baseband protocol engine (Enhanced Shock Burst). It has an inbuilt power amplifier and low noise amplifier to get more accurate results while transmitting or receiving.

The features for the same are listed below:

(a) Worldwide 2.4 GHz ISM Band Operation
(b) 250 kbps, 1 and 2 mbps on air data rates
(c) Ultralow power consumption
(d) On-Chip Voltage regulation
(e) 11.3 mA Tx at 0 dBm output power
(f) 13.5 mA Rx at 2 mbps data rate
(g) 900 nA in Power Drain
(h) Supply Voltage Range: 1.6–3.6 V
(i) Automatic Packet Handling
(j) Auto Packet Transaction Handling
(k) 6 Data Pipe Transceiver
(l) Low-Cost BOM
(m) 16 MHz Crystal

The device can be used for different purposes such as wireless PC peripherals, mouse, keyboards, VoIP headsets, game controllers, active RFID, watches and sensors, ultralow power sensor networks, home and commercial automation, etc. [1–4].

It can easily be configured using a microcontroller, through a Serial Peripheral Interface (SPI). The embedded baseband protocol engine (Enhanced Shock Burst) is based on a packet communication and supports various modes from manual operations to the advanced autonomous protocol operation. Internal FIFOs ensure a

Fig. 1 Image of nRF24L01 + PA + LNA

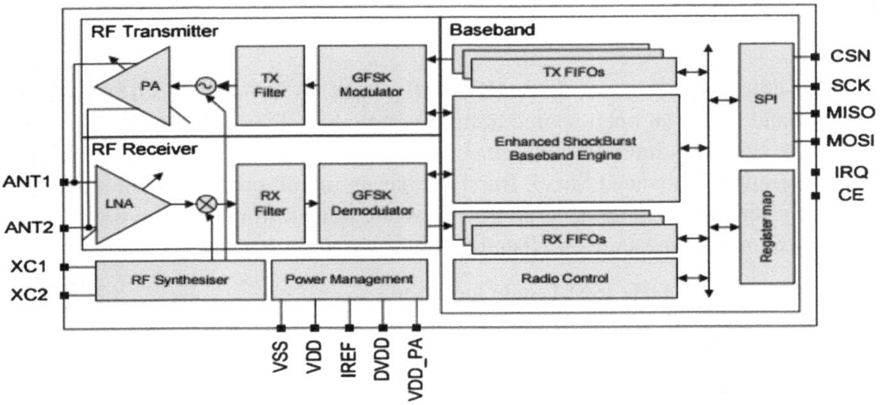

Fig. 2 Internal Block Diagram of nRF24L01

smooth data flow between radio front end and system's microcontroller. Complete specifications of nRF24L01 have been given in [5].

The radio front end uses the GFSK modulation scheme. It has user-configurable parameters such as frequency channel, output power, and air data rate. Internal voltage regulators ensure a high power supply rejection ratio and a wide power supply range.

2.2 *ATMega328P*

The Atmel picoPower ATMega328P is a low power CMOS 8-bit microcontroller based on the AVR-enhanced RISC (Reduced Instruction Set Computer) architecture. The block diagram of ATMega328P is shown in Fig. 3. By executing powerful instructions in a single clock cycle, ATMega328P achieves throughputs close to 1 MIPS (Million Instructions per Second) per MHz. This empowers system designers to optimize the device for power consumption versus processing speed.

The features of ATMega328P are listed below:

(a) Advanced RISC Architecture
(b) 131 Rich Instruction Set
(c) Most Single Clock Cycle Execution
(d) 32 × 8 General Purpose Registers
(e) Fully Station Operation
(f) Up to 20 MIPS (Million Instructions per Second) throughput for 20 MHz
(g) On-Chip 2 cycle Multiplier
(h) 32 KB of In-System Self-Programmable Flash Program Memory
(i) 1 KB EEPROM
(j) 2 KB Internal SRAM

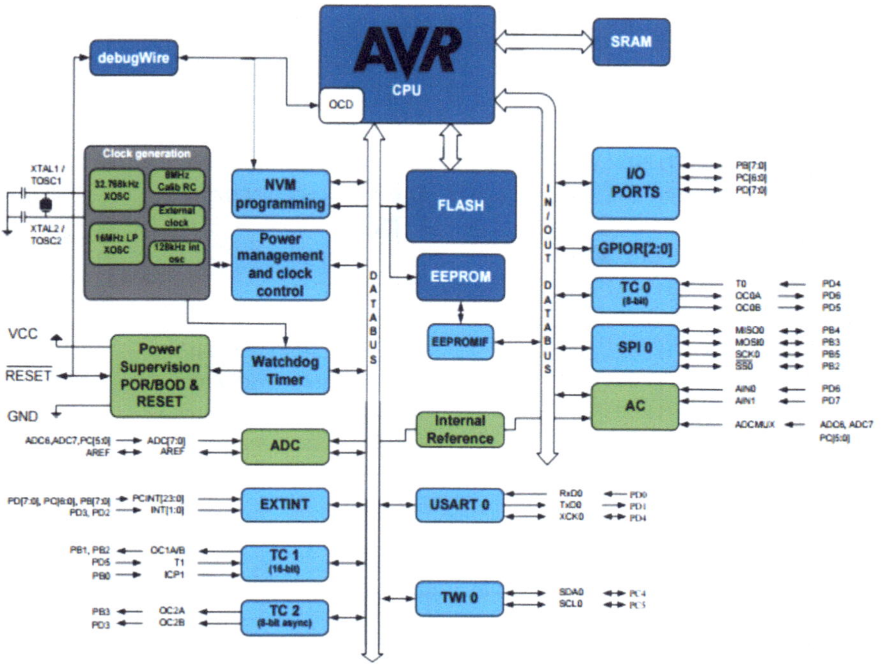

Fig. 3 Block diagram of ATMega328P

(k) Data Retention of 20 Years
(l) Programming Lock for Software Security
(m) Two 8 bit Timer/Counter with Separate Prescaler and Compare Mode
(n) One 16 bit Timer/Counter with Separate Prescaler and Compare Mode
(o) Real-Time Counter with Separate Oscillator
(p) Six PWM Channels
(q) 8-Channel 10 bit ADC (Analog to Digital Converter)
(r) Two Master/Slave SPI (Serial Peripheral Interface)
(s) One Programmable Serial UART
(t) One Byte-oriented 2 Wire Serial Interface
(u) One On-Chip Analog Comparator
(v) Interrupt and Wake Up on Pin Change
(w) Internal Calibrated Oscillator
(x) External and Internal Interrupt Sources
(y) 23 Programmable I/O Lines

The pin diagram for ATMega328P is shown in Fig. 4. It is a 28-Pin Integrated Circuit (IC). The microcontroller comes with 23 General Purpose Input Output (GPIO) Pins and has inbuilt features such as Serial Peripheral Interface, Universal Synchronous Asynchronous Receiver Transmitter (USART), Analog to Digital Converters, ADC Channels, 8 bit Timer/Counter, and 16 bit Timer/Counter.

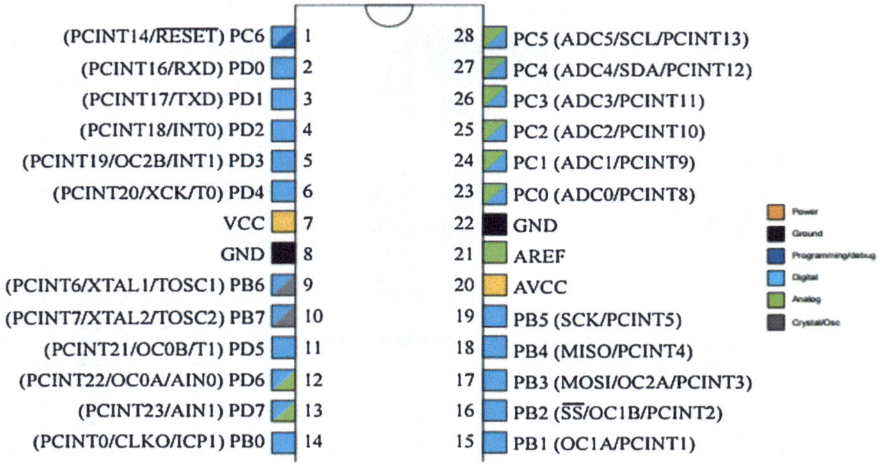

Fig. 4 Pin out diagram of Atmaga328P

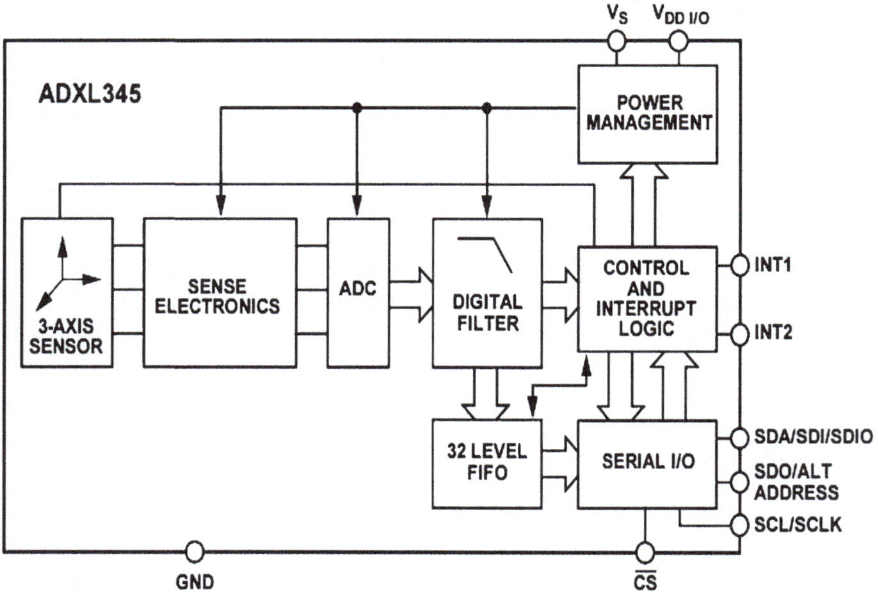

Fig. 5 Block diagram of ADXL345

2.3 *Accelerometer ADXL345*

An accelerometer ADXL345 is being used to measure the velocity of the vehicle on the road to measure congestion and traffic. Its block diagram is shown in Fig. 5.

Fig. 6 HC-05 module

Manufactured by Analog Devices, the ADXL345 is a small, thin, ultralow power, 3 axis accelerometer with high resolution (13 Bit) measurement up to 16 g. Digital output data are formatted as 16-bit twos complement and are accessible through either an SPI or I2C digital interface.

It is well suited for mobile device applications and measures the static acceleration of gravity in tilt sensing applications as well as dynamic acceleration resulting from motion or shock. Its high resolution enables measurement of inclination change less than 1.0°.

2.4 HC-05 Bluetooth Module

HC-05 module shown in Fig. 6 is an easy to use Bluetooth SPP (Serial Port Protocol) module, designed for transparent wireless serial connection setup. Serial Port Bluetooth module is a fully qualified Bluetooth V2.0 + EDR (Enhanced Data Rate) 3 Mbps modulation with complete 2.4 GHz radio transceiver and baseband.

It uses CSR Bluecore external single-chip Bluetooth system with CMOS technology and AFH (Adaptive Frequency Hopping) feature. Its operational voltage varies from 1.8 to 3.6 V.

3 Wireless Network Establishment

This section describes how the network is being established between multiple vehicles on road, and the action taken by the network in different practical situations.

As specified in the list of components and modules, a decentralized network of vehicles has been established using single-chip ultralow power-consuming module known as nRF24L01+ with inbuilt Power Amplifier and Low Noise Amplifier. It works on a frequency band of 2.4–2.4835 GHz and can be used by anyone. Being a self-network generator, it doesn't need any internet connectivity, and hence can easily be used in any state, city, or country.

Before discussing about the establishment procedure and special cases, firstly it is necessary to understand different parameters and/or variables that are being used in the vehicle for communication.

1. *Registration ID*: The Registration ID is the registration number of the vehicle provided by the manufacturer and is in format "DL4CN5166". This will be used to keep a record of vehicles connected in a network through a look-up table.
2. *Network ID*: Network ID is the special 6 digit ID generated randomly initially while establishing a network. Any vehicle getting connected to a network will transmit its message encapsulated within Network ID and its own Vehicle ID so that only those vehicles can decode the message who are in the network.
3. *Vehicle ID*: Vehicle ID is allotted by the primary vehicle to any other vehicle requesting for connection to any other vehicle. The vehicle ID is appended in the look-up table along with its registration ID.
4. *Generic Network ID*: Whenever a new vehicle comes on the road, it is instructed to transmit on the Generic Network ID to communicate and request for a network connection and its credentials. This generic network ID is set to "00001" and any vehicle initially transmits on this address only.

Some important terminologies that will be used further in the report are as follows:

1. *Primary Vehicle*: The vehicle that will be establishing the initial network is entitled as the primary vehicle. Only this vehicle will have permission to give permission to any other vehicle to connect to the network.
2. *Secondary Vehicle*: Vehicles other than the primary vehicle in the network are entitled as the secondary vehicles of that network.
3. *Special Vehicle*: Special vehicles involve emergency vehicles such as PCR van, ambulance, fire brigade, etc.

The process of establishment of the network is specified below with all the special cases.

3.1 Case 1: When Vehicle Enters Road and There Is No Other Vehicle on Road

Whenever a vehicle enters the road, it initially has its registration ID only. It broadcasts its registration ID with the request to attach to any network in the surrounding in the network pipe having the address "000001", i.e., the generic net ID.

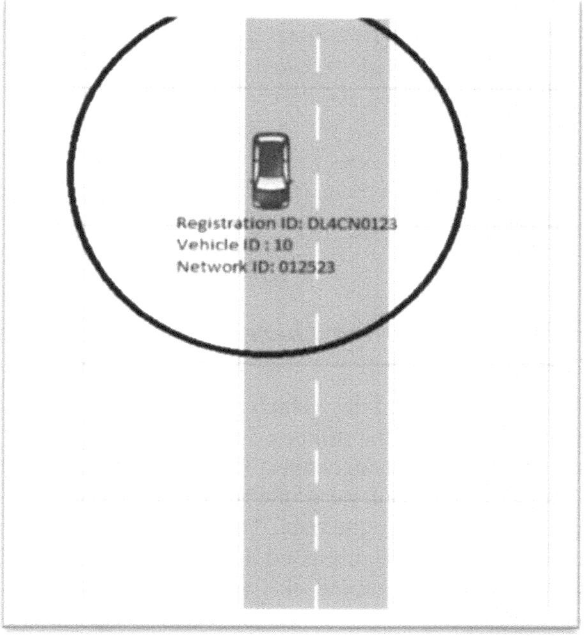

Fig. 7 Situation when no other vehicle is on road

Since there is no other vehicle in the surroundings as depicted in Fig. 7; the device will fail to get any response. Each vehicle makes 15 tries for connecting to a network at an interval of 15 ms. On failing, the vehicle is declared as the primary vehicle and the device generates the parameters to establish a network itself. A network ID is generated. A look-up table is established, and the first entry is made into it with the Registration ID of primary vehicle and the vehicle ID allotted to itself, i.e., 10, in case of primary vehicle. A special vehicle cannot establish its own network.

Consider a vehicle with registration ID DL4CN0123 that came on the road. It initially runs a pseudo-code as follows:

```
Start
SetTryLimit = 15;
SetTryTime = 15ms;
Registration_ID = "DL4CN0123";
Network_ID = "00001";
Radio.OpenWritingPipe (Network_ID);
Radio.Write ("Request_to_Connect"+ Registration_ID);
Radio.OpenReadingPipe (Network_ID);
If Radio.Available ()
```

```
     //Receive Network Credentials
If! Radio.Available ()
     //Establish Network
     Vehicle_ID => "10";
     Network_ID => "012523";
     Vehicle_Type => Primary;
     Radio.OpenReadingPipe (Network_ID);
Stop
```

3.2 Case 2: When There Are Other Vehicles on Road as Well

In case there are other vehicles on road as well when the vehicle transmits the request to attach, the primary vehicle will receive the signal and will transmit the network credentials to connect to the network. According to the vehicle ID allotted to the vehicle, it will be allotted a specific period time for the transmission. Else, it will keep receiving the data. There is a specified format of the message, i.e., the code word which is discussed in the next chapter. The vehicle will be able to communicate within a range of 1.5 km and will become a secondary vehicle of the network.

Below is the pseudo-code the device that has established the network will follow,

```
Start
SetTryLimit = 15;
SetTryTime = 15ms;
Registration_ID = "DL4CN0123";
Network_ID = "00001";
Radio.OpenWritingPipe (Network_ID);
Radio.Write ("Request_to_Connect"+ Registration_ID);
Radio.OpenReadingPipe (Network_ID);
If Radio.Available ()
     //Receive Network Credentials
If! Radio.Available ()
     //Establish Network
     Vehicle_ID => "10";
     Network_ID => "012523";
     Vehicle_Type => Primary;
Radio.OpenReadingPipe (Network_ID);

While (1) //Infinite Loop
If Radio.Available ()
RxString = Radio.read ();
If Flag == 0
```

```
     //Request for Connection Received
If Data_is_Valid ()
     Radio.OpenWritingPipe (Network_ID);
     Radio.Write (Network_Credentials);
     Radio.OpenReadingPipe ();
     Radio.Write (Regular_Data);
Stop
```

The vehicle that has transmitted the request to connect will follow the below specified pseudo-code,

```
Start
SetTryLimit = 15;
SetTryTime = 15ms;
Registration_ID = "DL4CN0123";
Network_ID = "000001";
Radio.OpenWritingPipe (Network_ID);
Radio.Write ("Request_to_Connect"+ Registration_ID);
Radio.OpenReadingPipe (Network_ID);
If Radio.Available ()
     //Received Network Credentials
     Radio.read (Network_Credentials);
If Data_is_Valid ()
     Decode_Received_Data ();
If! Radio.Available ()
     //Establish Network
     Vehicle_ID => "10";
     Network_ID => "012523";
     Vehicle_Type => Primary;
     Radio.OpenReadingPipe (Network_ID);
Stop
```

3.3 Case 3: When a Special Vehicle Enters the Road

In case a special vehicle such as a PCR van, an ambulance, or a fire brigade comes to the road, a priority message will be transmitted to all the vehicles in the network about their presence making a request to provide a path for them. This will help in increasing the efficiency of emergency vehicles on the road.

Parity	TxAdd	ToV	FLAG	SV	SOS	TS	DATA	CHKSM	Parity

Fig. 8 Code word used in the network

4 Data Transmission and Code Word

This section discusses the actual message being transmitted or received in the network. When a vehicle comes under a network; it is given different responsibilities such as keeping itself safe, detecting theft, informing other vehicles about traffic congestion, and a lot of other information. To do so using single transmission, there needs to be a specific code word that is transmitted every time the device gets permission to transmit the signal.

A code word as shown in Fig. 8 is a specific format of the message that other networks can understand within the network. It must be secure, safe, encoded, and easy to decode for the receiver. The below image depicts the general idea of the code word being used in the project work for a decentralized vehicular network.

The different parts of the code word seen above are Parity bits, Transmitter Address (TxAdd), Type of Vehicle (ToV), Flag, Stolen Vehicle Indication (SV), Traffic Status Indication (TS), Data, Checksum and Save of Ships (SOS).

All the bits are discussed below in detail:

4.1 Transmitter Address

Transmitter Address or TxAdd is an 8 bit address allotted to all the vehicles in the format of Vehicle ID which is issued in the beginning when the vehicle is connected to the network along with other required credentials. This helps other secondary and primary vehicles to keep a track of the source of the message. Also, if a vehicle with vehicle ID 74 (converted from 8 bit binary to decimal) transmits a message, the vehicle with ID 75 will come to know that the next permission for transmission is allotted to him, and hence can prepare its message and deliver it on time without causing any delay in the network.

4.2 Type of Vehicle

Type of vehicle specifies what class of vehicle is transmitting the message. By class, we mean the primary, secondary, or special class of the vehicles as shown in Table 1.

The primary vehicle is the one that has established the network and has the rights to grant permissions to all the incoming requests of connection from other vehicles falling in its range.

Table 1 Type of vehicles

Code word	Type of vehicle
01	Primary vehicle
10	Secondary vehicle
11	Special vehicle

The secondary vehicles are the incoming vehicles in the network which have a connection through a network established by a primary vehicle. It doesn't have the right to grant permissions to other incoming requests of connection and hence can only transmit or receive the required information at specified or allotted time periods.

Special vehicles are a class of vehicles reserved specifically for emergency services such as ambulances, police or patrolling vehicles, fire forces, VIP vehicles, or any other official vehicles in an emergency. These vehicles will stay primary or secondary depending on their state; unless their siren is turned on specifying that the vehicle is on official duty.

4.3 Flag

The flag is a special and most crucial bit in the code word. It carries one bit of data specifying the type of message.

There are basically two types of messages:

(a) A normal communication message when the device is connected to a network. In this state, the flag bit remains low, i.e., 0.

(b) A connect request message which is transmitted by any vehicle when it is not connected to any network. In this condition, the flag bit gets high, i.e., 1, and such messages can only be decoded by the primary vehicles. For the secondary vehicles, there won't be any relevance of this message.

4.4 Stolen Vehicle Indication

In case the vehicle detects itself to be stolen, it needs to inform the other vehicles about its theft so that the original owner of the vehicle can be informed. In this case, the SV or Stolen Vehicle Indicator bit gets high, i.e., 1 and the contact number of the owner is stored in the DATA bits and is transmitted to other vehicles.

The vehicles receiving the message with Stolen Vehicle Indication high extracts the contact number of the owner and the GPS location of the stolen vehicle, and using the Bluetooth protocol, informs the mobile phone to send an SMS (Short Messaging Service) to the original owner about the theft and the location of the vehicle.

4.5 Save Our Ship

Assume a condition that the vehicle has met with an accident, or there is some other problem and there is a patient in your vehicle, you can press the SOS button and ask for help from your neighboring vehicles in the network. The other vehicles in the network will receive this as a priority message and the driver will be asked if he/she wants to help that car or not. In case, he/she is willing to help, the GPS coordinates in the DATA section will help him reach the person in need.

4.6 Traffic Status

One of the most crucial applications of this device is to get real-time data about traffic congestion on the road. Inspiration for handling the traffic situation has been derived from [6, 7].

Two bit data has been used for explaining the traffic congestion on the road. The traffic congestion is detected using the speed of the vehicle which is calculated using the accelerometer installed in the device. If the vehicle is traveling with a very high speed, it will be considered that the road is completely free, and vice versa.

Table 2 explains the speed being undertaken and the data being transmitted accordingly.

4.7 Data

Data section has specific applications in different conditions specified below:

1. For stolen vehicles, it will carry the owner's contact number
2. In the case of SOS, it will carry the GPS location of the vehicle
3. In case of connection request, it will carry the vehicle's registration ID
4. In case of granting permission of connection, it will carry the network's credentials.

The parity and checksum need not be handled by the developer. Because of the Enhanced Shock Burst architecture in nRF24L01+, it has the capability to handle

Table 2 Speed of vehicle vs. traffic condition

Speed of vehicle	Condition of traffic incurred	Data bits
0–20 kmph	Extreme traffic jam	11
20–40 kmph	Normal traffic jam	10
40–50 kmph	Normal vehicles on road	01
50 kmph and above	Road completely free	00

both parity and the checksum. The device itself built the parity bits along with the checksum and encapsulates the data accordingly.

5 Practical Implementation

This section discusses the algorithm developed for the project. Each vehicle will be installed with a black box containing the controller, the transceiver section, the Bluetooth module, antenna, and the accelerometer. All such boxes follow a common algorithm.

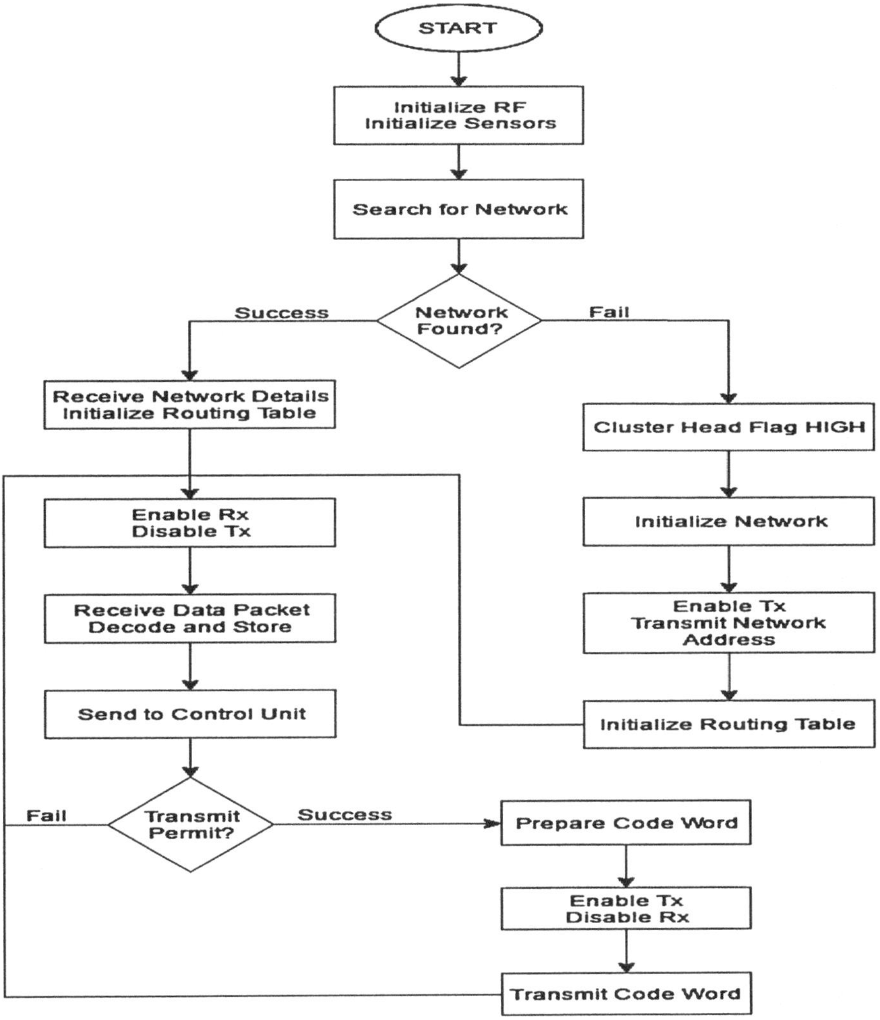

Fig. 9 Flow diagram

Figure 9 shows an approximate flow diagram of the algorithm used. Whenever a driver enters the vehicle, the very first thing the vehicle will do is recognizing the driver. This can be done either by facial recognition using image processing techniques or via the mobile that gets connected with the device via Bluetooth protocol. This recognition is used to detect whether the vehicle is being stolen or a registered driver is going to drive the vehicle. Another way to detect the same is analyzing the driving patterns, but since, in many cases, the driver is not always fixed for any vehicle, implementing the same will be difficult.

Once the driver is recognized and the vehicle comes up on road, it starts looking out for the nearby networks. For doing the same, the vehicle transmits a connection request on the address 00001. This address is known as the Generic Network ID and is reserved only for connections. If any device pings in the pipeline established for this particular network ID with Flag bit kept High and a valid registration ID in the Data section of the code word, the primary vehicle (if available) considers it as the connect request. A vehicle can make 15 tries of doing the same at an interval of 15 ms each for getting a response for the connect request sent by the device.

In case the device doesn't receive any connection from any other network in the surrounding, it infers that it is the only vehicle on the road, and it starts establishing its own network as shown in Fig. 10. While doing the same, it generates a network ID at which all communication will take place, a vehicle ID for itself which is fixed to be 10 for the primary vehicle, and type of vehicle being 0 that is primary vehicle.

With the establishment of the network, now this vehicle will be able to provide a network to other vehicles. If there is some other vehicle on road, and it sends a connection request on network address 000001, it will be received by the primary vehicle as shown in Fig. 11. To serve this connection request the primary vehicle will respond on the same address with flag 1, specifying all the network credentials required for the connection. Using the credentials, the other vehicle will also be able to communicate with the other vehicles in the network as specified in Fig. 12.

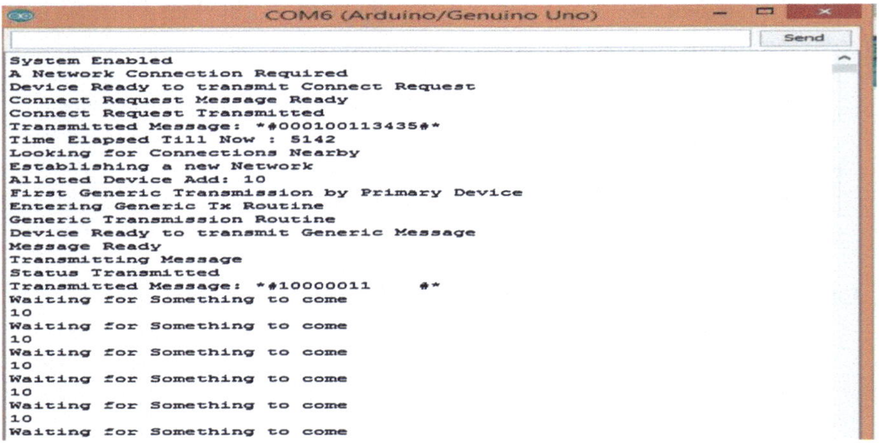

Fig. 10 Primary node getting established

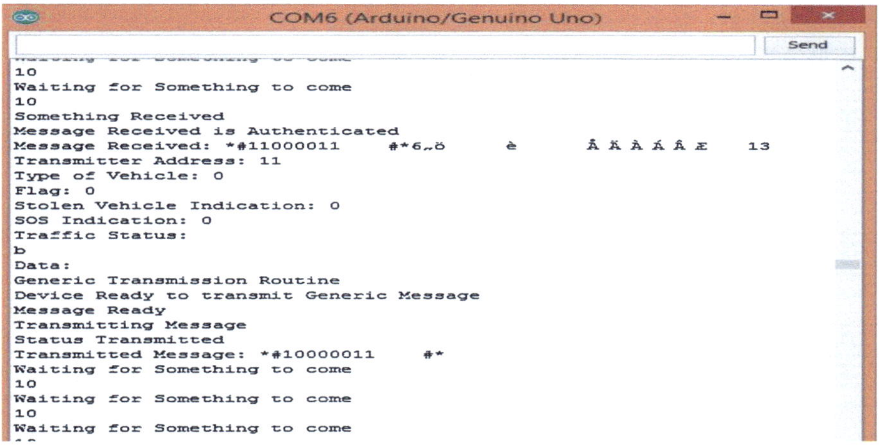

Fig. 11 Primary node receiving connection request

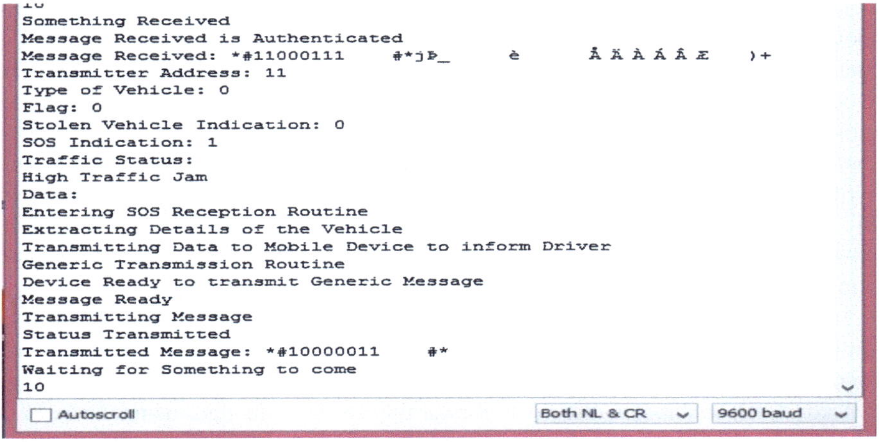

Fig. 12 Generic transmission

Now, after a network is established among multiple vehicles, in normal conditions, the first permit will be granted to the primary vehicle to transmit. It will only transmit its GPS location in the data section, and the traffic situation using TS bits. After its transmission being received by everyone, the next vehicle in the network being the one with vehicle ID 11 will get the transmission permission as shown in Fig. 13. It will transmit similar data according to its location, and a similar process will go on. The vehicles find out about their turn with the help of the transmitter address in the message received earlier. The vehicle having address incremented by one is the next having the permission of transmission in that particular network. Also, each vehicle is permitted only 30 ms of transmission and if the vehicle fails to transmit, the next vehicle will get permission for transmission.

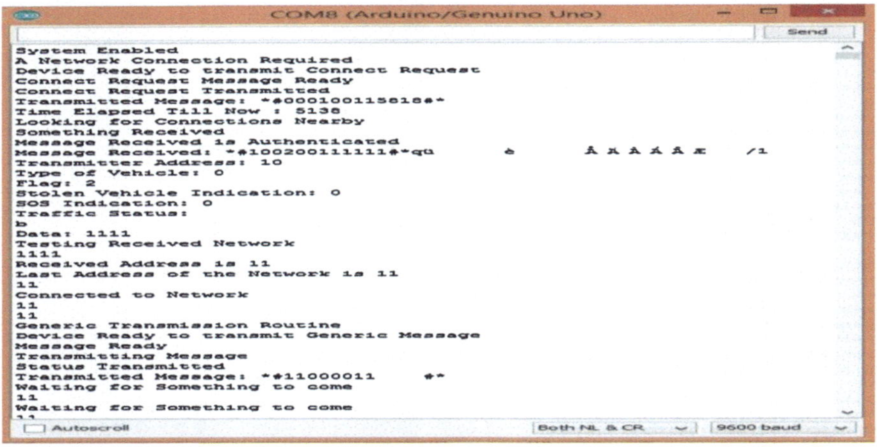

Fig. 13 Secondary node output

In the condition of a stolen vehicle, once the vehicle detects to be stolen by the above-stated methods, it gets the SV bit of the code word high, stores the owner's contact number in the data section, and transmits it in the network making a request to other vehicles to send an SMS to its owner about its present location. If any of the vehicles are connected to mobile via Bluetooth protocol, the device will feed the contact number, and using the mobile application, it will send an SMS to the owner of the vehicle. Since the stolen vehicle will keep updating the information the owner will keep getting updated information about the location of the vehicle. To keep it slow, the stolen vehicle will only transmit signals after losing ten turns of transmission so that the owner will get an update after every 15 min.

In case a person needs emergency service, in such a condition, the message transmitted with an SOS request will become a priority. In this state, the data section will contain the GPS location of the device. If any other driver accepts to help the person in need, it will be directed directly to the car through Google maps using the mobile application.

In case any special vehicle arrives on the road, it will be given special priority in case it satisfies the conditions for being a special vehicle. Such vehicles will make the type of vehicle bits to be '11' and will make it communicate with the network. In this state, it will keep transmitting a blank message which will be understood as a request to keep roads empty for emergency vehicles.

6 Results and Output

The device is initially tested using Serial Monitor Output. Assuming only one device is connected at a time, the following are the outputs received on the screen.

Outputs:

1. *Primary node establishing network as shown in Fig. 10.*
2. *The primary node receiving connect requests as shown in Fig. 11.*
3. *Code word transmission and reception as shown in Fig. 12.*
4. *Secondary device transmitting connect request as shown in Fig. 13.*

7 Future Prospects

The future direction in the presented work involves the implementation of the accelerometer to measure speed, including traffic signals for helping emergency vehicles and detecting violation of rules, easing out the work of traffic inspectors.

Further improvements could be made to implement image processing and at a very low-cost solution for detecting thefts.

A mobile application for smartphones needed to be developed to get the GPS location and SMS facilities. This mobile application will be working in collaboration with Google Maps making navigation easier for everyone.

References

1. Liu, Z.-P., Zhao, G.-L.: Short range wireless data transmission based on nRF24L01. Appl. Sci. Technol. **03**, (2008)
2. Zhurong, C., Chao, H., Jingsheng, L., Shoubin, L.: Protocol architecture for wireless body area network based on nRF24L01. In: Proceedings of IEEE International Conference on Automation and Logistics, ICAL 2008 (2008)
3. Weder, A.: An energy model of the ultralow power transceiver nRF24L01 for wireless body sensor networks. In: Second International Conference on Computational Intelligence, Communication Systems and Networks, Liverpool, pp. 118–123 2010
4. Nordic Semiconductors: nRF24L01 ultra low power 2.4GHz Transceiver IC. http://www.nordicsemi.com/eng/Products/2.4GHz-RF/nRF24L01
5. SparkFun. nRF24L01 Preliminary Product Specification. https://www.sparkfun.com/datasheets/Components/SMD/nRF24L01Pluss_Preliminary_Product_Specification_v1_0.pdf
6. Bando, M., Hasebe, K., Nakamaya, A.: Dynamical model of traffic congestion and numerical solution. Phys. Rev. E. **51**(2), 1035–1042 (1995)
7. Arnott, R.: ErenInci: an integrated model of downtown parking and traffic congestion. J. Urban Econ. **60**(3), 418–442 (2006)

Design of Auto-Braking System for Accident Prevention and Accident Detection System Using IoT

Gitanjali Mehta, Manoj Singh, Shubham Dubey, Uzair, and Yogesh Mishra

Abstract In recent times, frequencies of accidents have increased considerably. This is because of an increase in the number of vehicles, carelessness of drivers, and over speeding. Over speeding is the main reason for increase in the number of accidents. In this work, the primary concern is to decrease the impact of collision, and after that communicating with the nearby hospital for providing necessary support to the victims. According to data provided by the Ministry of National Highway, most of the deaths occurred because of not getting help in crucial times or not getting an ambulance service in time. Our main aim is to communicate with the nearest hospital through GPS and help the victims. Our work is divided into two main parts. One is sensing and communication part. Other is the braking part which has three steps. When the distance of the vehicle from the obstacle is more than 30 m then the system is disabled. If the distance becomes less than 30 m then a warning is generated by the system for the driver to apply brakes. If the distance is further reduced and becomes less than 4 m understanding that the driver has lost control over the vehicle, control is fully transferred to the braking system and plugging braking is used to stop the vehicle instantly to reduce the impact of a possible collision.

Keywords Accident prevention · Auto-braking system · Accident detection · Internet of Things

1 Introduction

1.1 Overall Description

India is one of the most populated countries, hence the population explosion has made a direct impact in the market of automobiles. The cause of the accidents is mainly due to the quality of roads and the unavailability of new technologies in vehicles. Transportation is the basic need for daily life and to reduce the severity of

G. Mehta (✉) · M. Singh · S. Dubey · Uzair · Y. Mishra
School of Electrical, Electronics and Communication Engineering, Galgotias University, Greater Noida, India

D. Gupta et al. (eds.), *Smart Sensors for Industrial Internet of Things*, Internet of Things, https://doi.org/10.1007/978-3-030-52624-5_7

accidental impact these technologies should be updated. Many people face various problems in their day-to-day life because of a lack of available traveling resources. The government is not taking the required decision and lacking the appropriate laws of traffic so that common men and children are not able to travel in secure vehicles [1–3]. So, for the betterment of the society, we came up with the research work "Design of Auto-Braking System for Accident Prevention and Accident Detection System using IoT".

There are many applications in the market to provide safety in vehicles but they are not up to the mark. In our work, we have developed such a technology which can prevent accidents and in case of failure which is evitable there is an alarming system built in the setup which will alarm the nearby hospital, police stations, and the relatives of the victim by sending messages to registered contact number [3, 4].

1.2 Survey

We use transportation to do many of our daily life works but it can create the worst scenarios and even kill people through accidents. In 2008, India ranked fourth in fatal injuries caused by road accidents and the age group which is more involved in these injuries is 15–29. In the absence of required actions, traffic crashes will reach the toll of death of around 1.9 million people annually by 2020.

1.3 Accident Detection

When an accident occurs, we are going to communicate with the nearby hospital to provide the necessary support to the victim. According to statistics from the Ministry of National Highway, accidents are classified into three types. In the first type, the accident is so severe and the victim dies on the spot. In the second type, the victim is seriously injured and can be safe if proper health facility is provided to him on time. In the third type, causality is not severe and takes time to recover but can be recovered. So our prime concern is on the second type and we are focused to provide them help on time thus providing a signal to the nearby hospital using IOTs [5, 6].

1.4 Motivation

We get the motivation for this work from fall detection technology in Apple watch. Similar technology is used by Apple watch, which detects when there is a hard fall or there is a severe impact. An emergency message is popped up in which we can ask for emergency help if we are seriously injured, or can click on "I M OK", to disable the system. If any option is not chosen then the watch will wait for a few seconds and automatically send the signal to relatives or nearby hospitals. Recently an incident occurred in San Francisco in which a person falls from the terrace and is not able to move, then watch sends signal automatically and the person is rescued.

1.5 Braking Mechanism

In this part, we are concerned about reducing the impact of collision, as by reducing the impact we can decrease the severity of accidents. According to statistics from the Ministry of Highway, Government of India, most of the accidents occurred due to over speeding, so by reducing the speed of vehicles at the last moment we can definitely reduce the casualties. According to statistics, many accidents occurred due to panic of the driver when an accident is going to occur, and he is unable to apply brakes or lose control over vehicles [7, 8]. So we also need to sense the distance between vehicle and obstacle and give warning to the driver to apply brakes (Fig. 1). When the distance is further reduced motor driver circuit comes into picture and decelerates the motor. When the distance is less than 4 m brakes are applied automatically.

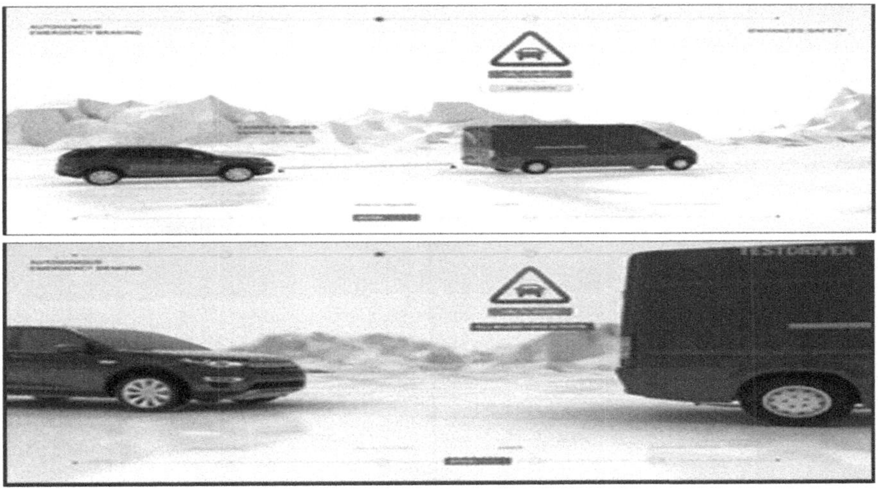

Fig. 1 Braking mechanism to prevent collision of two vehicles

2 Methodology

2.1 Block Diagram

Figure 2 shows the block diagram of the proposed system. The Arduino Uno is a microcontroller board based on the Microchip ATmega328P microcontroller. Arduino has made a huge impact in the electrical and electronics world. In this work, it is used to detect the input from the piezoelectric sensor and send the signal through GPS. A separate 9 V portable power supply powers the Arduino board or we can use the supply of the vehicle [8].

GPS is a global positioning system used to detect the location of anything on the earth. It uses longitude and latitudinal coordinates to find the position of the device. There is a transmitter and a receiver in the GPS module in which the transmitter transmits the signal to satellite and the receiver receives the coordinates. In this project, GPS is used to communicate with nearby hospitals.

A piezoelectric sensor is a type of transducer which takes pressure as input and converts it into electrical signal. In this system when there is a collision between two vehicles a piezoelectric sensor is used in the front part which detects the collision and sends a signal to Arduino.

Internet of Things is abbreviated as IoT. In the present time every device can be connected with each other with the help of IoT, data is stored in the cloud so that our time can be saved. IoT is almost used everywhere in today's world and after Industry 4.0 even large machines can be controlled from anywhere around the world [5, 6].

2.2 Flowchart

Figure 3 shows the flowchart of the process involved in the designed system.

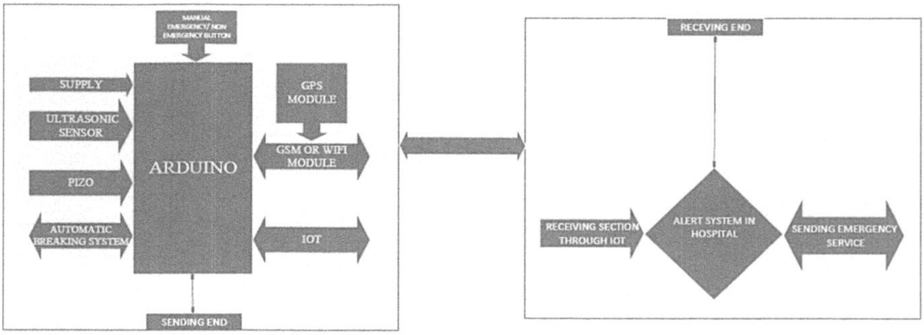

Fig. 2 Block diagram of the system

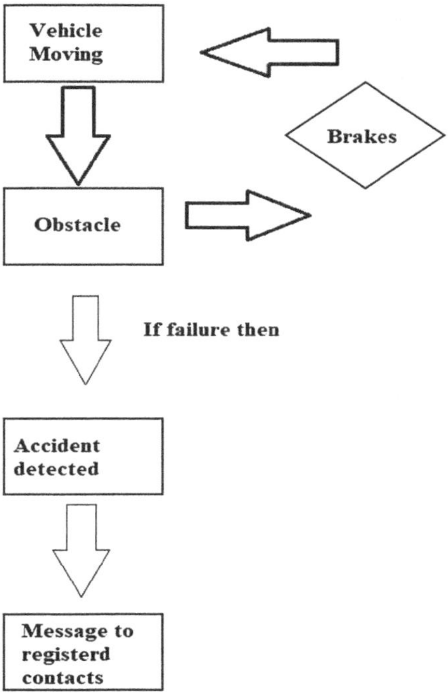

Fig. 3 Flowchart of the process involved in system design

2.3 *Software Tools*

MATLAB: Matrix Laboratory is a computing environment and mathematical programming language invented by MATHWORKS. It allows manipulation of data and functions. It is used to implement algorithms, create a user-friendly interface.

Multisim: It is developed on the BERKELEY SPICE software simulation. It is an electronic schematic maker and simulation program. It is used to make circuits and to do the simulation.

Proteus: It is a software tool suite developed for the designing of circuits. It generates digital blueprints of the required circuit.

3 Component Description

3.1 *Arduino*

Arduino is an open-source device, it has various types. In this we are using Arduino Uno. It uses its own Arduino language to compile the Arduino and works on that language to compile the Arduino. But it can be programmed in various other languages also like C, C++, Python, and Java. It consists of 28 pins, in the pin

Fig. 4 Arduino Uno Board

Fig. 5 Ultrasonic sensors

configuration it has 13 digital input-output pins which can be used for either input or output. It has six analog pins that can be also used for input and output port, and it works on 5 V supply, it consists of three GND terminals (Fig. 4) [8].

3.2 Ultrasonic Sensors

Ultrasonic sensor (Fig. 5) plays a major role in measuring the distance between two vehicles or any obstacles. Ultrasonic sensor uses high-frequency signals or waves which cannot be heard by normal human ears. Sound wave is emitted by ultrasonic sensors which strike with obstacles and revert, which calculates the distance between the vehicles and obstruction. Ultrasonic sensors are used in submarines, aircrafts, and radars. It consists of two parts, i.e., transmitter and receiver. The transmitter transmits the signal whereas the receiver collects the reflected signal to calculate the

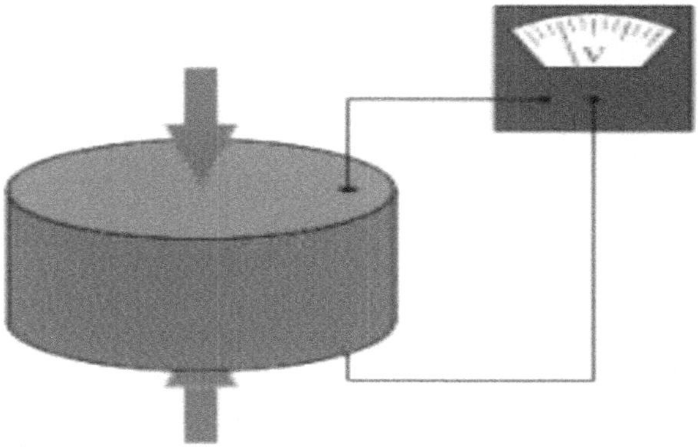

Fig. 6 Piezoelectric sensor

distance. The formula for calculating the distance is: Distance $d = 1/2 \times t \times v$ where d is the distance travelled, t is the time taken between the emission and reception, and v is the speed of sound.

3.3 *Piezoelectric Sensor*

A piezoelectric sensor (Fig. 6) is a type of transducer which takes pressure as input and converts it into electrical signal. In this system when there is a collision between two vehicles a piezoelectric sensor in front or back part of the vehicle detects the collision impact and sends signals to Arduino.

3.4 *Motor Driver IC L293D*

L293D Motor Driver IC shown in Fig. 7 is one of the important components used in braking. Motor driver circuit is used for controlling the speed of the motor. When the distance between vehicle and obstacle is less than 30 m then motor driver comes into picture and motor slightly decelerates and a warning is generated by Arduino. When the distance is further reduced final warning is generated and motor further decelerates. In the final step when the distance is less than 4 m then the driver circuit fully takes control and applies plugging brakes to stop the vehicle.

L293D Motor Driver IC consists of 16 pins: 4 output pins, 2 enable pins, 4 input pins, 2 Vss, and 4 GND. Input pins are connected with Arduino to take the signal from it. Output pins are used for controlling the speed of the motor. It works on 12 V and other terminal is used for providing voltage to motor (Fig. 8).

Fig. 7 L293D Motor Driver IC

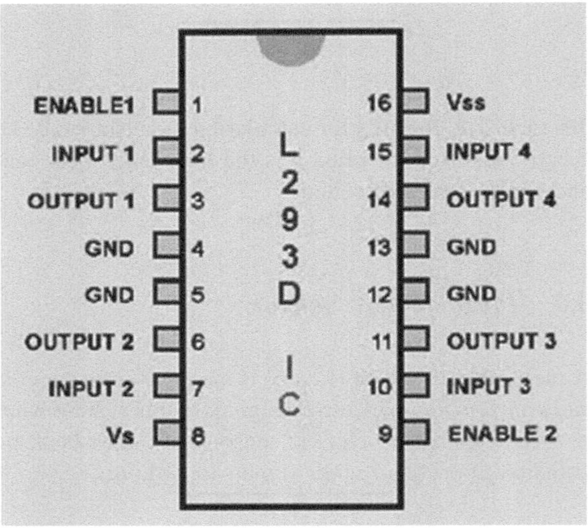

Fig. 8 Pin configuration of L293D Motor Driver IC

4 Design Process

4.1 Design Steps

While driving, the sensors monitor the distance between the vehicles. Figure 9 shows vehicle with sensors and auto-braking system. Working of the braking system can be divided into four parts:

1. When the distance between vehicles is more than 30 m, system is not activated or it is disabled.
2. When the distance is less than 30 m braking process will start. Motor starts decelerating or a little brake is applied by the system, and warning is given to the driver.
3. When the distance is less than 10 m further deceleration of motor takes place and extra warning is given by the system to the driver.

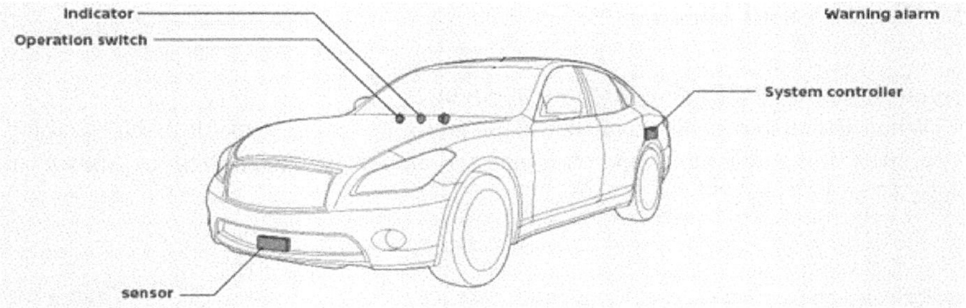

Fig. 9 Vehicle with sensors and auto-braking system

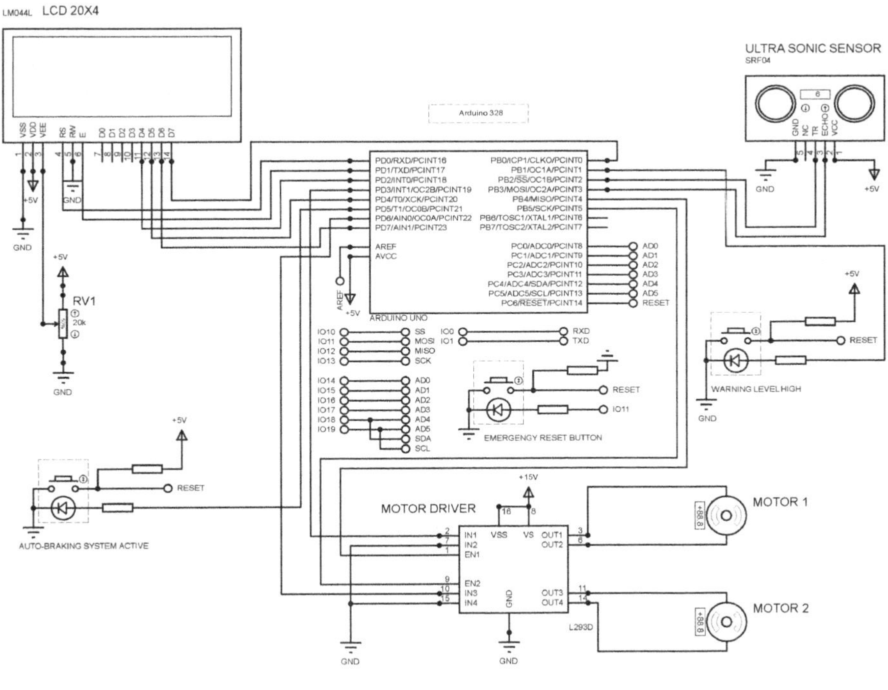

Fig. 10 System configuration

4. When the distance is less than 4 m whole system is fully automatic and there is no control of driver in vehicles. An emergency brake is applied by the system. This will reduce the damage caused by the collision of two vehicles.

4.2 System Configuration

Figure 10 shows the system configuration.

5 Results and Discussion

Figure 11 shows the Proteus simulation circuit.

When the distance between the obstacles is less than 4 m, the Auto-Braking System will be activated and automatic brakes will be applied as shown in Figs. 12 and 13.

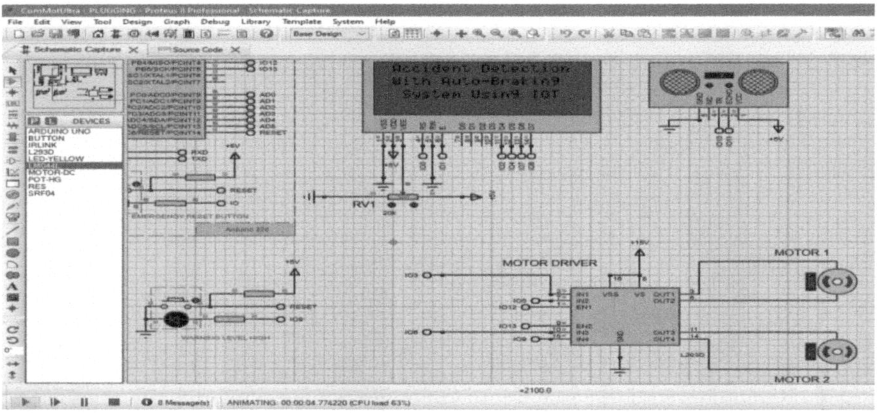

Fig. 11 Simulation circuit

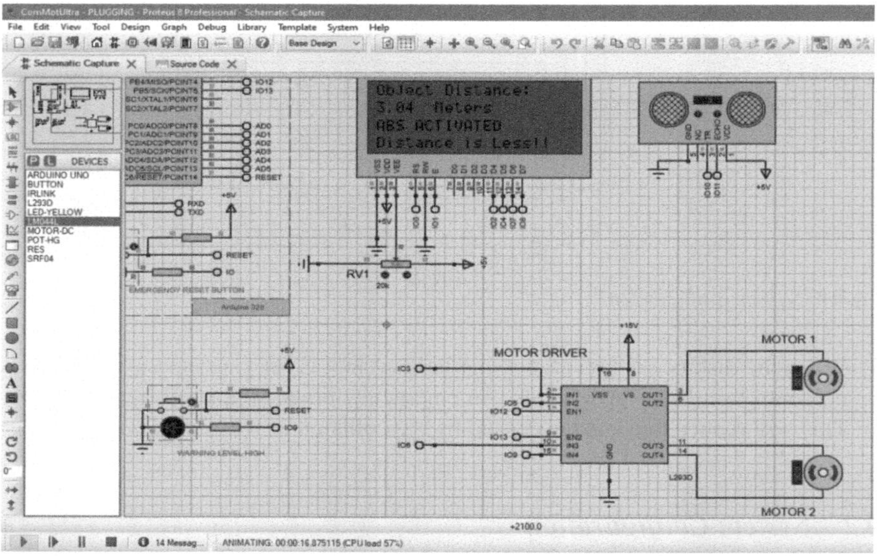

Fig. 12 Simulation result when obstacle is less than 4 m and automatic brakes applied

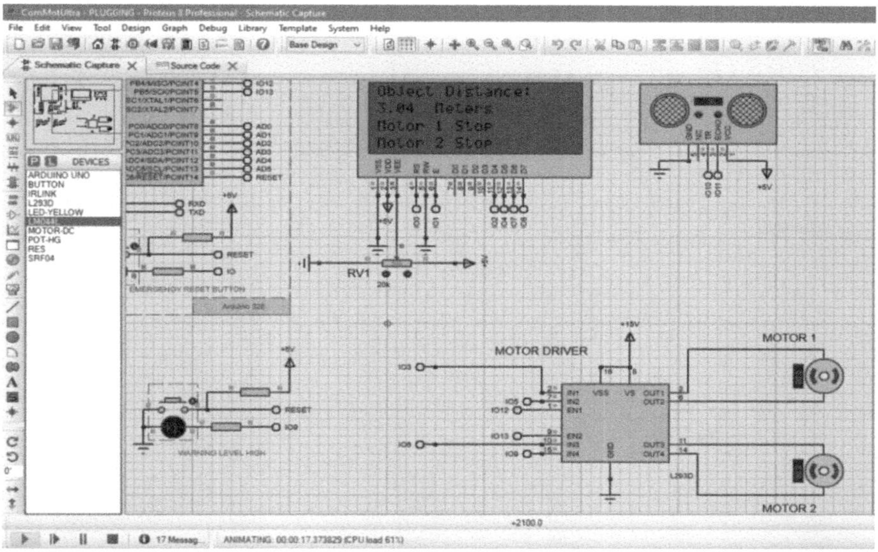

Fig. 13 Simulation result when automatic brakes applied to stop the car

When the distance between the obstacles is between greater than 4 m and less than 10 m, the Auto-Braking System will generate a signal to motor driver to limit the speed to the desired value and alert the driver of the car also as shown in Figs. 14 and 15.

When the distance between the obstacles is between greater than 10 m and less than 30 m, the Auto-Braking System will generate a signal to motor driver to limit the speed to desired value as shown in Fig. 16.

When the distance between the obstacles is greater than 30 m, the Auto-Braking System will be active but will not generate any signal to the motor driver (Fig. 17).

6 Review and Comparison

The objective of the proposed work is to prevent or reduce the severity of a collision. This safety feature will reduce those accidents which can be fatal or at least the impact can be reduced up to a maximum extent. The speed of the vehicle is automatically reduced so as to reduce the impact of collision. Even after automatic emergency brake if there is a collision, then the sensing and communication part takes place. Piezoelectric sensor senses the impact and Arduino detects the collision and sends the signal to the nearby hospital through GPS.

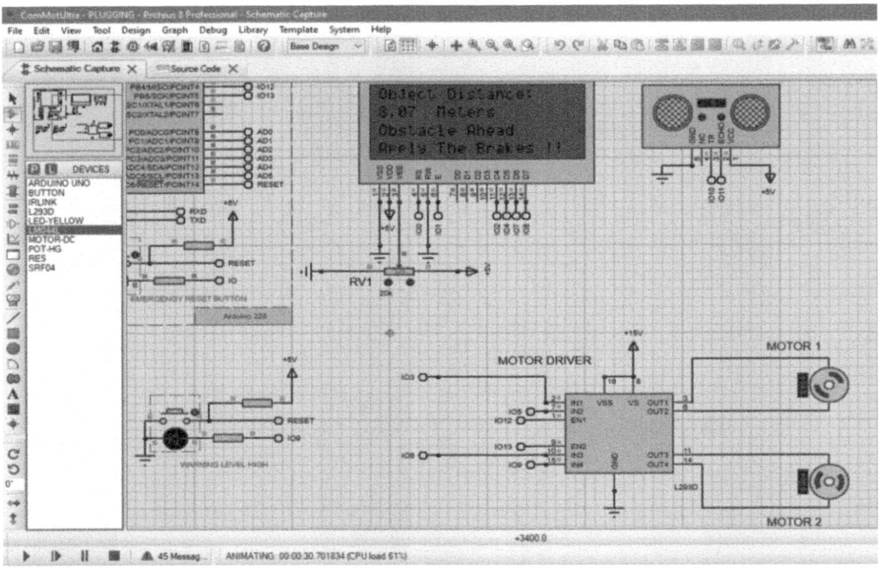

Fig. 14 Simulation showing alert signal when obstacle is between 4 and 10 m

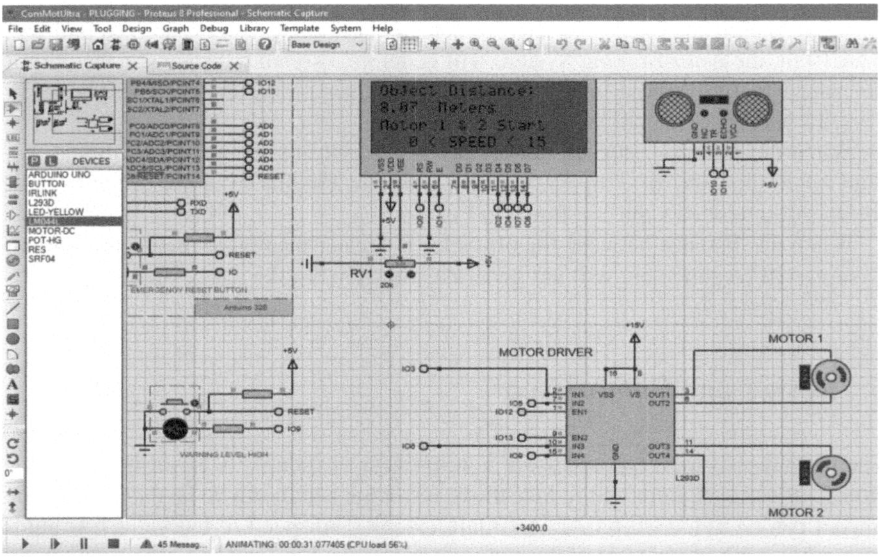

Fig. 15 Simulation signaling the driver to limit the speed to the desired value

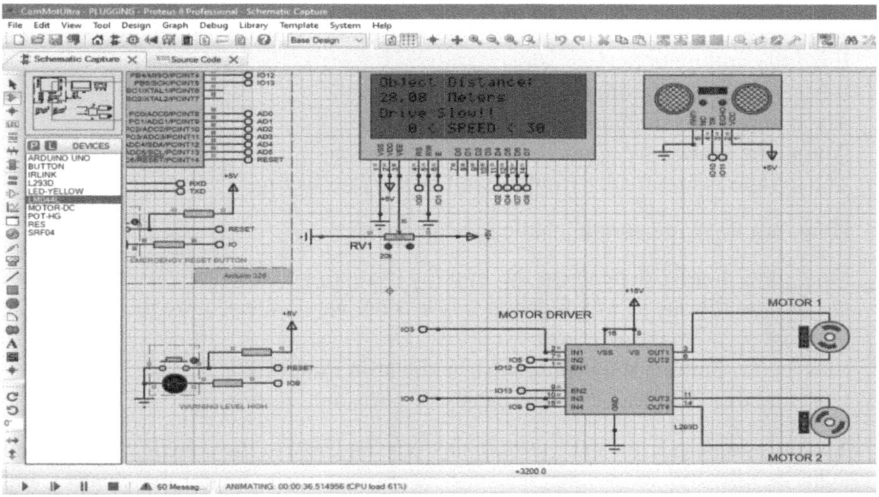

Fig. 16 When the distance between the obstacles is between 10 and 30 m

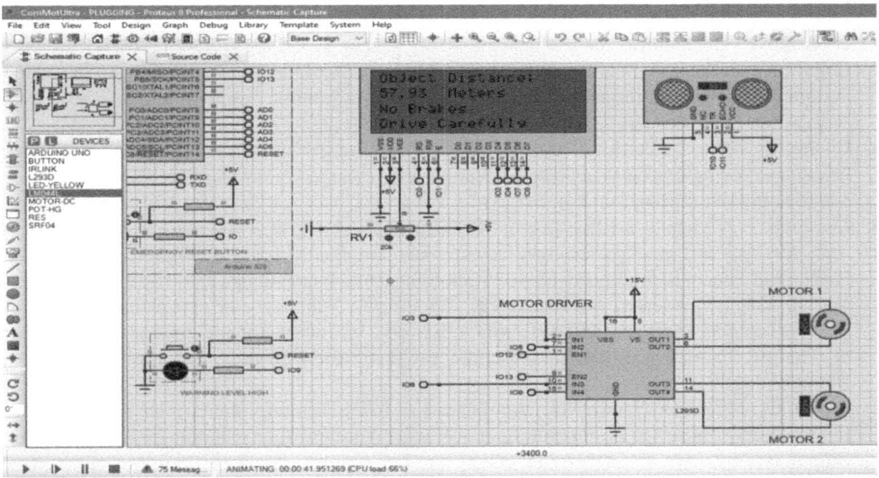

Fig. 17 When the distance between the obstacles is greater than 30 m

The advantages offered by the proposed system are:

- It is fully automatic.
- It has a simple electronic control unit.
- The components used are cost-effective.
- Automatic brakes are applied to prevent collision.
- The system is helpful for front as well as back collisions.

However, there are certain challenges involved:

- We cannot totally depend on the braking system, because sometimes there are various conditions in which distance is less but we do not require the emergency braking.
- Emergency braking may cause wear and tear in motor parts.
- Dust and dirt can reduce the efficiency of ultrasonic sensors.

The proposed system can be used in any type and size of the vehicle, AI-operated vehicles as effective assistance, and in concept cars for reducing the chances and impact of vehicle crash and providing immediate assistance in case of any casualty.

References

1. Tian, D., Zhang, C., Duan, X., Wang, X.: An automatic car accident detection method based on cooperative vehicle infrastructure systems. IEEE Access. **7**, 127453–127463 (2019)
2. Celesti, A., Galletta, A., Carnevale, L., Fazio, M., Ĺay-Ekuakille, A., Villari, M.: An IoT cloud system for traffic monitoring and vehicular accidents prevention based on mobile sensor data processing. IEEE Sensors J. **18**(12), 4795–4802 (2018)
3. Chang, W., Chen, L., Su, K.: DeepCrash: a deep learning-based internet of vehicles system for head-on and single-vehicle accident detection with emergency notification. IEEE Access. **7**, 148163–148175 (2019)
4. Coelingh, E., Eidehall, A., Bengtsson, M.: Collision warning with full auto brake and pedestrian detection – a practical example of automatic emergency braking. In: 13th International IEEE Conference on Intelligent Transportation Systems, Funchal, pp. 155–160 (2010)
5. Verma, M.K., Mukherjee, V., Yadav, V.K., Mehta, G.: Planning and optimizing the cost of DGs for stability of green field distribution network. In: International Conference on Internet of Things: Smart Innovation and Usages, pp. 1–6. Birla Institute of Applied Sciences, Bhimtal (2018)
6. Mehta, G., Mittra, G., Yadav, V.K.: Application of IoT to optimize data center operations. In: International conference on computing, power and communication technologies, pp. 738–742. Galgotias University, Greater Noida (2018)
7. Pagadala, V., Rani, S., Priya, B.K.: Design and implementation of the prevention and analysis of the accident for automobiles. In: International Conference on Advances in Computing, Communications and Informatics, Bangalore, pp. 2283–2289 (2018)
8. Mahamud, M.S., Monsur, M., Zishan, M.S.R.: An arduino based accident prevention and identification system for vehicles. In: IEEE Region 10 Humanitarian Technology Conference, Dhaka, pp. 555–559 (2017)

IoMT with Cloud-Based Disease Diagnosis Healthcare Framework for Heart Disease Prediction Using Simulated Annealing with SVM

Kishore Kumar Kamarajugadda, Pavani Movva, Manthena Narasimha Raju, S. Anup Kant, and Satish Thatavarti

Abstract Internet of Medical Things (IoMT) interlinks a collection of intelligent sensors on the patient's body to observe and interpret multimodal health data, including the patient's physiological and psychological signals. The large amount of data produced by IoMT devices in medical application is examined on cloud by replacing the restricted memory as well as processing resources of handheld tools. In this study, an IoMT-based healthcare diagnosis model is introduced by the use of intelligent techniques. This paper proposes a new IoMT-based disease diagnosis healthcare framework for heart disease prediction using the BBO-SVM model. The proposed model involves the parameter tuning of SVM using the BBO algorithm. The validation of the proposed model takes place using a Statlog Heart disease dataset. The detailed experimental analysis strongly pointed out that the proposed BBO-SVM model has shown excellent results by attaining a maximum precision of 88.33%, recall of 87.60%, accuracy of 89.26%, *F*-score of 87.96%, and kappa value of 78.27%.

Keywords IoT · Feature selection · Support vector machine · Healthcare

K. K. Kamarajugadda (✉) · P. Movva
Faculty of Science and Technology, ICFAI Foundation for Higher Education, Hyderabad, Telangana, India
e-mail: kkishore@ifheindia.org; pavanimovva@ifheindia.org

M. N. Raju
Shri Vishnu Engineering College for Women (Autonomous), Bhimavaram, Andhra Pradesh, India

S. A. Kant
CVR College of Engineering, Ibrahimpatnam, Telangana, India

S. Thatavarti
Professor in Department of CSE, NSRIT, Sontyam, Vishakapatnam, Andhra Pradesh, India

D. Gupta et al. (eds.), *Smart Sensors for Industrial Internet of Things*, Internet of Things, https://doi.org/10.1007/978-3-030-52624-5_8

1 Introduction

The current proliferation of data as well as communication models and integrated techniques has led to develop a novel method named as IoT. This IoT allows peoples and objects from external world, data and virtual platforms to communicate with one another [1]. Several domains applying IoT as the main source of data acquisition unit from modern environments like intelligent vehicles, intelligent buildings, telemedicine, etc., as the portion of a developing digital community. As an extension in IoT, supported medical tools are assumed to be the well-known potential research fields [2]. The increasing expense of healthcare and existence of diseases throughout the world requires the conversion of healthcare from a hospital-oriented system to a patient-oriented environment. With the aim of disease management as well as the personal problem, it is presented with a model that applies the ubiquitous sensing potential of IoT devices to forecast the probabilities of capable disease from a patient.

At the same time, cognitive IoT (CIoT) and cloud computing (CC) are based on one another. CIoT interlinks a collection of intelligent sensors on the patient's body to observe and interpret multimodal health data, including the patient's physiological and psychological signals. By the combination of these IoT and CC, it becomes more effective at the remote site by offering continuous health data for doctors. A typical IoT and cloud-based healthcare technology is depicted in Fig. 1. IoT has been oriented by a virtual unrestricted potential and source of cloud to replace the technical limitations like memory, functions, and power. Besides, a cloud obtains several benefits from IoT by expanding the value to resolve the real-time objects and dispatch the number of novel facilities from a distributed and dynamic fashion. But,

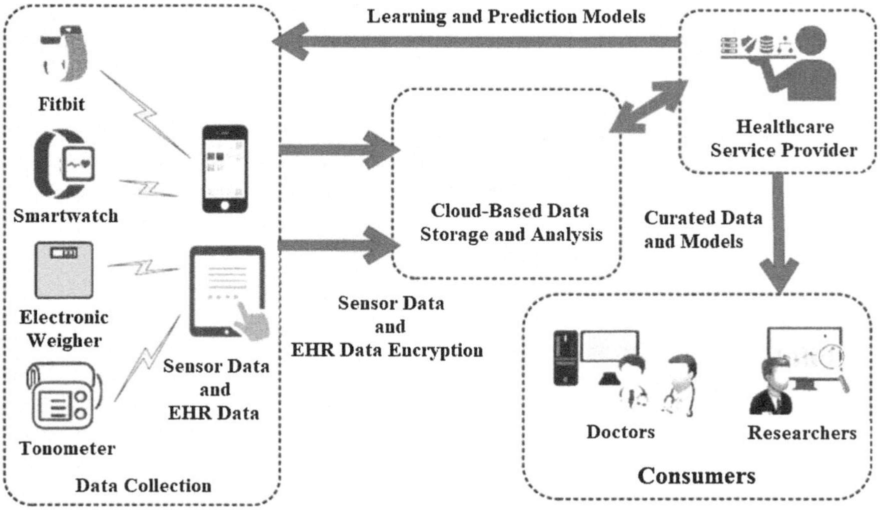

Fig. 1 General IoT and cloud-based healthcare framework

IoT centric-cloud structure could be advanced for developing novel domains and facilities in modern platforms [3]. In this method, cloud-centric IoT relied health diagnosing model has been projected with the application of computational science framework. There are few modern student communicative health models described for the IoT environment. With the application of the IoT medical system, the sequence of measurements is applied to gather the data in frequent modifications from health attributes across the given time and presence of anomalous state at the time of finite time. Furthermore, IoT devices as well as medical sensor values could be employed efficiently while analyzing the disease with a level of severity within a limited time interval.

Additionally, a fresh model which depends upon the IoT medical device's extensions is named as Body Sensor Network (BSN). In this model, the patient is observed with the help of diverse tiny-powdered as well as light-weight sensors. Furthermore, security needs in creating the BSN healthcare model are assumed in this work. And the background of IoT with its application form u-healthcare point of view. An ideal technique of IoT for u-healthcare has been developed. Several processes are evolved in solving the heterogeneity issues of data format in the IoT environment under the application of semantic data methods. Furthermore, resource-oriented data accessing technique (UDA-IoT) is developed for performing IoT data in a ubiquitous manner. Ref. [4] defined the advanced models to investigate the data gathered from wearable sensors in health monitoring. Ref. [5] explained the techniques to design m-health-based applications like website builder as well as applications developer to observe the patients' healthcare medical system. Thus, the IoT-based health observing system is to calculate the immediate health result like alcohol consumption and therapeutic effects.

Ref. [6] projected a Smart Hospital System (SHS) under the application of RFID, WSN, and smart phones. Such methods are combined with one another by applying IPv6 over the minimum-power wireless personal area network structure. Ref. [7] explained a model for important sign observing the human. This model estimates the pulse rate as well as body temperature from a distant location. Ref. [8] implied a Web middleware environment to link users to a physician using wearables, named as Eco Health. The main objective of the presented model is to enhance remote observations and patient diagnosis. Ref. [9] explained an interlinking approach for mobile health based on IoT. It has been established with technological creations to enhance the health monitoring system as well as patient tools with online facilities. Ref. [10] applied an adverse scenario to monitor the system with the help of context motion tracking for chronic disease individuals. This system is capable of analyzing the present condition of a patient with the use of contextual data and offers the required data by examining the regular habits of a patient. Ref. [11] developed a novel method termed as 4G health. It is depicted with multidisciplinary behavior of the dimension of a healthcare delivery model. Ref. [15] presented and executed a smart home relied environment, named as iHome Health-IoT. In this study, an IoMT-based healthcare diagnosis model is introduced by the use of intelligent techniques. This paper proposes a new IoMT-based disease diagnosis healthcare framework for heart disease prediction using the BBO-SVM model. The proposed model involves the parameter tuning of SVM using the BBO algorithm. The validation of the proposed model takes place using a Statlog Heart disease dataset.

2 The Proposed BBO-SVM Approach

2.1 SVM

Support Vector Machine (SVM) belongs to supervised classification or regression classification. But, it has been employed for classification problems. In the SVM method, every data is plotted as a point in *n*-dimensional space along with the measure of all features which is a specific coordinate. Followed by, the classification task is processed by exploring the hyper-plane which distinguishes two classes as depicted in Fig. 2.

2.2 BBO-SVM

Biogeography-Based Optimization (BBO) is defined as a population-based optimizing model which has been assumed as evolution and explored from predators and preys in diverse platforms. The experiments represent a simulation outcome attained with the help of BBO which may be a competitive with alternate population

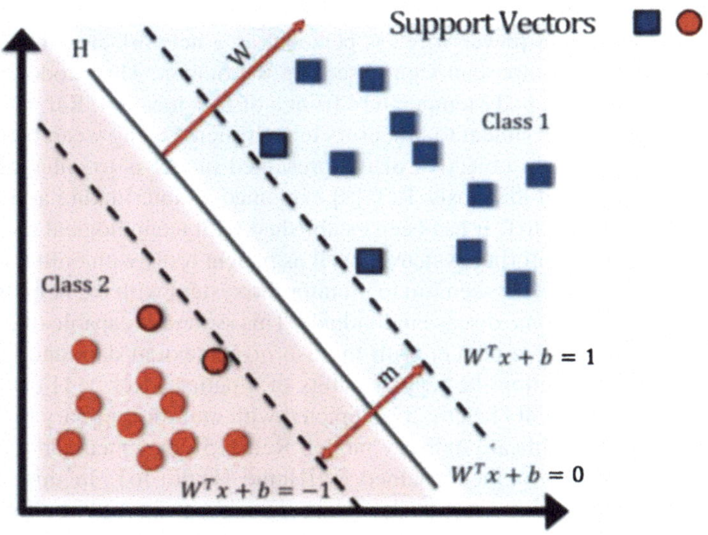

Fig. 2 Hyperplanes in SVM

relied models. It shows that an optimal performance is carried out by heuristic techniques like PSO, GA, and ACO on a few practical issues as well as standard functions. The phases of the BBO model could be defined in the following. Initially, BBO produces an arbitrary value for search agents termed as habitats, which denotes the vectors of variables in GA. Then, every agent is declared with emigration, immigration, and mutation values that promote features of a diverse environment. Also, a parameter named as HSI (Habitat Suitability Index) helps in calculating the fitness for all habitats. The maximum rate of HSI shows that habitat is applicable in the residence of biological species. Besides, a solution of BBO which has a higher value of HSI denotes that it is comprised of a result, whereas a solution with a minimum score of HSI implies a poor outcome. At the time of processing, a collection of solutions is retained for several rounds, and all habitat forwards and obtains habitants to and from various habitats which depend upon the immigration as well as emigration values that are adopted probabilistically. Random values of habitants undergo mutation. Finally, it develops a solution to get adapted by a learning process from neighbors in an algorithm format. The solution parameter is represented as Suitability Index Variable (SIV).

The function of BBO is comprised of two stages: migration and mutation. Initially, at the time of processing the migration phase, immigration (λ_k) and emigration (μ_k) values for all habitats apply the same method as shown in Fig. 3. Many numbers of habitants from a habitat improve the possibility of emigration and reduce the viability of immigration whereas in a mutation phase, the mutation operator in BBO distributes the habitants in a habitat. Unlike the mutation factor in GA, the mutation operator of BBO cannot be fixed in a random manner; it is based on a probability of species in all habitats.

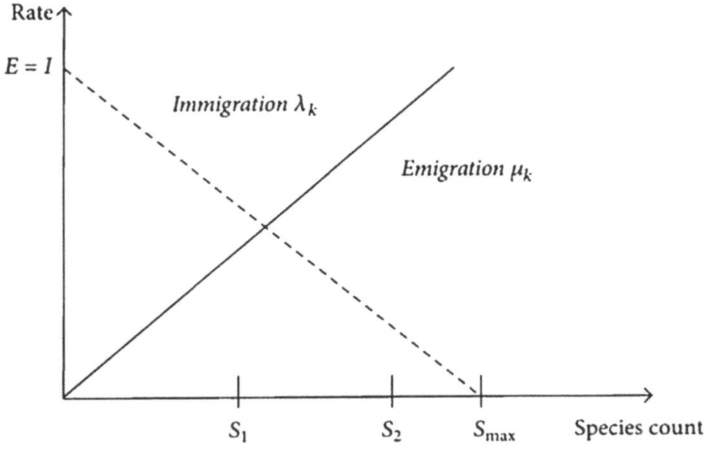

Fig. 3 Species model of a habitat

The numerical function of immigration (λ_k) and emigration (μ_k) is expressed as:

$$\lambda_k = I\left(1 - \frac{S_k}{S_{\max}}\right), \tag{1}$$

$$\mu_k = E\left(\frac{S_k}{S_{\max}}\right), \tag{2}$$

where I refers the high immigration value, E indicates the higher emigration score, $S_{\max}$ denotes the greater habitants, and S_k implies the habitant value of k.

The mutation for all habitats, which enhances the identification of BBO, is described as follows:

$$m(s) = m_{\max} \times \left(1 - \frac{P_n}{P_{\max}}\right). \tag{3}$$

Here $m_{\max}$ is assumed to be mutation explained by a user, $P_{\max}$ represents the higher mutation probability of every habitat, and P_n indicates the mutation possibility of nth habitat that is attained by

$$\dot{P}_n = \begin{cases} -(\lambda_n + \mu_n)P_n + \mu_{n+1}P_{n+1}, & n = 0; \\ -(\lambda_n + \mu_n)P_n + \mu_{n+1}P_{n+1} + \lambda_{n-1}P_{n-1}, & 1 \le n \le S_{\max} - 1 = 0; \\ -(\lambda_n + \mu_n)P_n + \lambda_{n-1}P_{n-1}, & n = S_{\max}. \end{cases} \tag{4}$$

The overall performance of the BBO method is mentioned in Algorithm 1; here $I : \phi \rightarrow \{H^n, \text{HSI}^n\}$ initiates an eco-system of habitats and determines every adjacent HSI and $\Gamma = (n, m, \lambda, \tau, \Omega, M)$ which is a transition operator that alters an ecosystem from a single optimizing process. The units of the six-tuple could be described in the following: n refers the count of habitats; m denotes a number of SIVs; λ signifies an immigration value; τ is the emigration score; Ω refers a migration factor; and M implies a mutation operator.

This model presents by applying the BBO model to find optimal measures for SVM and compute the applicable subset at the same time. Also, it is projected with a BBO method for parameter determination in SVM, called as BBO-SVM. The main aim of BBO-SVM is while searching an optimal parameter rates for SVM, to increase the accuracy value in dividing the testing dataset. It is similar to the optimization issue which seeks for higher solution. While resolving the maximization issue, when possible solutions increase the measure of an objective function, then it is approved in the form of recent viable solution, which has been represented as a starting point for a novel search for upcoming possible solution. When a future solution has minimum accuracy value in classifying recent solution, then a Metropolis acceptance rule has been applied to select a replacement of present solution with the future possible solution. The process of the presented BBO-SVM technique is explained briefly in the following. The primary solution X has been produced randomly. For all iterations, applying X as initial point, a random vector is used to choose the future solution Y. Suppose obj(X) is an estimation of objective function rate of X and ΔE is the variation from obj(X) and obj(Y), which defines that $\Delta E = \text{obj}(Y) - \text{obj}(X)$. A probability of interchanging X with Y,where X is a recent solution and Y implies an upcoming solution, is provided as $\Delta E < 0$, by $e^{\Delta E/T}$. This probability can be achieved by producing an arbitrary value $r \in [0, 1]$, and substituting the solution X with Y if $r < e^{\Delta E/T}$. Simultaneously, when $\Delta E \geq 0$, then a probability for replacing X with Y is 1 whereas $\text{obj}(X) \geq \text{obj}(X_{opt})$, then $X_{opt} = X$, and temperature T is cooled. The BBO-SVM stops while iterations I_{iter} are attained. The best parameter rates for SVM as well as an advantageous subset of features which is acquired from X_{opt}.

The BBO-SVM model needs a primary possible solution. When the viable solution has been produced arbitrarily, massive search process might be required to enhance the solution, due to the ineffective primary solution. In order to eliminate this issue, it is applied with a greedy model. The value spaces of attributes C and Υ are classified into 6 areas, correspondingly. The edges of the regions are applied as viable solutions to compute the primary solution, by generating partial solutions to be sampled. Optimal solutions from other solutions are applied with possible solution from BBO-SVM. When diverse solutions are composed with the same classifying accuracy values, then a minimum valued C has been selected. While C values are similar, then a lower Υ score is selected, respectively.

3 Performance Validation

3.1 Dataset Used

Table 1 shows the details of the dataset used to assess the results of the BBO-SVM model. The dataset is a Statlog heart disease dataset, which comprises a set of 270 instances under the presence of 13 features. Besides, a set of 44.5% of samples comes under the present class and 55.5% of samples comes under the absent class. The frequency distribution of the attributes that exist in this dataset is shown in Fig. 4.

Table 1 Dataset description

Description	Values
Number of instances	270
Number of features	13
Number of class	2
Percentage of present samples	44.50%
Percentage of absent samples	55.50%
Data sources	[11]

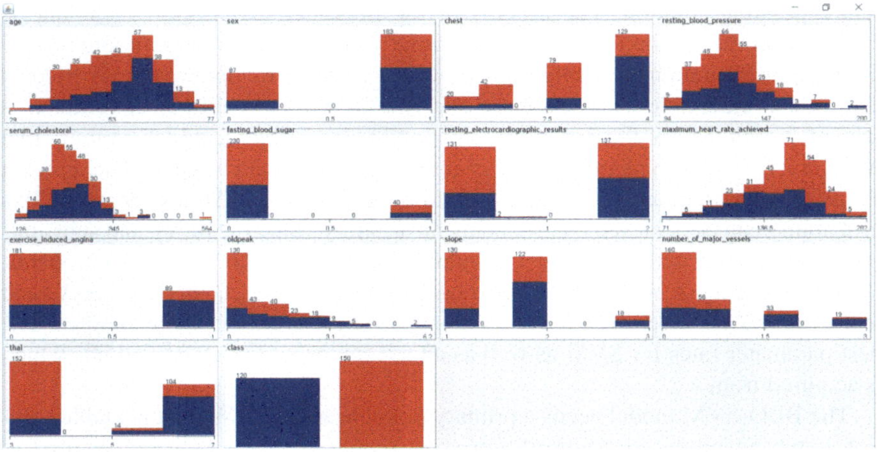

Fig. 4 Frequency distribution of heart disease dataset for all attributes

Table 2 Performance evaluation of different classifiers on heart disease dataset

Classifiers	Precision	Recall	Accuracy	*F*-Score	Kappa
BBO-SVM	88.33	87.60	89.26	87.96	78.27
CFS+J48	81.11	81.11	81.11	81.00	61.52
J48	73.33	73.95	76.66	73.64	52.71
Random Tree	74.16	72.95	76.29	73.55	52.08
RBF Network	80.33	82.91	84.07	81.86	67.67

3.2 *Results Analysis*

Table 2 and Figs. 5 and 6 show the classifier results analysis of diverse models on the applied dataset. On assessing the results in terms of precision, it is shown that the J48 model has shown the least classification outcome with a minimum precision value of 73.33%. Next to that, the RT model has exhibited slightly better performance with a precision value of 74.16%. Along with that, the RBF Network model has provided a moderate classifier outcome with a precision value of 80.33%. Simultaneously, manageable classifier outcome has been provided by the CFS+J48 model with a precision value of 81.11%. On all the above, the BBO-SVM model has attained a maximum precision value of 88.33%.

On assessing the results in terms of recall, it is shown that the Random Tree model has shown the least classification outcome with a minimum precision value of 72.95%. Next to that, the RT model has exhibited slightly better performance with a precision value of 73.95%. Along with that, the CFS+J48 model has provided a moderate classifier outcome with a precision value of 81.11%. Simultaneously, manageable classifier outcome has been provided by the RBF Network model with a precision value of 82.91%.

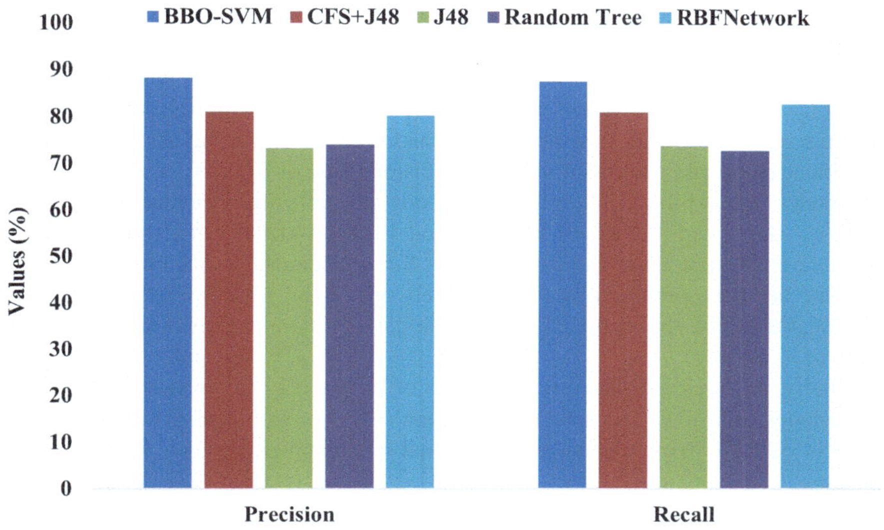

Fig. 5 Precision and recall analysis of diverse models

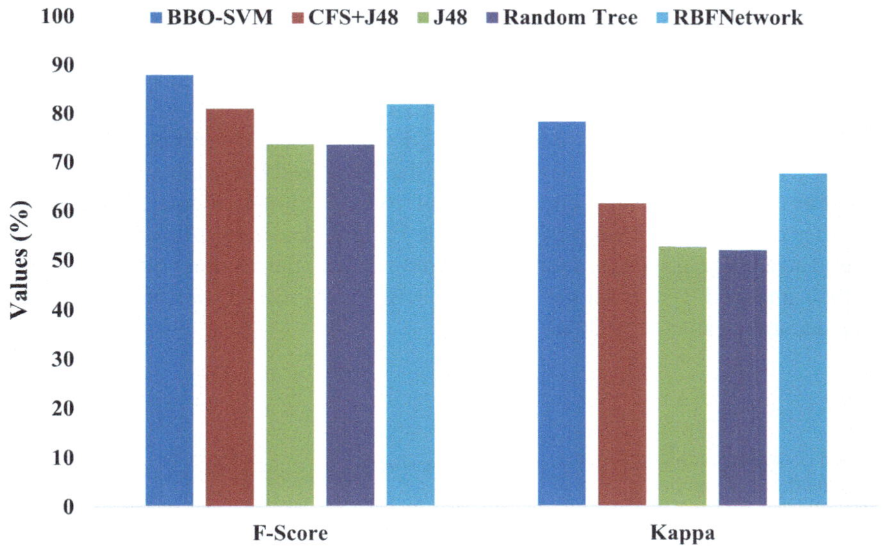

Fig. 6 *F*-score, Kappa value analysis of diverse models

On all the above, the BBO-SVM model has attained a maximum precision value of 87.60%. On assessing the results in terms of accuracy, it is shown that the Random Tree model has shown the least classification outcome with a minimum precision value of 76.29%. Next to that, the RT model has exhibited slightly better performance with a precision value of 76.66%. Along with that, the CFS+J48 model has provided a moderate classifier outcome with a precision value of 81.11%. Simultaneously, manageable classifier outcome has been provided by the RBF Network model with a precision value of 84.07%. On all the above, the BBO-SVM model has attained a maximum precision value of 89.26%.On assessing the results in terms of *F*-score, it is shown that the Random Tree model has shown the least classification outcome with a minimum precision value of 73.55%. Next to that, the RT model has exhibited slightly better performance with a precision value of 73.64%. Along with that, the CFS+J48 model has provided a moderate classifier outcome with a precision value of 81.00%. Simultaneously, manageable classifier outcome has been provided by the RBF Network model with a precision value of 81.86%. On all the above, the BBO-SVM model has attained a maximum precision value of 87.96%.

Finally, on assessing the results in terms of kappa, it is shown that the Random Tree model has shown the least classification outcome with a minimum precision value of 52.08%. Next to that, the RT model has exhibited slightly better performance with a precision value of 52.71%. Along with that, the CFS+J48 model has provided a moderate classifier outcome with a precision value of 61.52%. Simultaneously, manageable classifier outcome has been provided by the RBF Network model with a precision value of 67.67%. On all the above, the BBO-SVM model has attained a maximum precision value of 78.27%.

Table 3 and Fig. 7 have undergone a comparative study of the proposed method with the recently proposed models. On assessing the results in terms of accuracy, it is shown that the Pruned J48 Decision Tree model has shown the least classification outcome with a minimum accuracy value of 73.79%.

Next to that, the RT model has exhibited slightly better performance with an accuracy value of 74.29%. Along with that, the J48 model has provided a moderate classifier outcome with an accuracy value of 76.66%. Simultaneously, manageable classifier outcome has been provided by the Extreme Learning Machine, Rotation Forest Ensemble Classifiers, and CFS+J48 models with the accuracy values of 80%, 80.49%, and 81.11%. On all the above, the BBO-SVM model has attained a maximum accuracy value of 89.26%.

Table 3 Comparison with recent methods for applied dataset in terms of accuracy

Classifiers	Accuracy
BBO-SVM	*89.26*
CFS+J48	81.11
J48	76.66
Random Tree	76.29
Rotation Forest Ensemble Classifiers	80.49
Pruned J48 Decision Tree	73.79
Extreme Learning Machine	80.00

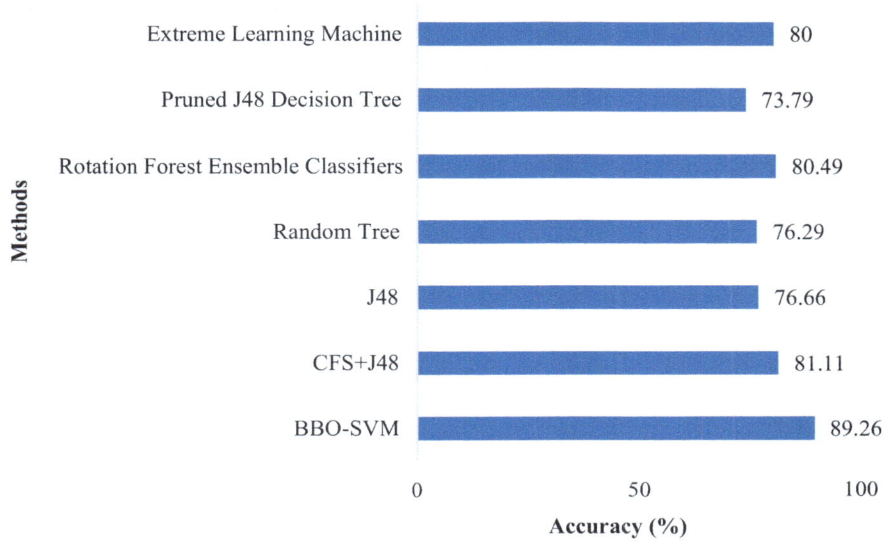

Fig. 7 Comparison of recent classifiers on heart disease dataset in terms of accuracy

4 Conclusion

In this study, an IoMT-based healthcare diagnosis model is introduced by the use of intelligent techniques. This paper has introduced an IoMT with Cloud-Based disease diagnosis healthcare framework for heart disease prediction using the BBO-SVM model. The proposed model involves the parameter tuning of SVM using the BBO algorithm. The validation of the proposed model takes place using a Statlog Heart disease dataset. The detailed experimental analysis strongly pointed out that the proposed BBO-SVM model has shown excellent results by attaining a maximum precision of 88.33%, recall of 87.60%, accuracy of 89.26%, *F*-score of 87.96%, and kappa value of 78.27%.

References

1. Atzori, L., Iera, A., Morabito, G.: The Internet of Things: a survey. Comput. Netw. **54**(15), 2787–2805 (2010)
2. Baig, M.M., Gholamhosseini, H.: Smart health monitoring systems: an overview of design and modeling. J. Med. Syst. **37**(2), 1–14 (2013)
3. Melillo, P., Orrico, A., Scala, P., Crispino, A., Pecchia, L.: Cloud-based smart health monitoring system for automatic cardiovascular and fall risk assessment in hypertensive patients. J. Med. Syst. **39**(10), 1–7 (2015)
4. Banaee, H., Ahmed, M.U., Loutfi, A.: Data mining for wearable sensors in health monitoring systems: a review of recent trends and challenges. Sensors. **13**(12), 17472–17500 (2013)

5. Zhang, M.W., Ward, J., Ying, J.J., Pan, F., Ho, R.C.: The alcohol tracker application: an initial evaluation of user preferences. BMJ Innov. **2**, 8–13 (2016)
6. Catarinucci, L., De Donno, D., Mainetti, L., Palano, L., Patrono, L., Stefanizzi, M.L., Tarricone, L.: An IoT-aware architecture for smart healthcare systems. IEEE Internet Things J. **2**(6), 515–526 (2015)
7. Kakria, P., Tripathi, N.K., Kitipawang, P.: A real-time health monitoring system for remote cardiac patients using smartphone and wearable sensors. Int. J. Telemed. Appl. **2015**, 373474 (2015)
8. Maia, P., Batista, T., Cavalcante, E., Baffa, A., Delicato, F.C., Pires, P.F., Zomaya, A.: A web platform for interconnecting body sensors and improving health care. Procedia Comput. Sci. **40**, 135–142 (2014)
9. Jara, J., Zamora-Izquierdo, M., Skarmeta, F.: Interconnection framework for m-health and remote monitoring based on the Internet of Things. IEEE J. Sel. Areas Commun. **31**(9), 47–63 (2013)
10. Kim, S.H., Chung, K.: Emergency situation monitoring service using context motion tracking of chronic disease patients. Cluster Comput. **18**(2), 747–759 (2015)
11. Istepanaian, R.S.H., Zhang, Y.T.: Guest editorial introduction to the special section: 4G Health The long-term evolution of m-Health. IEEE Trans. Inf. Technol. Biomed. **16**(1), 1–5 (2012)

Hyperparameter Optimization of Deep Neural Network in Multimodality Fused Medical Image Classification for Medical and Industrial IoT

Velmurugan Subbiah Parvathy, Sivakumar Pothiraj, and Jenyfal Sampson

Abstract Industrial Internet of Things (IIoT) refers to the extension and utilization of the Internet of Things (IoT) in industrial sectors and applications. Medical image fusion and classification has been utilized to get valuable significant multimodality medical image data. Recently, deep learning methods offer an effective way to design an end-to-end model that can determine the final classification labels. This chapter introduces a Multimodality Image Fusion Classification (MMIFC) by the incorporation of image fusion, feature extraction, and classification techniques. Initially, the input medical images were fused by the use of optimal shearlet transform where the coefficients of shearlets are optimized by the use of Enhanced Monarch Butterfly Optimization (EMBO) algorithm. In the next stage, the fused images were classified based on the feature extractor and deep learning model to check the test image is malignant and benign. To further optimize the performance of the deep learning model, hyperparameter tuning process takes place by the use of Bayesian optimization model to optimize the weights of structure. Finally, this ODNN model classifies the MMFI as class 1 or 0. A detailed simulation result takes place to ensure the effective performance of the proposed method. From the experimental results, the proposed strategy accomplishes better fusion rate and classification results compared with other supervised and non-supervised learning techniques.

Keywords Medical image · Image fusion · Industrial IoT · Classification · Transform · Hyperparameter · Deep neural network

---

V. S. Parvathy (✉) · S. Pothiraj · J. Sampson
Department of Electronics and Communication Engineering, Kalasalingam Academy of Research and Education, Krishnankoil, India
e-mail: s.p.velmurugan@klu.ac.in; siva@klu.ac.in

D. Gupta et al. (eds.), *Smart Sensors for Industrial Internet of Things*, Internet of Things, https://doi.org/10.1007/978-3-030-52624-5_9

1 Introduction

Industrial Internet of Things or IIoT is defined as a collection of machines, computers, and people permitting intelligent industrial operations by the use of latest data analytics for transformational business outcomes. A main benefit of the IIoT is the minimization of the human error, manual labor, and time by maximizing the efficiency. Image fusion is the technique for consolidating data from numerous images of a similar scene taken from different sensors, various positions, or distinctive time intervals [1]. Two fundamental sorts of classification were considered in this examination, namely, supervised and unsupervised classification. It fuses numerous images gotten by the image sensors with various wavelengths, and diverse imaging standards were present in a similar domain into a new image through a particular fusion algorithm [2]. Multimodal medical image fusion (MMIF) process intends to join data from various medical images of same methodology or of various modalities, to give a solitary intertwined image [3]. MMIF has wide scope because of its conservative and exact portrayal of data [4]. For example, fusion of SPECT and CT images does not encourage oncologists for arranging and screening cancer operation yet additionally helps cardiologists to study the nearness and degree of coronary artery disease [5]. Further, selected-level algorithms are something simple but difficult to implement. Moreover, it is computationally efficient and favored for MMIF combination by wavelet changes [6].

Notwithstanding the quick advancement of the imaging strategies, there have been countless methodologies in the field of MMIF [7]. Some scientists proposed some novel techniques consolidating multistate transform and SR [8], to put the data structure of the images source into the intertwined images, which create techniques substantially more astounding and time-devouring [9]. Enhancing the accuracy, time affectability, and energy of image registration are issues that should be settled in the current examination by optimization process [10]. In a decade ago, different procedures of delicate computing were connected in multimodal image combination applications; at long last the fused multimodality images are classified by machine or deep learning methodology. To start with, deep models may overfit the preparation information, which is a long way from satisfactory, as there is typically a little dataset in medical image examination and this identifies with the job required in obtaining the data image and afterward in image comment [11, 12].

Primarily, images of medical are deteriorated into numerous layer scales, and along these lines assorted visual subtleties can be isolated from various layer scales. Furthermore, for every scale layer, the non-negative scanty model of coding with Fisher discriminant examination is built to get the discriminative meager portrayal of fused medical images [13]. There are many classification methods like KNN, naive Bayes classifier, machine learning, deep learning model, and some optimization methodologies like particle swarm optimization (PSO), genetic algorithm (GA), etc. The greater part of current methodologies uses pre-trained DNN to concentrate feature descriptors from MMIF images [14]. There have been numerous examinations of various classification strategies; anyway no single classification strategy has been observed to be better overall datasets. Accordingly, the issue stays an examination subject [15]. Additionally, the immense measure of medical image data available to the overall public calls for growing new devices and classification strategies to viably diagnosis, dissect, and classify medical data [16].

This chapter introduces a Multimodality Image Fusion Classification (MMIFC) by the incorporation of image fusion, feature extraction, and classification techniques. Initially, the input medical images were fused by the use of optimal shearlet transform where the coefficients of shearlets are optimized by the use of Enhanced Monarch Butterfly Optimization (EMBO) algorithm. In the next stage, the fused images were classified based on the feature extractor and deep learning model to check the test image is malignant and benign. To further optimize the performance of the deep learning model, hyperparameter tuning process takes place by the use of Bayesian optimization model to optimize the weights of structure. Finally, this ODNN model classifies the MMFI as class 1 or 0. A detailed simulation result takes place to ensure the effective performance of the proposed method. From the experimental results, the proposed strategy accomplishes better fusion rate and classification results compared with other supervised and non-supervised learning techniques.

The remainder of this chapter is composed as follows: Section 2 briefs the related works and Section 3 presented a detailed explanation of the proposed method. Section 4 examined the execution results of the proposed framework and conclusion is presented in Sect. 5.

2 Review of Recent Literature

In Shahdoosti and Mehrabi [17], scanty unblending by factor parting and extended Lagrangian (SUnSAL) classifier has been developed. Coefficients to be fused into the melded image are picked by the classifier. Coefficients to be melded into the fused image are selected by the classifier. At last, the spinning cycle procedure is misused to stay away from ancient rarities. Test outcomes on third sets of medical images approve the unwavering quality and believability of the proposed strategy in medical applications when contrasted with four best in class fusion techniques. The practical limitation of convolutional neural network (CNN) algorithm is dependent on the pattern information and neuron pathway pruning recuperating algorithms by Feng-Ping and Zhi-Wen [18]. Here, the learning of features are carried out via unlabeled image block samples, and the initial parameters of all the layers of the network are trained. In addition, the labeled image block samples are further tuned by the use of convolutional neural network (CNN), which has a classification function. Examinations demonstrated that the proposed division technique has high division precision, yet in addition has amazingly high versatile segmentation capacity for different medical images.

In 2019, El-Hoseny et al. [19] have proposed the Modified Central Force Optimization (MCFO) and the Dual-Tree Complex Wavelet Transform (DT-CWT) methods. The initial phase in the proposed framework is the histogram matching of one of the images to the next to permit a similar powerful extension for the two images. The Dual-Tree Complex Wavelet Transform was utilized after that to break down the images to be melded into their coefficients. The proposed fusion framework was emotionally and equitably tried and assessed with various fusion quality measurements. Zhang et al. [20] developed a Deep Convolutional Neural Network (DCNN) using a pair of DCNNs. Every pair of DCNNs has their learned image representation concatenated as the input of a synergic network, which has a fully connected structure which identifies whether the pair of input images belonged to an identical class. In this way, in the event that one DCNN creates a right classification, a slip-up created by another DCNN prompts a synergic mistake that fills in as an additional power to refresh the method.

Another Convolutional Sparse Kernel Network (CSKN) is developed in [21] for learning the visual features of the medical images. From the results, CSKN model has shown effective outcom over the classical moldes like supervised convolutional neural networks (CNNs), Fully Convolutional Layer Fusion Network (FCLFN), and Side Output Decision Fusion Network (SODFN). In fact, SODFN upgrades each convolutional layer in the system to make a side characterization map. From that point onward, the dominant part casting a ballot is connected on these maps to acquire the fused outcome by Zhao et al. [7]. In the FCLFN, one fully convolutional layer (FCL) is first made for every convolutional layer, and afterward FCLs from various convolutional layers are joined to shape one FCL in the previous layer. At long last, the new FCL can be utilized to acquire the last arrangement map.

The feature is removed from the combination of the Gray Level Co-occurrence Matrix (GLCM) and the Log-Mel Spectrogram (LMS) for the acoustic scene characterization by Oo and Oo [22]. Multi-class Support Vector Machine (SVM) trains the highlight so as to classify the sort of condition for audio information record. The fundamental commitment of this chapter is to extricate the viable highlight from the blend of signal preparing approach and image handling approach. The inspiration behind this feature is to lessen computational time for characterization. Strange kidney images were grouped through grayscale conversion, district of premium, multi-scale wavelet-based Gabor extraction, probabilistic principal component investigation-based feature choice, and adaptive artificial neural network system by Chaitanya and Kumar [23]. Results demonstrated that the proposed methodology had 100% specificity and 94% accuracy. What's more, it's false acknowledgment rate is 0%, while that of existing techniques isn't <27%.

3 Overview of Framework

Figure 1 shows the working principle of the presented MMIFC model. At the beginning, the input medical images were fused by the use of optimal shearlet transform where the coefficients of shearlets are optimized by the use of EMBO algorithm. In the next stage, the fused images were classified based on the feature extractor and deep learning model to check the test image is malignant and benign. To further

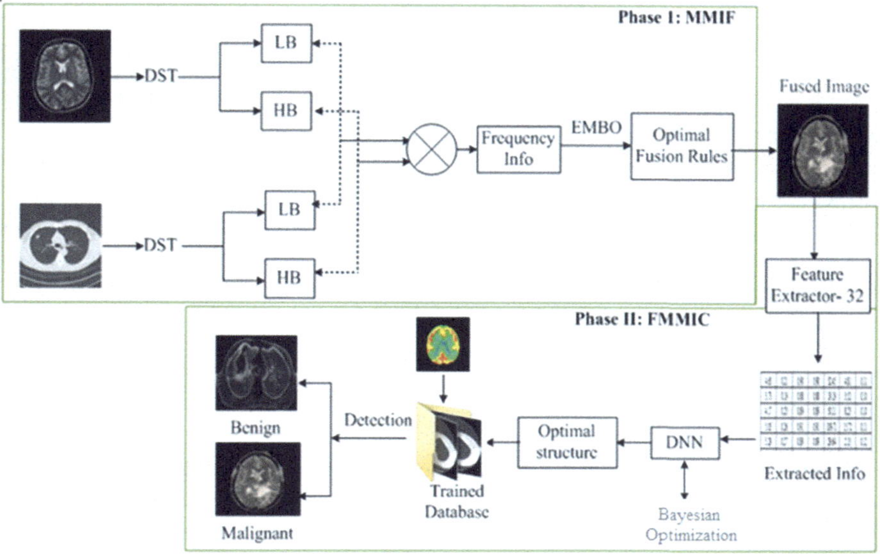

Fig. 1 Working process of proposed model

optimize the performance of the deep learning model, hyperparameter tuning process takes place by the use of Discrete Gravitational Search (Bayesian) to optimize the weights of structure. Finally, this ODNN model classifies the MMFI as class 1 or 0.

3.1 Multimodality Image Fusion Process

In this phase, the MMIF process takes place by the use of discrete shearlet transform, and the coefficients of shearlets are optimized by the use of EMBO algorithm. The image fusion technique comprises image decomposition and ideal rules are significant. Some significant points to consider for MMIF examination are as follows:

- MMIF has been fused which is handled before Gaussian filters.
- The image combination guidelines are utilized to unite the different features into sub-images at various goals.
- Image registration process takes place for the fusion process through the preprocessed diverse methodology images.
- If both or any of the images are not in grayscale, then it is converted to grayscale by histogram equalization process.
- Finally, enlisted MMI is fused by the use of optimal DST algorithm to next level examination.

3.1.1 MMI Decomposition by DST

The shearlet transform has the capacity to detect directionality and thus contrasting the conventional wavelet transform. Shearlet has one of the most critical properties as the information that they give ideally sparse approximations. The proposed image pixel values are classified into two bands like higher and lower band information for further fusion investigation. To play out the determination of shearlet low-frequency coefficients, the human feature visibility fusion plan is utilized. The idea of human feature deceivability is acquainted as a strategy to assess the nature of an image. In the coefficients' investigation process, few rules are removed from the MMI as mentioned in Eq. (1).

$$\mathrm{MMI}_{(A,B)} = \sum_{m,n} \mathrm{DSTCoeff}^{(A,B)}(x,y) \tag{1}$$

From condition (1), m and n fulfills the normalization rule. The fused images have some energy information of unique distinctive modality images. This transform remains unaltered as it stood out from that of the source images, which makes it easier to develop a communication between the sub-band images, and along these lines it is helpful in making fusion rules.

3.1.2 EMBO-Based Optimization for Image Fusion

Frequency wavelet coefficients include the tremendous amount of energy of an image with the objective that it is continuously important for the incredible appearance of an image by EMBO algorithm. During the optimization issue through the conjugate inclination approach, the image fusion process gets acquired. During this optimization, the optimal threshold is selected by the foraging behavior of butterflies. The important processes of MBO algorithm are migration operator and butterfly adjusting operator. MBO plans to override the relative more awful arrangements by the new better solutions and keep it for the next generation of the fusion procedure [24]. To make an answer, ideal thresholds are picked individually and are delegated to one of the cells in light of the control parameter and sporadic walk. Optimal high- and low-frequency coefficients from different images are joined by a standard subject to nearby energy while upon intensive examinations of the duties from different components, for instance, variance, normal slope, and spatial frequency to image quality.

3.2 FMMI Classification Model

3.2.1 Feature Extraction for FMMI

Fused multimodality images were considered for further classification process by the use of texture feature extractor and deep learning classifier. In any case, a particular kind of feature is effective and increasingly viable for specific sorts of areas.

It is useful to treat regions diversely as per their nearby qualities to classify the image. The various features of these areas are removed. The suitable features from all the image source are converged to get the melded image. They are less receptive to noise and progressively strong [25–29]. Our proposed model has extensive features, for example, GLCM and LGXP, which are texture-based feature models. To improve results, most existing techniques utilize numerous features like a consolidated vector for matching, which infers that individual features contribute similarly to region matching over the whole image.

This programmed arrangement of feature extractors is a sort of portrayal learning. It permits to naturally building an appropriate information portrayal to be utilized together with a classifier and does not require extensive space skill. The utilization of feature extractor that does not fit to a specific classifier, and the other way around, frequently yields exceptionally poor performance. To simplify this procedure, we propose an uncomplicated way to utilize structure that figures out how to concentrate low-dimensional features from raw estimations. For the surface feature extraction alongside well-known GLCM and texture strategy, algorithm has been acquainted which concentrates on the textural inconsistency of the lesion zone. Alongside the shape and texture features, diverse statistical measures have been extricated from FMMI.

$$\text{MMIFeature} = \left\{ f_{11}, f_{12}, \ldots f_{24} \; \& \; f_{21}, f_{22}, \ldots f_{32} \right\} \tag{2}$$

3.2.2 Texture Feature

Texture feature extraction assumes a significant job in low-level image processing and computer vision.

It has been utilized in a wide number of utilizations, for example, automated inspection medical image investigation. Our proposed model, gray level-based features, is removed for classification, and the GLCM technique, one of the most prominent texture strategies, has been reached out to a three-dimensional space also [30]. The GLCM defines the texture of the image by detemining how frequently the pair of pixels with particular values and in a predefined spatial relationship exist in the image, generating a GLCM and extracts the statistical measure from the matrix. The matrix segment signifies the image features are confined by the deliberate pixel esteem the whole features model shows in Fig. 2; the spatial connection between two neighboring pixels can be determined from multiple points of view with various counterbalances and edges, the default one being between a pixel and its quick neighbor on its right side to classify the medical images.

As an option in contrast to the bottom-up examination of image pixels to decide texture properties for classification, model-based techniques to integrate texture have been considered. In these strategies a model for texture is first accepted and its parameters are then assessed from the image region with the end goal that an image orchestrated utilizing the model intently looks like the input image region.

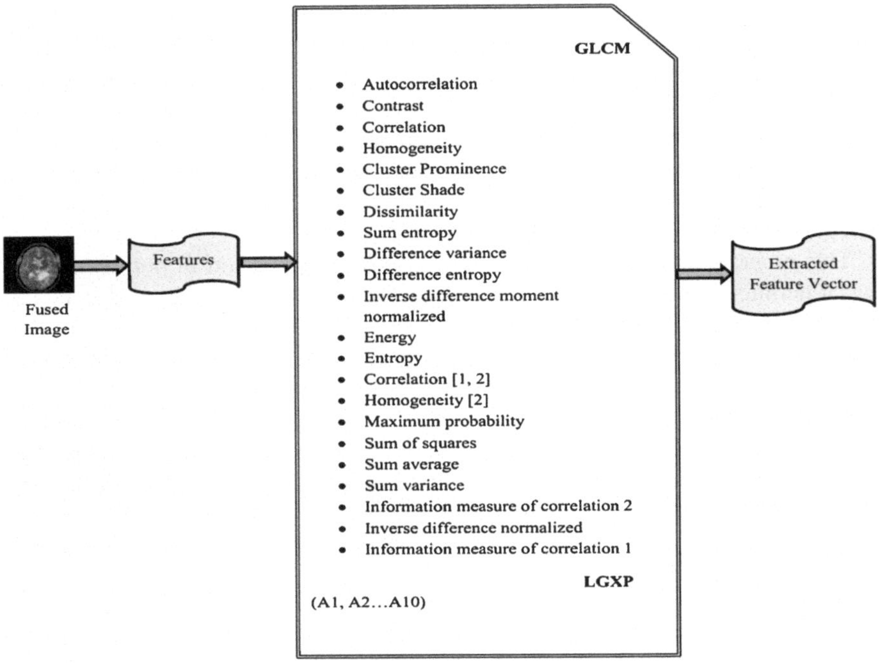

Fig. 2 Features of GLCM

3.2.3 LGXP Feature

From the start, LGXP descriptor phase is quantized into different reaches. The quantity of phase reaches is made in such a way to create the examples vigorous to the varieties of Gabor stage; henceforth it can't be excessively high. LGXP is utilized for feature extraction process and is named as local Gabor XOR patterns. In our work, feature extraction is performed by following advances. At first, input image is sectioned into different blocks and each block is isolated into 3 × 3 matrix structure. This feature has absolutely ten trait esteems for FFMMI classification process, and it's scientifically described in below condition;

$$\mathrm{LGXP}_{x,y}^{i} = \mathrm{O}(\phi_{x,y}(F_c)\,\mathrm{XOR}\,\mathrm{O}\big(\phi_{x,y}(F_i)\big) \tag{3}$$

Here, $\phi_{x,\,y}(F_c)$ denotes the stage and the estimation of (0, 1), (1, 0), (1, 1)... (1, 10). $p\,(\phi_{x,\,y}(F_i))$ speaks the quantized estimation of the stage where $\varphi_{x,\,y}(k)z_c$ delineates the focal pixel position in Gabor phase map with scale ω and orientation y, p represents the element of the neighborhood.

3.2.4 Local Binary Pattern (LBP)

The LBP is characterized depending on the connection between the middle pixel and its encompassing neighbors in an image and the sample process of LBP is shown in Fig. 3. The relationship which is gathered between the middle pixel and its neighbors depends on the edge between the neighbor and the middle pixel.

3.2.5 Normalization Process of LBP

The normalization process of LBP takes place to make all the score levels of different classification models to a certain identical range. Yet, in a vector sense, the normalization algorithm is all the more of vector transform starting with one vector space then onto the next. Subsequently, the decision for a perfect normalizer winds up significant as this may twist the vectors thereby improving or diminishing the class distinctness between two particular class vectors. These LBP codes encode the local structure around every pixel. The different steps engaged with creating LBP features are expressed by certain conditions. The relating pixel is allotted the value "0," and in the event that it is positive, the pixel is allocated a value "1." All the attained binary codes are connected in a clockwise way starting from the left one at the top or bottom to frame a binary number. A ternary pattern is gotten by consolidating the neighboring pixels subsequent

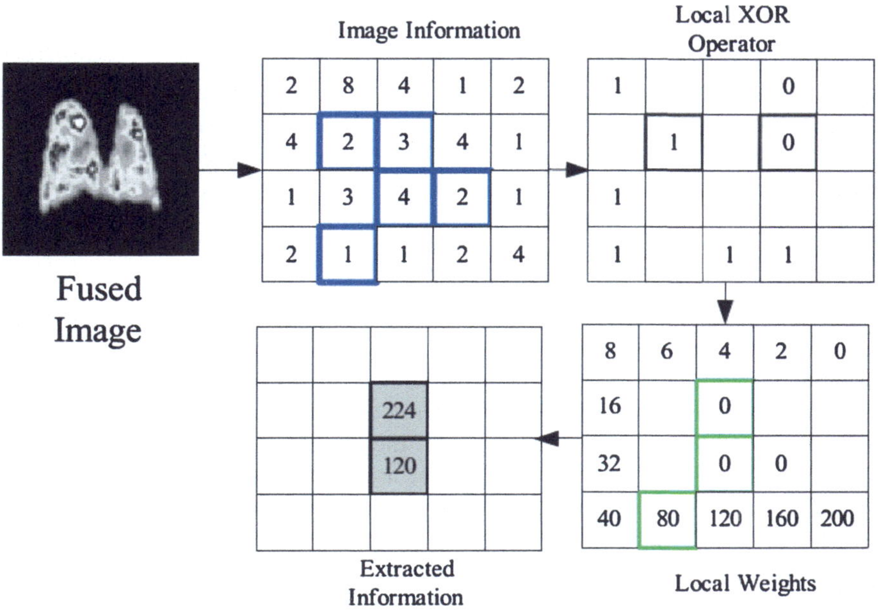

Fig. 3 Procedure of LBP

to thresholding. In the event that the histogram of these ternary values is registered, it will bring about an enormous range, so the ternary pattern is part into two binary patterns.

Expression of final feature vectors of classification is

$$\text{Extracted Features} = \begin{cases} \text{GLCM}_{11}\text{GLCM}_{12},\ldots\ldots\text{GLCM}_{22},\text{LGXP}_{11},\text{LGXP}_{12},\ldots\text{LGXP}_{10} \\ \text{GLCM}_{21},\text{GLCM}_{22},\ldots\ldots\text{GLCM}_{22},\text{LGXP}_{21},\text{LGXP}_{22},\ldots\text{LGXP}_{20} \\ \ldots\ldots\ldots\ldots\ldots\ldots\ldots\ldots\ldots\ldots\ldots\ldots \\ \ldots\ldots\ldots\ldots\ldots\ldots\ldots\ldots\ldots\ldots\ldots\ldots \\ \text{GLCM}_{1n},\text{GLCM}_{1n}\ldots\ldots\text{GLCM}_{2n},\text{LGXP}_{1n},\text{LGXP}_{1n},\ldots\text{LGXP}_{nn} \end{cases} \tag{4}$$

3.2.6 FMMI Classification via Extracted Feature Information

Consequent to choosing optimal features from the extracted signal, the FMMI is used for predicting the abnormal and normal signal using DNN. An artificial neural network method with the various layers of the input and output layers is named as DNN. Also, it contains both pre-preparing (utilizing creative Deep Belief Network (DBN) and adjusting phases in its parameter learning. To improve the level of classification, Bayesian optimization is used to streamline the highlights DNN structure appears in Fig. 4. Here, we utilized DNN based on restricted Boltzmann machine (RBM) which is shallow, two-layer neural nets that comprise the structure squares of deep-belief systems.

The primary layer of the RBM is known as the obvious information layer, and the second is the concealed layer. At every hidden node, each input x is multiplied by its separate weight w. The weights between two layers will consistently frame a matrix where the lines are equivalent to the input nodes and the segments are equivalent to the yield nodes. The standard sort of RBM has binary-esteemed (Boolean/Bernoulli) x hidden and y visible neurons and comprises a matrix of weights and the function speaks to by utilizing the following condition

$$\text{Weight}_{i,j} = -\sum_{i=1}^{i} Q_{ij}, E_j * p_j - \sum_{j=1}^{i} \alpha_j * p_j - \sum_{j=1}^{i} \delta_j p_j \tag{5}$$

where E_{ij} symbolizes the symmetric association term among the unmistakable unit Q_i and the hidden unit p_j, α, β is the bias term, and i, j is the number of noticeable and concealed units [31]. The subsidiary of the likelihood of a preparation vector concerning a weight is inconsistently simple. Between shrouded units in an RBM, there are no immediate impacts; it is tremendously simple to select an unprejudiced example of$(I_i, u_j)_{\text{data}}$.

Since the neurons are binary-valued, the probabilities of the conditions of the visible and hidden neurons can be gotten by means of the sigmoid function:

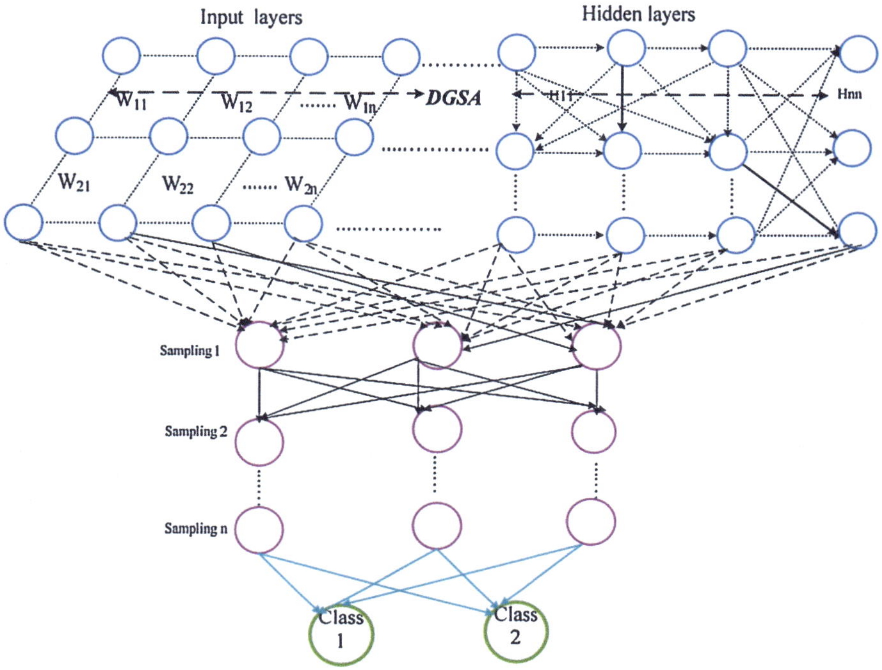

Fig. 4 Optimal DNN structure

$$Q_{I_i} = q\left(I_i = 1\right) = \frac{1}{1 + \exp\left(-\sum_i \text{weight}_{i,j} p_j\right)} \tag{6}$$

To refresh the covered up and noticeable units are parallel in the given obvious and shrouded units. For executing the stochastic steepest rising in the log likelihood of the preparation information, this demonstrates the route a much-uncomplicated learning rule as

$$Hw_{ij} = \theta\left(k_{si} y_j\right)_{\text{data}} \left(y_{si} g_j\right)_{\text{reconstruction}} \tag{7}$$

where HW_{ij} speaks to the updated weight achieved from the changed weight in hidden layer; at the point when the RBM is prepared, a different RBM can be "stacked" over it to frame a multilayer model. The output layer of the effectively prepared layers is locked in as contribution to the novel RBM. The deep system loads are occupied with preparing an adjusting stage.

3.2.7 Hyperparameter Optimization of DNN Using Bayesian Optimization

Hyperparameters are settings which can be tuned for controlling the actions of a machine learning (ML) technique. The procedure of choosing hyperparameters is the main dimension of any deep learning solutions. Many deep learning models clearly describe particular hyperparameters which manage diverse dimensions like storage or computation cost. But, the extra hyperparameters are required to be tuned by a specific technique for adapting with a particular scenario. The practitioners classically spend quite a bit of time tuning hyperparameters in order to achieve the best performance for a particular model.

Upon the selection and optimization of the hyperparameters, there are two basic approaches: namely, manual and automatic selection. All these methods are officially feasible and the decision normally defines a trade-off among the deep interpretation of a machine learning model needed for choosing hyperparameters manually versus the maximum computation cost needed by automatic selection models. The hyperparameters in the DNN model are tuned by the use of Bayesian optimization model. The Bayesian optimization is a widely employed model for the global optimization of noisy, expensive black-box functions. It is based on the generation of a probabilistic technique which describes the distribution over objective functions from the input space to the objective of interest. Conditioned on a prior over the functional form and a set of N observations of input-target pairs $D = \{(xn, yn)\}, N,n = 1$, the relatively cheap posterior over functions is then queried to reason about where to seek the optimum of the expensive function of interest. The promise of a new experiment is quantified using an acquisition function, which, applied to the posterior mean and variance, expresses a trade-off between exploration and exploitation. Bayesian optimization proceeds by performing a proxy optimization over this acquisition function in order to determine the next input to evaluate. The primary commitment of our ODNN is to classify the given test image is benign or malignant based fused multimodality images. Moreover, the prepared dataset is capable until the enhanced load is gotten, or most extreme accuracy is practiced with the help of objective function. To organize the exhibition of the structure, an output layer is normal in the most astounding purpose of the DNN, moreover the DNN system assessment; it finds the presence or absence of medical image prediction process.

Algorithm 1 Overall Steps Involved in MMIFC Process

```
Training and testing process
Input Images: Medical images (CT, MRI, and SPECT)
Fusion Model: Apply DST
Fusion Rules and Threshold determine by EMBO
    Updating by Migration operator and Butterfly adjust-
    ing operator
                Optimal parameters for MMIF
```

```
Extract GLCM and LGXP in FMMI by Extractor
Classification DNN model with Bayesian Optimization
Testing
Input: Trained database
Optimal Classifier
Output: Fused image as Class 1 or 0 (Benign or Malignant)
```

4 Results and Discussion

For validating the performance of the proposed MMIFC framework, it is implemented in MATLAB2018a with i7 processor and 4GB RAM. For this result analysis, four types of database images are used, all database image containing more than 50 images, with 512 * 512 matrix size. This section discussed the results about some test images with different metrics like fusion factor and spatial frequency of the MMIF process and then the classification evaluation by confusion matrix-based metrics like accuracy, sensitivity, and specificity; and finally some existing techniques are considered for the comparison model.

The presentation of combination algorithms is for the most part assessed abstractly. The emotional assessment includes various human observers to pass judgment on the nature of the combined image shown in Fig. 6a and b and Table 1. Figure 5 shows some of the sample multimodality images like CT, MRI and SPECT images with fused images. The proposed techniques safeguard better both nearby edge and surface data, which is the essential data for diagnosis. Therefore, the proposed method of medical image fusion is an effective method in both subjective and objective evaluation criteria as shown in Table 1 compared with DST, DST + SSO (Social Spider Optimization), and DST + EMBO model. For test image evaluation fusion factor and spectral frequency are considered, and from that four databases are evaluated.

It is noted that the fusion factor is found to be high with the value of 6.55 in the database II. CT, SPECT, and MR images are chosen as the information source images. It tends to be seen that as a result of various imaging gauges, the source images with different modalities contain the necessary data.

At that point Fig. 6 demonstrates the similar investigation of MMIF process, and this examination is dependent on source and fused images. Hence, the image fusion procedure can be exceptionally powerful with all the required information. Image fusion is a strategy which can be utilized to combine integral images, different fleeting images, and multifocal images into one medical image with all the essential required data from input medical images. From Fig. 6, the fusion factor 6.55 in source image 1 (S1) compared with source image 2 (S2) is minimal, comparatively spatial frequency parameters. In many cases, the fusion factor is found to be high compared to the source image. At that point spatial frequency is determined dependent on the DST transform low band and high band coefficients.

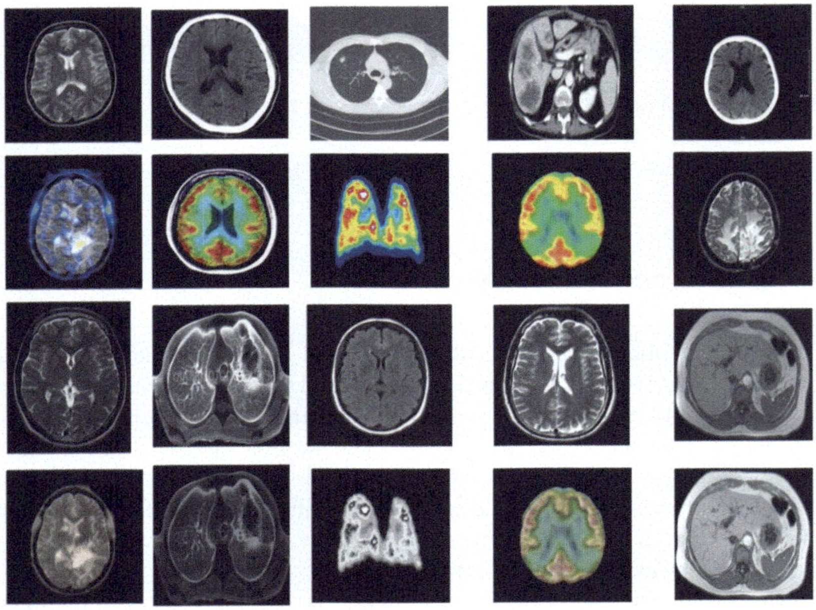

Fig. 5 (1) CT, (2) SPECT, (3) MRI, and (4) fused image

Table 1 MMIF result analysis

	Fusion factor			Spatial frequency		
Dataset	DST + EMBO	DST	DST + SSO	DST + EMBO	DST + SSO	DST
I	6.45	1.58	5.56	23.56	19.55	23.45
II	6.55	2.54	4.26	28	23.78	20.25
III	7.22	1.89	6.55	35.666	31.24	17.56
IV	5.89	3.75	3.78	20.15	18.04	18.89
Average	6.52	2.44	5.03	26.8	23.15	20.03

Tables 2 and 3 examined the proposed MMIF classification dependent on separated feature fusion for image classification. Feature vector of GLCM and LGXP test esteems appears in Table 2, and it's needful for image classification. This outcome confirms prevalence of proposed image fusion method regarding visual and factual and basic information achieved in the fused image dependent on sources and fused images that are clearer and contained higher differentiation. The proposed model achieves a better fusion factor for various features such as 0.14, 0.77, 0.52, 0.98, 0.87, 8.19, 3.63, 28.64, 0.50, and 0.94, etc. At the same time, the fused image is classified as benign type with the accuracy of 92.22%. For exploratory outcomes, the framework performs approvals and ten times cross approvals to demonstrate that

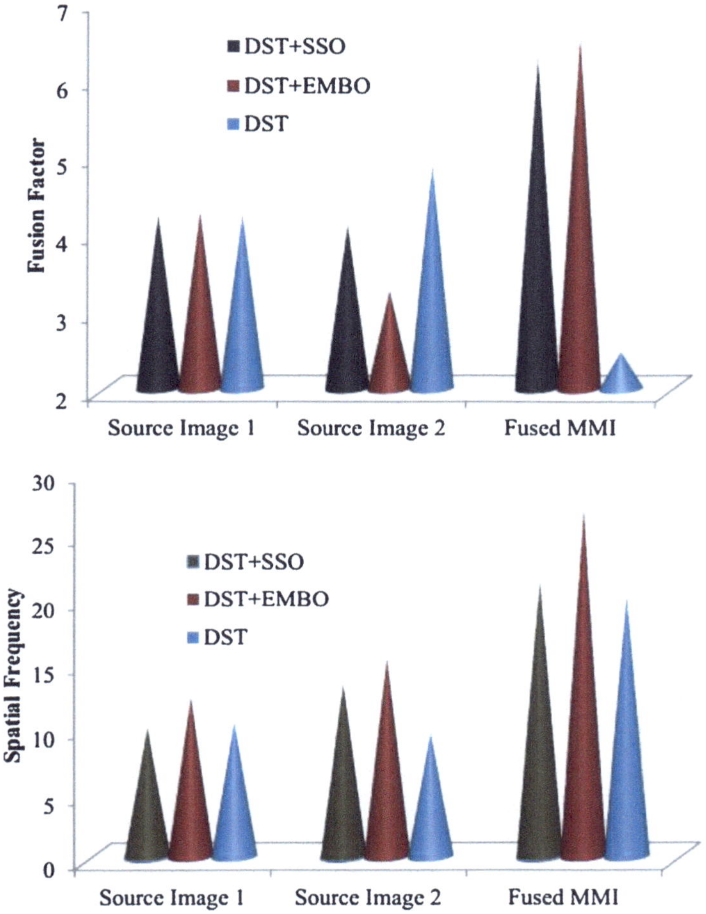

Fig. 6 Comparative analysis of MMIF

the proposed highlight does not have information inclination. An adjusted arrangement of features is produced from random by working straight blend of the first factors to classification process.

Table 4 and Fig. 7 talked about comparative examination of FMMI classification and the considerable comparison methods like convolutional neural network (CNN), artificial neural network (ANN), SVM, DNN, and proposed model with different images like I, II, III, IV, V, VI, and VII with precision, accuracy, sensitivity, and specificity. For instance, Table 4a demonstrates the comparison estimation of accuracy in a different image, for example, ANN is 56.22%, SVM is 79.22%, CNN is 71.22%, and our proposed methodology is 100, its most extreme worth. The proposed method has attained better classification results with minimum computation. Table 4b portrayed the sensitivity examination of FMMI, likewise different

Table 2 MMIF result analysis

Feature	Images						
	I	II	III	IV	V	VI	VII
F1	7.77	7.56	24.38	16.78	17.60	21.06	22.17
F2	0.94	0.39	0.82	0.90	0.46	0.57	0.71
F3	0.90	0.96	0.96	0.95	0.98	0.97	0.97
F4	0.90	0.96	0.96	0.95	0.98	0.97	0.97
F5	2392.87	2403.39	2607.77	3025.58	3051.94	2651.43	2745.33
F6	193.71	193.09	153.55	245.25	245.70	191.30	189.91
F7	0.14	0.06	0.12	0.13	0.07	0.09	0.10
F8	0.77	0.79	0.51	0.60	0.60	0.52	0.53
F9	0.52	0.46	0.75	0.66	0.63	0.87	0.74
F10	0.98	0.99	0.99	0.98	0.99	0.98	0.99
F11	0.98	0.99	0.98	0.98	0.99	0.98	0.98
F12	0.87	0.88	0.61	0.73	0.73	0.66	0.65
F13	8.19	7.70	24.74	17.26	17.74	21.25	22.42
F14	3.63	3.52	7.29	5.61	5.74	6.58	6.79
F15	3.50	3.82	3.49	3.61	3.85	4.10	4.67
F16	8.62	10.07	7.75	8.83	11.28	13.25	19.47
F17	1.27	1.39	1.29	1.39	1.46	1.16	1.23
F18	0.09	0.21	0.19	0.23	0.26	0.27	0.32
F19	0.25	0.32	0.31	0.28	0.29	0.31	0.31
F20	−0.79	−0.74	−0.73	−0.79	−0.78	−0.72	−0.74
F21	0.91	0.91	0.90	0.92	0.93	0.87	0.89
F22	0.99	0.99	0.99	0.99	0.99	0.99	0.99
F23	1.00	1.00	1.00	1.00	1.00	1.00	1.00
F24	3.26	3.60	2.85	3.31	3.92	4.44	5.76
F25	0.08	0.13	0.13	0.17	0.17	0.15	0.25
F26	0.94	0.90	0.85	0.87	0.90	0.92	0.91
F27	0.07	0.07	0.06	0.08	0.06	0.04	0.04
F28	0.43	0.46	0.45	0.41	0.35	0.50	0.83
F29	1.08	0.94	1.03	1.13	1.29	0.89	0.49
F30	0.97	0.97	0.97	0.96	0.97	0.98	0.98
F31	0.97	0.97	0.97	0.96	0.97	0.98	0.98
F32	0.55	0.58	0.57	0.53	0.50	0.64	0.91

Table 3 FMMI classification results

Fused medical image	I	II	III	IV	V	VI	VII
	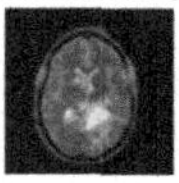	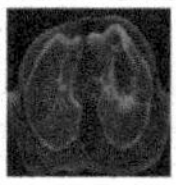		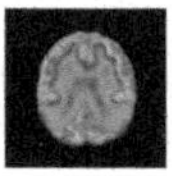	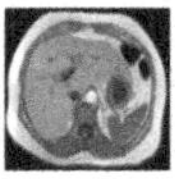	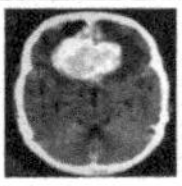	
Class	Malignant	Benign	Benign	Malignant	Malignant	Malignant	Benign
Accuracy	92.22	96.22	98.41	89.56	96.58	92.1	93.89

Table 4 FMMI classification results

Features	32						
Classified images	I	II	III	IV	V	VI	VII
(a) Accuracy							
ANN	70.52	71.58	69.78	75.45	83.66	71.22	70.89
CNN	87.48	78.45	84.22	79.22	90.45	90.75	89.22
SVM	72.25	65.89	76.22	89.45	75.14	83.52	74.22
DNN	90.14	86.45	89.22	90.41	86.22	79.22	86.89
Proposed model (DNN + Bayesian)	92.22	96.22	98.41	89.56	96.58	92.1	93.89
(b) Sensitivity							
ANN	75.22	76.558	81.45	83.22	65.22	63.45	59.22
CNN	69.52	82.22	76.55	72.22	89.45	78.44	86.48
SVM	70.12	92.22	0	86.22	76.55	92.11	96.3
DNN	90.45	92.22	87.12	100	0	78.14	93.22
Proposed model (DNN + Bayesian)	100	99.56	100	96.89	100	98.55	100
(c) Specificity							
ANN	75.22	62.22	100	86.45	72.22	92.22	69.45
CNN	100	96.22	92.22	84.22	100	75.22	76.22
SVM	56.22	65.22	69.45	76.45	82.22	91.22	86.55
DNN	78.44	82.22	76.22	82.22	0	92.11	96.22
Proposed model (DNN + Bayesian)	100	88.45	86.22	100	96.58	89.77	0
(d) Precision							
ANN	56.22	56.85	62.22	72.22	68.22	69.78	51.08
CNN	71.22	73.22	82.22	76.22	73.22	82.22	81.11
SVM	79.22	76.22	56.22	89.22	100	76.22	91.22
DNN	90.11	76.55	100	79.22	59.22	82.22	100
Proposed model (DNN + Bayesian)	100	99.45	96.45	89.47	82.45	0	100

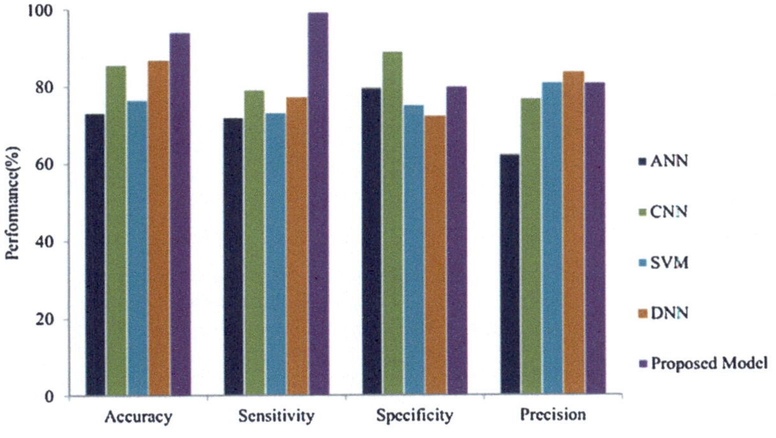

Fig. 7 Classifier-based comparison

measurements in Table 4c and d. To improve the classification model the weights are enhanced in DNN utilizing Bayesian optimization approach. From the figure and table, the proposed model exactness, affectability, explicitness, and accuracy normal worth is 94.14%, 99.28%, 80.45%, and 81.11%, respectively, its most extreme worth compared with different techniques. This is a better execution compared to various conventional systems. It provides significantly more image subtleties, bigger MMIFC quality, the briefest preparation time, and a superior visual investigation.

5 Conclusion

IIoT makes a connected enterprise through the integration of information and operational department of the industry. In numerous significant imaging applications, medical images are very important, and so in this chapter, we have talked about the MMIF and classification process through optimization and deep learning methodology. Particularly in image combination, edge conservation is significant in acquiring the integral subtleties of the information images. This chapter has introduced an MMIFC method by incorporating image fusion, feature extraction, and classification techniques. At first, the input medical images were fused utilizing optimal shearlet transform where the coefficients of shearlets are optimized by the use of EMBO algorithm. In the next stage, the fused images were classified based on the feature extractor and deep learning model to check the test image is malignant and benign. To further optimize the performance of the deep learning model, hyperparameter tuning process takes place by the use of Bayesian optimization model. Finally, this ODNN model classifies the MMFI as class 1 or 0. A detailed simulation result takes place to ensure the effective performance of the proposed method. From the experimental results, the proposed strategy accomplishes better fusion rate and classification results compared with other supervised and non-supervised learning techniques. To further enhance the fusion and classification results, the presented model can be improved by the use of optimization techniques such as AlexNet- or ResNet-based CNN model for classification process.

References

1. Constantinos, S.P., Pattichis, M.S., Micheli-Tzanakou, E.: Medical imaging fusion applications: an overview. In: Conference Record of Thirty-Fifth Asilomar Conference on Signals, Systems and Computers (Cat. No. 01CH37256), vol. 2, pp. 1263–1267. IEEE, Washington, DC (2001)
2. Zong, J.J., Qiu, T.S.: Medical image fusion based on sparse representation of classified image patches. Biomed. Signal Process. Control. **34**, 195–205 (2017)
3. Xu, X., Wang, Y., Chen, S.: Medical image fusion using discrete fractional wavelet transform. Biomed. Signal Process. Control. **27**, 103–111 (2016)

4. Sreeja, G., Saraniya, O.: Image fusion through deep convolutional neural network. In: Deep learning and parallel computing environment for bioengineering systems, pp. 37–52. Academic Press, Cambridge (2019)
5. Manchanda, M., Sharma, R.: A novel method of multimodal medical image fusion using fuzzy transform. J. Vis. Commun. Image Represent. **40**, 197–217 (2016)
6. Daniel, E., Anitha, J., Kamaleshwaran, K.K., Rani, I.: Optimum spectrum mask based medical image fusion using Gray Wolf Optimization. Biomed. Signal Process. Control. **34**, 36–43 (2017)
7. Zhao, G., Liu, G., Fang, L., Tu, B., Ghamisi, P.: Multiple convolutional layers fusion framework for hyperspectral image classification. Neurocomputing. ***339***, 149–160 (2019)
8. Elkholy, M., Hosny, M.M., El-Habrouk, H.M.F.: Studying the effect of lossy compression and image fusion on image classification. Alex. Eng. J. **58**(1), 143–149 (2019)
9. Razzak, M.I., Naz, S., Zaib, A.: Deep learning for medical image processing: overview, challenges and the future. In: Classification in BioApps, pp. 323–350. Springer, Cham (2018)
10. Yang, W., Liu, J.R.: Research and development of medical image fusion. In: 2013 IEEE International Conference on Medical Imaging Physics and Engineering, pp. 307–309. IEEE, Washington, DC (2013)
11. Padmavathi, N.B.: Fusion of multimodal abdominal cancerous images and classification using support vector machine. In: 2017 International Conference on Intelligent Sustainable Systems (ICISS), pp. 266–269. IEEE, Washington, DC (2017)
12. Liu, Y., Yang, J., Sun, J.: PET/CT medical image fusion algorithm based on multiwavelet transform. In: 2010 2nd International Conference on Advanced Computer Control, vol. 2, pp. 264–268. IEEE, Washington, DC (2010)
13. Song, Y., Cai, W., Huang, H., Zhou, Y., Feng, D.D., Wang, Y., Fulham, M.J., Chen, M.: Large margin local estimate with applications to medical image classification. IEEE Trans. Med. Imaging. **34**(6), 1362–1377 (2015)
14. Valavanis, L., Stathopoulos, S., Kalamboukis, T.: Fusion of bag-of-words models for image classification in the medical domain. In: European Conference on Information Retrieval, pp. 134–145. Springer, Cham (2017, April)
15. Rahman, M.M., You, D., Simpson, M.S., Antani, S.K., Demner-Fushman, D., Thoma, G.R.: Multimodal biomedical image retrieval using hierarchical classification and modality fusion. Int. J. Multimed. Inf. Retr. **2**(3), 159–173 (2013)
16. Bhatnagar, G., Wu, Q.J., Liu, Z.: Directive contrast based multimodal medical image fusion in NSCT domain. IEEE Trans. Multimed. **15**(5), 1014–1024 (2013)
17. Shahdoosti, H.R., Mehrabi, A.: Multimodal image fusion using sparse representation classification in tetrolet domain. Digit. Signal Process. **79**, 9–22 (2018)
18. Feng-Ping, A., Zhi-Wen, L.: Medical image segmentation algorithm based on feedback mechanism convolutional neural network. Biomed. Signal Process. Control. **53**, 101589 (2019)
19. El-Hoseny, H.M., El-Rahman, W.A., El-Rabaie, E.S.M., El-Samie, F.E.A., Faragallah, O.S.: An efficient DT-CWT medical image fusion system based on modified central force optimization and histogram matching. Infrared Phys. Technol. **94**, 223–231 (2018)
20. Zhang, J., Xie, Y., Wu, Q., Xia, Y.: Medical image classification using synergic deep learning. Med. Image Anal. **54**, 10–19 (2019)
21. Ahn, E., Kumar, A., Fulham, M., Feng, D., Kim, J.: Convolutional sparse kernel network for unsupervised medical image analysis. Med. Image Anal. **56**, 140–151 (2019)
22. Oo, M.M., Oo, L.L.: Fusion of Log-Mel Spectrogram and GLCM feature in acoustic scene classification. In: International Conference on Software Engineering Research, Management and Applications, pp. 175–187. Springer, Cham (2019)
23. Chaitanya, S.M.K., Kumar, P.R.: Oppositional gravitational search algorithm and artificial neural network-based classification of kidney images. J. Intell. Syst. **29**(1), 485–496 (2018)
24. Kumar, R., Quan, C.: Asymmetric multi-user optical cryptosystem based on polar decomposition and Shearlet transform. Opt. Lasers Eng. **120**, 118–126 (2019)

25. Shankar, K., Elhoseny, M., Lakshmanaprabu, S.K., Ilayaraja, M., Vidhyavathi, R.M., Elsoud, M.A., Alkhambashi, M.: Optimal feature level fusion based ANFIS classifier for brain MRI image classification. Concurr. Comput. Pract. Exp. (2018, First Online: 27 Aug 2018). https://doi.org/10.1002/cpe.4887
26. Mohamed, E., Shankar, K., Uthayakumar, J.: Intelligent diagnostic prediction and classification system for chronic kidney disease. Nat. Sci. Rep. (2019). https://doi.org/10.1038/s41598-019-46074-2
27. Shankar, K., Lakshmanaprabu, S.K., Khanna, A., Tanwar, S., Rodrigues, J.J.P.C., Roy, N.R.: Alzheimer detection using Group Grey Wolf Optimization based features with convolutional classifier. Comput. Electr. Eng. **77**, 230–243 (2019)
28. Elhoseny, M., Bian, G.-B., Lakshmanaprabu, S.K., Shankar, K., Singh, A.K., Wu, W.: Effective features to classify ovarian cancer data in internet of medical things. Comput. Netw. **159**, 147–156 (2019)
29. Elhoseny, M., Shankar, K.: Optimal bilateral filter and convolutional neural network based denoising method of medical image measurements. Measurement. **143**, 125–135 (2019)
30. Malegori, C., Franzetti, L., Guidetti, R., Casiraghi, E., Rossi, R.: GLCM, an image analysis technique for early detection of biofilm. J. Food Eng. **185**, 48–55 (2016)
31. Liu, C., Gardner, S.J., Wen, N., Elshaikh, M.A., Siddiqui, F., Movsas, B., Chetty, I.J.: Automatic segmentation of the prostate on CT images using deep neural networks (DNN). Int. J. Radiat. Oncol. Biol. Phys. **104**(4), 924–932 (2019)
32. de Moura Oliveira, P.B., Oliveira, J., Cunha, J.B.: Trends in gravitational search algorithm. In: International Symposium on Distributed Computing and Artificial Intelligence, pp. 270–277. Springer, Cham (2017)
33. Velmurugan, S.P., Sivakumar, P., Rajasekaran, M.P.: Multimodality image fusion using centre-based genetic algorithm and fuzzy logic. Int. J. Biomed. Eng. Technol. **28**, 322–348 (2018)
34. Parvath, S., Velmurugan, Pothiraj, S.: Multi-modality medical image fusion using hybridization of binary crow search optimization. Health Care Manag. Sci. (2019). https://doi.org/10.1007/s10729-019-09429-2

Cognitive IoT-Based Smart Fitness Diagnosis and Recommendation System Using a Three-Dimensional CNN with Hierarchical Particle Swarm Optimization

Chalumuru Suresh, M. Ravikanth, B. Srivani, and Thatavarti Satish

Abstract Due to the worldwide increase in economic development, fitness centers are rising quickly all over the globe. In this view, people require methodical and realistic supervision for building their body. At the same time, Cognitive Internet of Things (CIoT) is capable of learning, sensing, decision-making, and transferring intelligence to humans. Recently, several methods have been evolved to address the fitness and health, where the wearable fitness models let down the market rate. In this paper, a Cognitive IoT-Based Smart Fitness Diagnosis and Recommendation System using a three-dimensional convolutional neural network (CNN) with hierarchical particle swarm optimization (PSO) algorithm has been presented. The proposed model is applied to observe the health statuses of exercisers. The system offers proper supervision to the exercisers. A set of experiments takes place, and the simulation results highlight the betterment of the proposed model.

Keywords Cognitive IoT · CNN · Recommendation system · Hierarchical PSO

1 Introduction

Hip osteoarthritis (HO) The faster growth of fitness world tends peoples to spend maximum energy to maintain a healthy body state [1]. Recently, several methods have been evolved to address the fitness and health, where the wearable fitness

C. Suresh (✉) · M. Ravikanth
Department of CSE, VNR VJIET, Hyderabad, Telangana, India
e-mail: ravikanth_m@vnrvjiet.in

B. Srivani
Department of IT, VNR VJIET, Hyderabad, Telangana, India
e-mail: srivani_b@vnrvjiet.in

T. Satish
Department of CSE, NSRIT, Sontyam, Visakhapatnam, Andhra Pradesh, India

D. Gupta et al. (eds.), *Smart Sensors for Industrial Internet of Things*, Internet of Things, https://doi.org/10.1007/978-3-030-52624-5_10

models let down the market rate. Nowadays, many devices have been evolved with a more number of features that helps to maintain the fitness [2]. Some of the devices are smart watches, heart rate monitors, as well as GPS tracking machines that may be discovered in the future. According to the report from China, the fitness industry has been enhanced partially from a maximum annual growth rate. Simultaneously, many fitness clubs have been developed mostly in China. Besides, the evolved models have led to initialize several fitness organizations by providing the required services to observe and modify the physical activity. In addition, Internet of Things (IoT) is assumed to be a smart technique that is applied by many individuals in everyday life use of wireless sensor network (WSN) [3]. There are numerous sensor nodes and devices that help people to assist in fitness exercise. Hence, the above mentioned results in developing fitness club with IoT and web-based methodologies in this study [4].

Here, the cognition of IoT is an extended model of human. At this point, it is essential to incorporate IoT with artificial intelligence (AI). The two models are capable of reducing the line from sci-fi and a novel high-tech method. But, AI is not effective as IoT. AI could include the extended features of the modern world. The AI across common machine intelligence for cognitive reasoning as well as rationalizing issues such as human intelligence may be a cognitive AI [5]. The system has been uploaded with the thought process of human that is mainly used to divide the space from human and machine. Cognitive computing is defined as the process of real-time enactment that includes a novel layer functionality to the previous IoT structure. Hence, it is named as Cognitive IoT (CIoT) which is capable of learning, sensing, decision-making, and transferring to humans. CIoT enables the complete ability of IoT to make a bonding among humans to external objects in order to develop an effective, intelligent, intuitive, as well as interactive model. Also, CIoT augments the present IoT along with included cognitive potential which is more familiar with human cognition.

In general, IoT is based on the extended version of the Internet. It is comprised only of the Internet, where some tools can be applied only with the help of the Internet [6]. Some of the devices are mobile units, wired and wireless sensors, and many other interlinking tools. IoT can be employed to observe the remote devices with the application of the Internet. Nowadays, the Internet can be accessed from mobile phones, PCs, and other electronic devices. By the application of IoT, several IoT-based applications are used in business and commercial purposes. At the same time, a massive IoT application has been worked in several domains [7].

At the same time, recommender systems (RS) have gotten one of the significant utilizations of examples removed from the use conduct of web clients and assets. With regard to web use, all real web clients show some particular navigational conduct. Likewise, examples can likewise be seen in the use of web assets. The route information of a web client and web asset which is put away in web logs give an approach to anticipate future moves of the web client and use of a web asset. RS misuse this data to anticipate the potential interests of web clients dependent on their history. Certain RS are intricate to work as contrasted with unequivocal RS. This is on the grounds that of two reasons, the nature of the certain information which contains unimportant web demands and the huge amount of understood information. These both compromise the productivity of the mining strategies for producing precise and approach continuous suggestions. The main issue of handling the nature of information is vital as studies have indicated that 60–80% of web utilization information contains unimportant information, for example, picture demands and web-bot demands.

The information should be gone through various investigation and preprocessing stages to clean it enough to be utilized by investigation methods for design extraction. The subsequent issue is the handling of the gigantic measure of information. Utilizing standard information investigation devices for separating designs from such information isn't effective. One arrangement is to utilize information mining strategies [8]; however, existing information mining techniques should be improved and altered so as to produce close to constant suggestions, give adaptability in the quantity of proposals, and create precise suggestions that are important to the client. Information grouping is one of the information mining methods which decrease the objective inquiry space for suggestions by separating the inquiry space into related pieces of information. Be that as it may, information grouping methods need to be altered and improved to adapt to the necessity of RS.

Ref. [9] presented that, applying practical information collected from infrared sensors from unwanted bins for normalizing the collection task. Ref. [10] deployed a method which employed sensors to gather and examine data from city-relied IoT as well as big data models. According to WSN models, an IoT method for investigating the traffic data from sensors results in the reduction of congestion vehicular traffic [11]. Ref. [12] executed a healthcare technique to observe the health state of peoples in hospitals. Ref. [13] employed radio-frequency identification (RFID) as well as IoT models to study the use of medical nursing system that may be more applicable. According to the IoT method, Ref. [14] studied a data accessing model in case of requiring emergency medical facilities.

In this study, an innovative development of fitness system has been integrated with IoT and ML methods. A data acquirement model has been developed. Also, it is proposed with the techniques to predict the heart rate and to design a RS. Under the application of DL method, this technology is capable of observing and reminding the correctness of user events. In this paper, a Cognitive IoT-Based Smart Fitness Diagnosis and Recommendation System using a three-dimensional CNN with hierarchical particle swarm optimization (PSO) algorithm has been presented. The proposed model is applied to observe the health statuses of exercisers. The system offers proper supervision to the exercisers. A set of experiments takes place, and the simulation result highlights the betterment of the proposed model.

2 Proposed Model

The structure of the proposed system is depicted in Fig. 1. This model is comprised of the points given below:

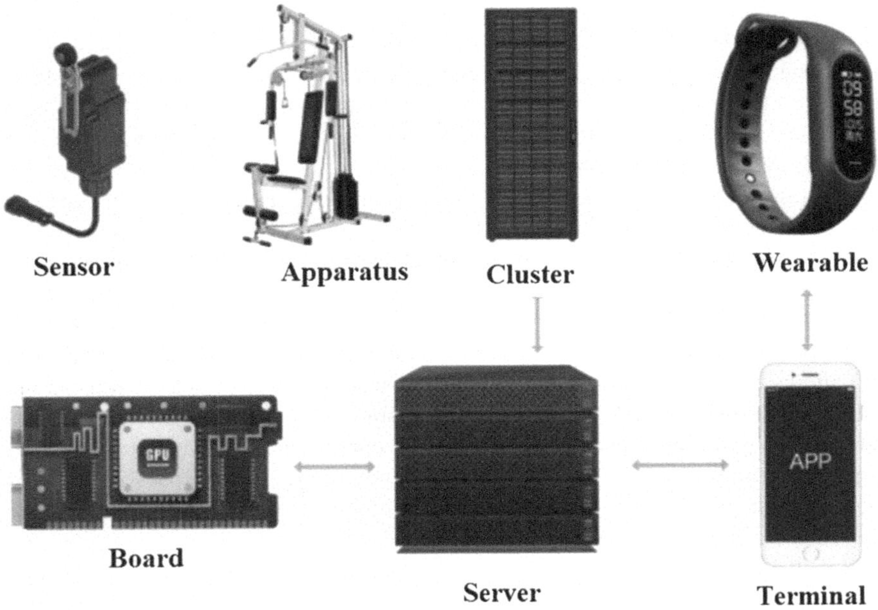

Fig. 1 Framework of intelligent devices in healthcare

1. Fitness apparatus: Generally, there are a more number of fitness devices available in gym. Such tools are reconstructed by embedding sensors. For instance, a distance sensor has been included in a pulling device to estimate the pulling distance. Then, it is calculated with the pulling distance.
2. Sensor: It is mainly applied in fitness tools to estimate the value of workload for users. Typically, it has been employed with the WSN to eliminate the cable connection issue. It is a specified issue which is based on the apparatus.
3. Wearable device: It is a device which helps to save the regular exercises, like real-time heart rate, number of times the exercise to be repeated, and so on. This device can be majorly utilized in recording the routine function whenever an individual attends the fitness class. The main purpose of gathering this information is to determine the complete health situation of a user.
4. Development board: This objective is helpful in collecting data from sensors. Usually, it is modeled with a touch screen which might be easy for clients to operate. A purpose of embedded in development board is same as mobile terminal. Also, it is applied to forward data from sensors to server. At the same time, it is considered to be an additional device for mobile users.
5. System server: The intermediate portion of a system is said to be a server that helps to gather and investigate the data sent transmitted from sensors to terminal. It offers a web application interface (API) to stimulate the mobile terminal.

6. Computing cluster: This computation is mostly applied to functions greater computational operations. Such kinds of computing nodes are delivered with system server.
7. Mobile terminal: The terminal displays the physical data or exercise. It is implemented on iOS or android platform.

2.1 The Use of Sensors

Here, it is applied with the distance sensors to calculate the distances of data. Based on the workload function, its used calories might be computed using $F \times S$. F is a pulling force that is often provided with bodybuilding devices. S denotes the distance that has been estimated by a distance sensor. Hence, a measure of applied calories is a significant variable for fitness. Therefore, distance sensors are applied in an IoT fitness system.

2.2 SDL

The customer's information can be determined on a system with the application of two models as given in the following:

- Sensor acquirement
- Fitness band acquirement

2.2.1 Sensor Acquirement

Usually, Information regarding the fitness exercises is gathering utilize of sensors. These sensors are linked with the system by either wired or wireless connections. By the assumption of cost and satisfaction, it is applied with WIFI model ESP82662 that has been linked with sensors through general-purpose input/output (GPIO) interface. Thus, the data obtained from sensors is acquired from GPIO that is an easy implementation. This Wi-Fi offers a sequence of API functions to read or write the measure of GPIO. In order to activate the Wi-Fi technique, it requires fixing the Wi-Fi mode in AP mode with the help of AT command.

The availability of Wi-Fi usage points has been screened. Later, the Wi-Fi model has been linked with Wi-Fi network. Then, the variable of AP mode is fixed. Once these procedures are completed, Wi-Fi method is capable of connecting to the system. Therefore, a TCP connection is developed for data transmission. Typically, data could be read with the application of a loop and forward data by TCP connection. Alternatively, an interrupt function for GPIO has been developed. Figure 2 shows the flowchart process of data acquirement. Once the weights

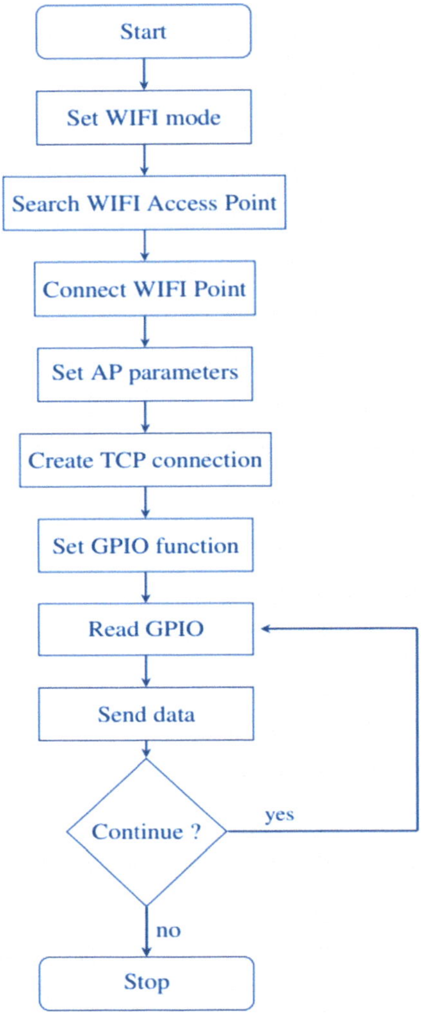

Fig. 2 Process involved in data acquirement

are declared, the energy cost can be evaluated by a work formula. Finally, it has to be measured with distance of motion; thus, it provides the result of energy consumed for an exercise.

2.2.2 Fitness Band Acquirement

Recently, fitness band has become familiar and a smart device. Many users use this fitness band in daily life. The fitness band is mainly applied to observe the state of a user's body. In addition, fitness band can be employed for authentication. If an individual enters a fitness class, fitness band maps the user's account. Simultaneously, the models examine the regular data from fitness band and recommend or advise to follow a suitable exercise regularly.

2.3 *Burnt Calories*

The burnt calories are divided into two steps: exercise and walk part.

The calories consumed with the exercise are measured with work notation. Initially, it collects the distance as well as weight data from sensors. Then, it undergoes a combination to obtain the exercise portion of applied calories.

Generally, to estimate the consumed calories either by walking or running, few experiments were carried out to attain the optimal solution [2]:

$$d = \mathrm{Ht} \times 0.45 \times \mathrm{step} \tag{1}$$

$$\mathrm{cal} = \mathrm{Wt} \times d \times 1.036 \tag{2}$$

In Eqs. (1) and (2), Ht is a height of a user, "step" refers the count of steps, and d represents a distance (meter). Upon the validation of several experiments, it has been found that phase length is proportional to the height of an individual and a scale factor is 0.45. Wt represents a weight of a user (kg); cal denotes a rate of calories burnt. Also, the burnt calories are approximately proportional to a user's weight as well as walking distance. The scale coefficient has been fixed to 1.036. In this model, a step is given by the fitness band. Practically, the calorie formula is said to be an empirical formula which is a fairy rough method to estimate the burnt calories. But, it is more applicable for customers to follow the exercise and also utilized to refer the effects of exercise.

2.4 *Heart Rate Detection*

Heart rate is defined as the count of times the heart beats per minute (bpm). Different types of models are used to calculate the heart rate based on the physical environment. In case a user applies a fitness band, the heart rate is captured in an automated manner. Several well-known fitness bands offer API to attain the original heart rate. Usually, heart rate could be calculated from smart phone like iPhone, android, and so on.

The process of predicting the heart rate with smart phone exploits a smart phone camera to predict the capillary differences of user's forefingers. If a user places the forefinger on a camera, it feels with few modifications in light intensity of a forefinger as the blood gets pressed inside a skin of the forefinger. Therefore, it collects the bpm by regular variation in brightness. Figure 3 depicts the heart rate prediction on

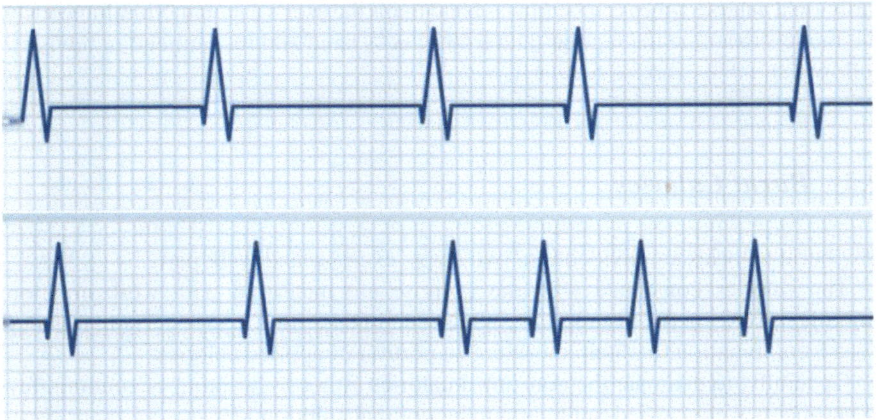

Fig. 3 Cardiograph on mobile phone

mobile phone. Detecting the heart rate is more helpful to many people for keeping the health condition in a balanced manner. In Fig. 3, it demonstrates a kind of arrhythmias which is an irregular heartbeat while sitting without any interruption. It is caused due to the existence of palpitation. Therefore, a novel technique helps to analyze the ailments of heart disease. While gathering the heart rate, the user may be in a similar state, like sitting quietly. Next, it calculates the bpm of every beat with 60/t and obtains a sequence of continuous bpm rates. Later, the mean values as well as SD of 30 bpm rates are measured. Based on the values of simulation outcome, ordinary mean values of bpm exist from 50 to 100. A SD value should be higher when compared with 1.0 to maintain a normal heartbeat.

2.5 Recommendation System

HPSO clustering applies PSO for clustering data into same groups on the basis of distance affinity from data elements. This model helps to integrate the features of hierarchical clustering as well as partitional clustering models along with PSO. As there are more flexible cluster granularity, expandability, and efficiency, HPSO clustering designs the web application data to cluster the RS. In this model, every particle denotes a centroid of a cluster; the whole swarm is a solution for clustering issue. Also, it is retained with maximum particles, as the complete search space has been covered and disjoint small clusters could be found. Simultaneously, all particles are initiated systematically that tends to distribute input data space with the help of

$$\mathrm{loc}\left(X\left(i\right)\right)=i\times\left(\frac{N}{k}-1\right) \tag{3}$$

where loc(X) is a position of a particle; i represent the index of a particle, which exist from 0 to the higher number of particles k; and N denotes the overall count of data vectors. After completing the initialization process, first-generation swarm has been iterated to explore optimal positions. In order to attain a better position, particles apply the corresponding cognitive, social, as well as self-organizing units.

Here, a cognitive unit of a model undergoes encoding as (pBest − $X_i(t)$), where $X_i(t)$ refers a present location of a particle and pBest is measured as

$$\mathrm{pBest}_i\left(t+1\right)=\begin{cases}\mathrm{pBest}_i\left(t\right) & \text{if } f\left(X_i\left(t+1\right)\right) \text{is not better than } f\left(\mathrm{pBest}_i\left(t\right)\right)\\ X_i\left(t+1\right) & \text{if } f\left(X_i\left(t+1\right)\right) \text{is better than } f\left(\mathrm{pBest}_i\left(t\right)\right)\end{cases} \tag{4}$$

where personal best position (pBest) is a position of closer data vector. It can be attained by maintaining the location of best position which is accomplished by a particle from particular generation. Every particle is comprised of a corresponding pBest value that has been measured with minimum difference in PSO pBest equation to find minima.

2.6 *Action Recognition*

In a fitness process, the physical events are observed, mostly in the situation of wrong action that leads to physical injury on equipment. To resolve these issues, four models have been developed to analyze the fitness actions and behaviors. There are few webcams introduced in fitness class to gather realistic video frames. Recently, a deep neural network (DNN) is an evident to be efficient on image analysis. CNN is another kind of DNN.

CNN might be a specialized feed-forward NN. It is enclosed with input layer, conv layer, subsampling layer, and fully connected (FC) layer. These layers are based on the robust formation of CNN technique. Usually, the input layer is in the vector format of an image matrix, and the simulation outcome is a type of probability. Action recognition is the same as image identification. The only difference is that the input of action recognition is not provided as an image; it is provided only as continuous image frames. Thus, it is employed with 3D CNN rather to analyze the fitness events, as given in Fig. 4. The variation from 3D CNN and CNN is that 3D CNN filters space and time features with the help of 3D conv kernels. A common KTH dataset 3 has been applied to sample the test event of 3D CNN.

There are 20 matrices derived from continuous frames of 20 × 16 × 16 vector. This vector can be applied as input vector. In a conv layer, it is employed with 32 conv filters of size 5 × 5 × 5 named as conv kernels. Hence, the size of a

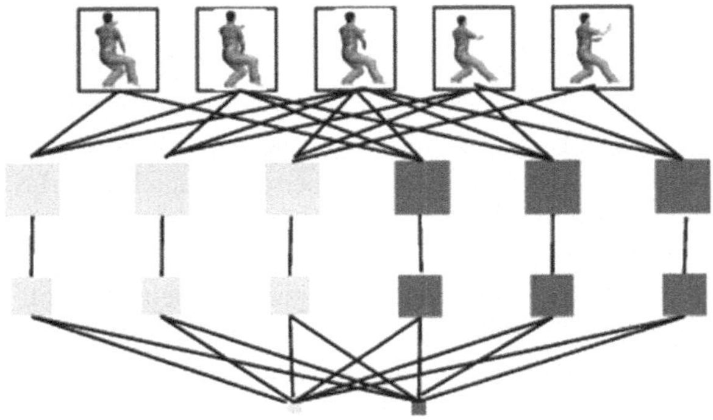

Fig. 4 The structure of 3D CNN

max-pooling in subsampling layer is fixed as 3 × 3 × 3. Later, FC is used to differentiate the actions. The neuron unit is assigned as 500 with a dropping measure as 0.5. Finally, the resultant layer is used with softmax function to represent the possibilities of various events.

3 Performance Validation

Figure 5 shows the training and validation accuracy of the 3D CNN for action recognition under varying epoch counts. From the figure, it is shown that the training and validation accuracy gets increased with increasing epoch count. It is shown that the training accuracy reached to a maximum of 95% and 97% under the validation process.

Table 1 and Fig. 6 provide a detailed comparative analysis of diverse models in terms of precision and recall. It is shown that the precision and recall values get increased with an increase in particles. Under the presence of particle 1, it is shown that the CNN-HPSO model achieves a maximum precision of 0.78. Similarly, under the presence of particle 2, it is shown that the CNN-HPSO model achieves a maximum precision of 0.80. Likewise, under the presence of particle 3, it is shown that the CNN-HPSO model achieves a maximum precision of 0.82. Likewise, under the presence of particle 4, it is shown that the CNN-HPSO model achieves a maximum precision of 0.84. Equally, under the presence of particle 5, it is shown that the CNN-HPSO model achieves a maximum precision of 0.87. In the same way, under the presence of particle 6, it is shown that the CNN-HPSO model achieves a maximum precision of 0.89. Likewise, under the presence of particle 7, it is shown that the CNN-HPSO model achieves a maximum precision of 0.91. In the same way, under the presence of particle 8, it is shown that the CNN-HPSO model achieves a maximum precision of 0.93. Afterward, under the presence of particle 9, it is shown that the CNN-HPSO model achieves a maximum precision of 0.94. In the same

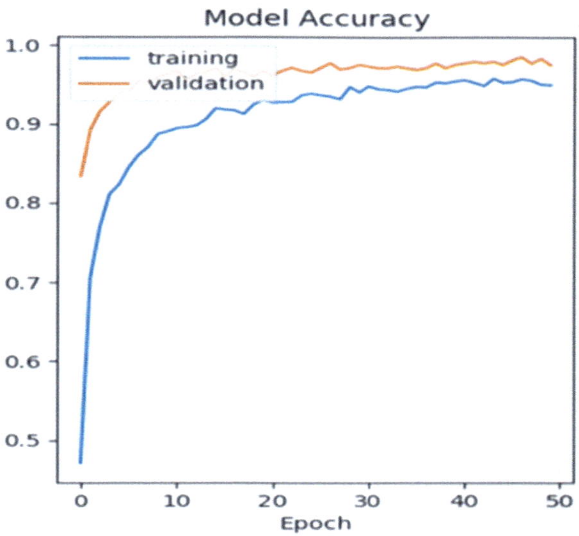

Fig. 5 Training and validation accuracy of 3D CNN for action recognition

Table 1 Precision and recall analysis of CNN-HPSO model

No. of particles	Precision	Recall
1	0.78	0.65
2	0.80	0.68
3	0.82	0.72
4	0.84	0.75
5	0.87	0.77
6	0.89	0.81
7	0.91	0.83
8	0.93	0.84
9	0.94	0.85
10	0.96	0.89
Average	*0.87*	*0.78*

way, under the presence of particle 10, it is shown that the CNN-HPSO model achieves a maximum precision of 0.96.

Under the presence of particle 1, it is shown that the CNN-HPSO model achieves a maximum recall of 0.65. Similarly, under the presence of particle 2, it is shown that the CNN-HPSO model achieves a maximum recall of 0.68. Likewise, under the presence of particle 3, it is shown that the CNN-HPSO model achieves a maximum recall of 0.72. Likewise, under the presence of particle 4, it is shown that the CNN-HPSO model achieves a maximum recall of 0.75. Equally, under the presence of particle 5, it is shown that the CNN-HPSO model achieves a maximum recall of 0.77. In the same way, under the presence of particle 6, it is shown that the CNN-HPSO model achieves a maximum recall of 0.81. Likewise, under the presence of particle 7, it is shown that the CNN-HPSO model achieves a maximum recall of 0.83. In the same way, under the presence of particle 8, it is shown that the

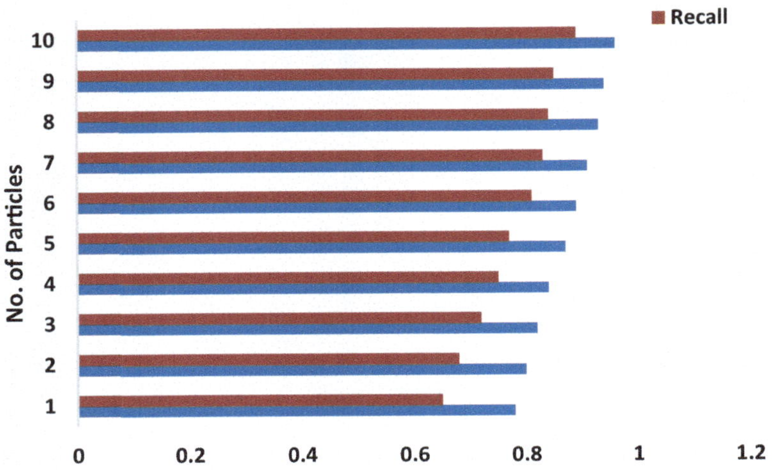

Fig. 6 Precision and recall analysis of CNN-HPSO model

Table 2 Comparative analysis of average precision and recall

Methods	Precision	Recall
CNN-HPSO	0.87	0.78
PSO	0.82	0.75
ACO	0.84	0.76
GA	0.79	0.72

CNN-HPSO model achieves a maximum recall of 0.84. Afterward, under the presence of particle 9, it is shown that the CNN-HPSO model achieves a maximum recall of 0.85. In the same way, under the presence of particle 10, it is shown that the CNN-HPSO model achieves a maximum recall of 0.89.

Table 2 and Fig. 7 offer a brief comparative study of the performance offered by the CNN-HPSO model with PSO, ACO, and GA models. On measuring the results in terms of precision, it is shown that the GA offers the least performance by attaining a minimum value of 0.79. Besides, it is shown that the PSO algorithm tried to show manageable results and ended up with a slightly higher precision of 0.82. Moreover, the ACO algorithm has exhibited near optimal results and offered a manageable precision value of 0.84. However, the CNN-HPSO model shows excellent results by achieving a maximum precision value of 0.87. On measuring the results in terms of recall, it is shown that the GA offers the least performance by attaining a minimum value of 0.72. Besides, it is shown that the PSO algorithm tried to show manageable results and ended up with a slightly higher recall of 0.75. Moreover, the ACO algorithm has exhibited near optimal results and offered a manageable recall value of 0.76. However, the CNN-HPSO model shows excellent results by achieving a maximum recall value of 0.78. The abovementioned tables and figures clearly

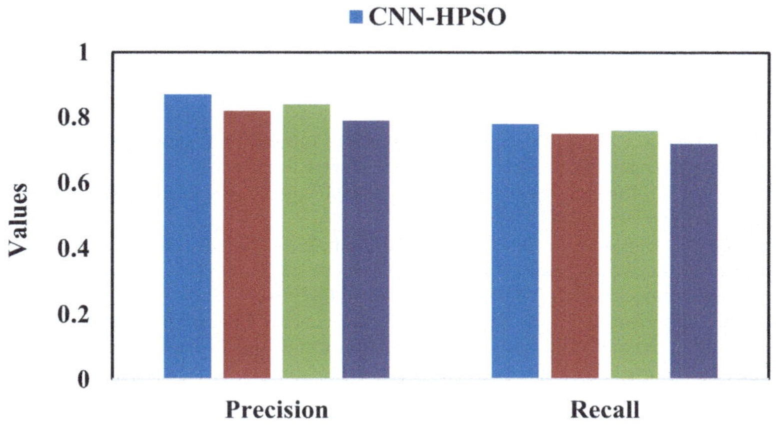

Fig. 7 Average precision and recall analysis

notified the superior results of the CNN-HPSO model with the precision and recall values of 0.87 and 0.78, respectively.

4 Conclusion

In this paper, a Cognitive IoT-Based Smart Fitness Diagnosis and Recommendation System using CNN-HPSO model has been presented. The proposed model is applied to observe the health statuses of exercisers. The system offers proper supervision to the exercisers. An innovative development of fitness system has been integrated with IoT and ML methods. Data acquirement model has been developed. Also, it is proposed with the techniques to predict the heart rate and to design a RS. Under the application of DL method, this technology is capable of observing and reminding the correctness of user events. The experimental results clearly notified the superior results of the CNN-HPSO model with the precision and recall values of 0.87 and 0.78 respectively.

References

1. Thompson, R.: Walter, worldwide survey of fitness trends for 2016: 10th anniversary edition. ACSMs Health Fit. J. **19**(6), 9–18 (2016)
2. Yong, B., Xu, Z., Wang, X., Cheng, L., Li, X., Wu, X., Zhou, Q.: IoT-based intelligent fitness system. J. Parallel Distrib. Comput. **118**, 14–21 (2018)
3. Kim, J., Yun, J., Choi, S.C., Seed, D.N.: Standard-based IoT platforms interworking: implementation, experiences, and lessons learned. IEEE Commun. Mag. **54**, 48–54 (2016)

4. Parviainen, V., Yin, A., Romu, A., Virkkala, R.: Efiiot: an efficient software architecture for internet of things. In: 2012 Seventh IEEE Conference on Industrial Electronics and Applications (ICIEA), pp. 709–712 (2012)
5. Duffy, B.R.: Anthropomorphism and the social robot. Robot. Auton. Syst. **42**(3–4), 177–190 (2003)
6. Gubbi, J., Buyya, R., Marusic, S., Palaniswami, M.: Internet of Things (IoT): a vision, architectural elements, and future directions. Futur. Gener. Comput. Syst. **29**(7), 1645–1660 (2013)
7. Lee, I., Lee, K.: The Internet of Things (IoT): applications, investments, and challenges for enterprises. Bus. Horiz. **58**(4), 431–440 (2015)
8. Krishnaraj, N., Elhoseny, M., Thenmozhi, M., Selim, M.M., Shankar, K.: Deep learning model for real-time image compression in Internet of Underwater Things (IoUT). J. Real-Time Image Proc. (2019). https://doi.org/10.1007/s11554-019-00879-6
9. Singh, A., Aggarwal, P., Arora, R.: IoT based waste collection system using infrared sensors. In: 2016 Fifth International Conference on Reliability, Infocom Technologies and Optimization (Trends and Future Directions) (ICRITO) (2016)
10. Rathore, M.M., Ahmad, A., Paul, A.: IoT-based smart city development using big data analytical approach. In: 2016 IEEE International Conference on Automatica (ICA-ACCA) (2016)
11. Nugra, H., Abad, A., Fuertes, W., Galarraga, F., Aules, H., Villacis, C., Toulkeridis, T.: A low-cost IoT application for the urban traffic of vehicles, based on wireless sensors using GSM technology. In: 2016 IEEE/ACM 20th International Symposium on Distributed Simulation and Real Time Applications (DS-RT) (2016)
12. Ghosh, A.M., Halder, D., Hossain, S.K.A.: Remote health monitoring system through IoT. In: 2016 Fifth International Conference on Informatics, Electronics and Vision (ICIEV) (2016)
13. Huang, C.H., Cheng, K.W.: RFID technology combined with IoT application in medical nursing system. Chem. Eng. News. **88-18** (2014)
14. Xu, B., Xu, L.D., Cai, H., Xie, C.: Ubiquitous data accessing method in IoT based information system for emergency medical services. IEEE Trans. Ind. Informat. **10**(2), 1578–1586 (2014)

Industrial Internet of Things (IIoT) with Cloud Teleophthalmology-Based Age-Related Macular Degeneration (AMD) Disease Prediction Model

R. J. Kavitha, T. Avudaiyappan, T. Jayasankar, and J. Arputha Vijaya Selvi

Abstract Industrial Internet of things (IIoT) utilizes smart sensors and actuators for enhancing manufacturing and industrial processes. Due to the advanced technological developments in healthcare industry, it has been proved that the primary detection of chronic diseases, namely, diabetic retinopathy (DR) as well as age-related macular degeneration (AMD), is capable of preventing loss of vision. In this study, a scalable cloud-oriented teleophthalmology structure by an Internet of medical things (IoMT) to detect AMD is projected. In the presented system, patient's wearable camera for transmitting the retinal fundus photographs for a secured cloud drive storage for diagnosing the severity of disease as well as predictive progression examination. A projected optimal generative adversarial network (OGAN) helps to investigate the images to find as well as to compute AMD disease severity. The GAN would be optimized with the application of a bat method. The performance of the proposed OGAN model has been validated using a set of benchmark images. A set of three measures used to examine the results are sensitivity, specificity, and accuracy. The experimental outcome showed the superior performance of the proposed model over the compared methods by attaining a maximum accuracy of 98.03%.

Keywords Age-related macular degeneration · Cloud teleophthalmology · IIoT · Deep learning

---

R. J. Kavitha (✉)
University College of Engineering, Panruti, Panruti, Tamil Nadu, India
e-mail: drrjk@ucep.edu.in

T. Avudaiyappan
K.Ramakrishnan College of Technology, Tiruchirappalli, Tamil Nadu, India

T. Jayasankar
University College of Engineering, BIT Campus, Anna University, Tiruchirappalli, Tamil Nadu, India

J. A. V. Selvi
Kings College of Engineering, Punalkulam, Tamil Nadu, India

D. Gupta et al. (eds.), *Smart Sensors for Industrial Internet of Things*, Internet of Things, https://doi.org/10.1007/978-3-030-52624-5_11

1 Introduction

Industrial Internet of things (IIoT) influences the power of smart machines and real-time analytics for taking the benefits of the data which dumb machines have produced in industrial settings for years. The driving viewpoint behind IIoT is that smart machines are not only better than humans at capturing and investigating data in real time, but they are improved at communicating essential data which could be employed for driving business decisions faster and more precisely. Age-related macular degeneration (AMD) is said to be an typical eye state which leads to loss of vision mostly for old peoples [1], along with a blinding prevalence all over the world that causes vision impairment such as uncorrected refractive faults, unoperated cataracts, as well as DR. Recently, the existence of AMD is around 1.1%. AMD is not capable of self-simulating which does not provide blindness, but vision loss occurred by AMD could interfere with elegant activities, like potential to observe faces, drive, read, and so on. Hence, it can be vital to many patients for vitreoretinal centers at the initial phase of AMD to conserve the visual applications. Several advancements in studies, medicine, awareness, as well as communications result in a major reduction of dangerous eye diseases like trachoma and onchocerciasis for the past decades. Therefore, in maximum countries, the minimization of vision impairment in people has become small as the absence of hospitals, physicians, geographical position, and so on. As the analysis of vision-impairing diseases has to be carried out at the initial stage, accurate and effective rates must be obtained to report the diseases. Recently, models and related works like cloud computing and Internet of thing (IoT) components have led to modify the type of communication, exchange, use, and save data from devices.

Telemedicine and telescreening have become more possible models of healthcare delivery, with respect to avoidance and organization. Telemedicine is a conversion of different medical information, namely, electronic health records (EHR), higher-definition images as well as videos of patients, and the remote locations, to alternate geographical position by accessing the professional physicians and clinics [2]. It is described in a brief manner as the application of telecommunication as well as processing models to offer medicinal data, remote facilities, and medical events like analysis, remedy, elimination of disease, etc. The current research of teleophthalmology activated maximum emergency diagnosis and recovery options for patients lagging in sufficient medical facilities. Hence, many numbers of doctors are in need to process teleophthalmology solution effectively and accurately which produces several barriers for doctors. It has been considered to be a portion as many cases have been recovered by doctors regularly. Also, many numbers of benign cases are treated by physicians. So the proper robust screening models to precede the function of doctors and to stimulate the teleophthalmology pipeline.

By applied the extended models like Artificial Intelligence (AI). In addition, by the applications of big data as well as the structure of maximum computation, different issues are handled with the help of AI spanning healthcare, academics, vehicle, and R1C5. Deep learning (DL), the class of ML, is defined as AI revolution that is applied in many numbers of applications such as medical research. AI can be used in medical application which enables doctors to perform the best imaging, analysis, risk prediction, and assisted care at the time of giving accurate medicine, drug discovery, lifestyle, observing, remote patient care, and so on [3]. Many studies have reported various models which are predefined samples in clinics to predict breast cancers, minor lung tumors on chest X-rays, fractures, and so on [4]. The DL-oriented AI engines are employed for analyzing diverse retinal infections like DR [5] as well as macular degeneration [6] from the color fundus images. These models enable doctors to analyze retinal diseases in primary phases and eliminate loss of vision. A different study in mobile teleophthalmology is a great extent for [7, 8]. concentrated on deploying mobile environments that depends upon depression ontology. By the applications of image sensors and smartphone models, it is feasible to make possible images with no cost of camera. Exploring the efficiency of automatic diagnosis when relate by human labels from a fundus image screening center. The employment of smartphone cameras in retinal cameras is found to be an inspiring study managed by ophthalmologists and developers identical to the recent years.

Ref. [9] exploit the smartphone cameras for capturing fundus photographs by a qualified and diagnostic component of conventional fundus camera images. The survey that addressed the medical verification of smartphone-oriented adapters in optic disk (OD) imaging is carried out by [10]. Consequently, nonclinical images with the application of a minimum cost smartphone adapter are captured using a reputed image under the application of a desktop retinal camera processed with an ophthalmic assistant. Ref. [11] associated a smart mobile ophthalmoscope through a slit-lamp biomicroscope to rank DR in patients with diabetes mellitus (DM). The reasonable arrangement from a smartphone ophthalmoscope through a dilated retinal biomicroscopy to rank DR has been initiated, and developers advised to utilize smartphones as a portable, suitable method of capturing images and distributing to doctors by a domain of smartphone.

Ref. [12] deployed smartphone, non-mydriatic, and mydriatic images that have been associated with the potential to forecast the rank of DR. As a result, it has been stated that smartphone and non-mydriatic fundus images are capable of predicting DR and the non-mydriatic cameras are more sensible in predicting DR when compared with smartphone. Regardless of the benefits smartphone offers of conventional retinal imaging models, many researchers advised in maintaining classical techniques due to maximum sensitivity. But, it proves that smartphone ophthalmology continues to enhance sensors and models like image stabilization. In this study, a scalable cloud-oriented teleophthalmology structure by an IoMT to detect AMD is projected. In the presented system, patient's wearable camera for transmitting the retinal fundus photographs for a secured cloud drive storage for diagnosing the severity of disease as well as predictive progression examination. A projected OGAN helps to investigate the images to find as well as to compute AMD disease severity. The GAN would be optimized with the application of the bat method.

2 The Proposed OGAN Model

The presented structure is comprised of three primary units to activate ophthalmologist in a loop telemedicine platform at the time of integrating along with AI-based health assistive professional as depicted in Fig. 1. The wearable head-embedded camera by AMD-ResNet deep convolutional neural network (CNN) is applied for AMD detecting the severity of disease. This domain enables the customers to obtain novel retinal fundus images of the eyes after completing the encryption process, securely forwarded through the Internet to a middle position safe in a firewall. A headset provides a local inference under the application of incorporated-AI model in the cloud back end by applying AI and data analytics services. The patient information could be recorded locally by encrypting with the secured storage. The cloud computing (CC) approach, which handles as well as serves a link to every Ophtho AI IoMT headset, protects the broadcast of fundus images, secures patient cloud memory drive, and observes as well as meters machine resources, AI engines to disease detection, and so on. These examinations are controlled by a directory as well as system manager. The doctor dashboard including a protected multiple tenant cloud backend offers privacy-aware role-oriented accessing management for patient details as well as resources. Hence, a secure multi-tenancy guarantees with respect to data loss, misuse, privacy violation, and so on.

Generative adversarial networks (GANs) are the effective class of NN which has been applied for unsupervised learning. GANs are configured with the system of two competing NN methods that compete with one another as well as the capability of analyzing, capturing, and duplicating the difference inside a dataset. Unexpectedly, the technique is included with maximum confidence in an ineffective detection process. The cause for this ML method from a minimized data is comprised of major demerit, as it is more vulnerable for overfitting. In addition, mapping from input to

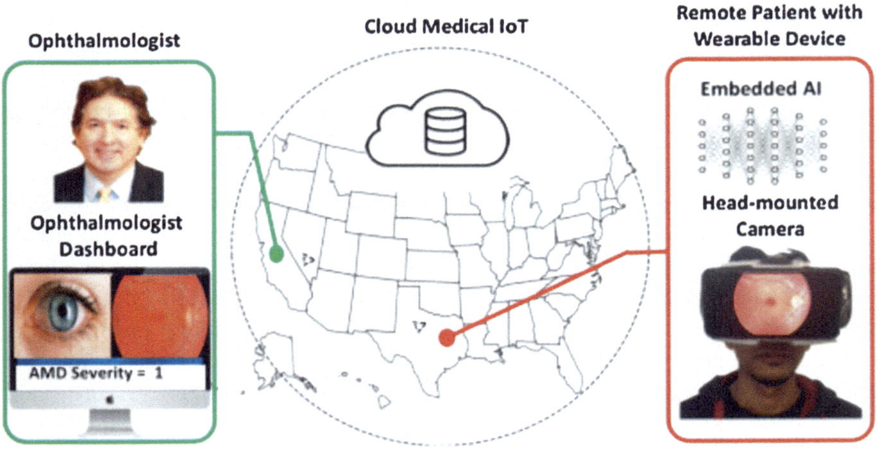

Fig. 1 Basic architecture of the proposed method

output is mostly in a linear fashion, even though it seeks for a division among different categories linearly. Actually, it is constrained with linearity's and minimum variations in a point from a feature space which tend to misclassify the data.

The GANs could be divided into three portions:

- Generative: For learning a generative method that defines the production of data with respect to probabilistic technique.
- Adversarial: A model can be trained by an adversarial setting.
- Networks: Apply DNN as the AI model for training the applied methods.

The GAN is a generator and discriminator; a generator creates forged instances of information (like image, audio, and so on.) and tries to idiot the discriminator. A discriminator conversely attempts to differentiate among original and forged instances. A generator and discriminator are called NNs, and they both run in competition with each other in the training stage. This phase is frequent many times in this generator as well as discriminator obtains optimal and optimal in their particular jobs following all replication. Now, generative methods capture the sharing of information and are trained in such a method which attempts to the highest possibility of the discriminator in creating the fault. A discriminator conversely depends on a concept that estimations the possibility that the instance it gets are obtained from the training information and not from the generator.

GAN is created as a minimax game, if the discriminator is annoying to diminish its return $V(D, G)$ and the generator is annoying to diminish the discriminator's return or in other words highest its failure. It is mathematically explained with the formula under

$$\underset{G}{\mathrm{mn}}\,\underset{D}{\mathrm{mx}}\,V(D,G) \tag{1}$$

$$V(D,G) = \mathbb{E}_{l\sim p_{\mathrm{data}}(l)}\left[\log D(l)\right] + \mathbb{E}_{z\sim p_z(z)}\left[\log\left(1 - D(G(l))\right)\right] \tag{2}$$

where

G = generator
D = discriminator
$P_{\mathrm{data}}(l)$ = shared of original information
$P(z)$ = shared of generator
l = instance from $P_{\mathrm{data}}(l)$
z = instance from $P(z)$
$D(l)$ = discriminator network
$G(z)$ = generator network

Thus, essentially, training GANs have two parts:

- *Part 1*: The discriminator is trained as the generator is unused. During this phase, the network is only forward-propagated and no backpropagations are finished. The discriminator is trained on actual information to n epochs and sees when it

is accurately forecasted them as original. Moreover this phase, the discriminator is beside to train the forged created information from the generator and sees when it is accurately forecasted them as forged.
- *Part 2*: The generator is trained as the discriminator is unused. Behind the Discriminator is trained with created as forged information of the Generator, it is obtain its forecast and utilize the outputs to trained the Generator and obtain optimal from the preceding condition to attempt and fool the discriminator.

The above technique is frequent to some epochs; next manually ensure the fake information when it seems genuine. When it seems suitable and training is closed, it's permitted for continuing to some more epochs. BA is introduced as a meta-heuristic search technique which utilizes the interesting capabilities of bats like catching the prey and distinguishing various kinds of insects still in without any light. The higher echolocation ability of bats turns them very interesting. These capabilities motivated several researchers to employ these several domains.

Bats employ the conventional sonar known as echolocation to discover preys and eliminate the obstacles. Bats, particularly microbats, have the ability to identify the object's location using the distribution of high and short audio signals with collision as well as reflection of these broadened signals. In BA, the echolocation features are used in the proposed method by taking the advantageous of such features of bats:

- Every bat uses its echolocation procedure to compute distance and also determine the distance between food/preys and obstacles.
- A bat flies in a random manner with velocity v_x at position l_x with a frequency f_{mn}, by altering the wavelength and loudness A_0 for searching the prey. It can regulate the wavelength (or frequency) of its emitted pulses in an automatic manner and also alters the rate of pulse emission $r_x[0, 1]$, based on the proximity of its goal.
- Though the intensity is different in various ways, it is considered that the loudness ranges between large (positive) A_0 and a minimum constant value A_{mn}.

At the beginning, the bats with random positions and velocities are initialized in the search space. The newly generated solutions l_x^t and velocities v_x^t at time step t are indicated in Eqs. (3)–(5):

$$f_x = f_{mn} + \left(f_{mx} - f_{mn}\right)\beta \tag{3}$$

$$v_x^t = v_x^{t-1} + \left(l_x^t - l^*\right) f_x \tag{4}$$

$$l_x^t = l_x^{t-1} + v_x^t \tag{5}$$

where $\beta \in [0, 1]$ is an arbitrary vector drawn from a uniform distribution and l^* is an existing global optimal location that is positioned once the results between every n bat are compared at every round. Primarily, every bat is arbitrarily allocated to a frequency that is drawn uniformly from $[f_{mn}, f_{mx}]$. Figure 2 shows the flowchart of the BA model.

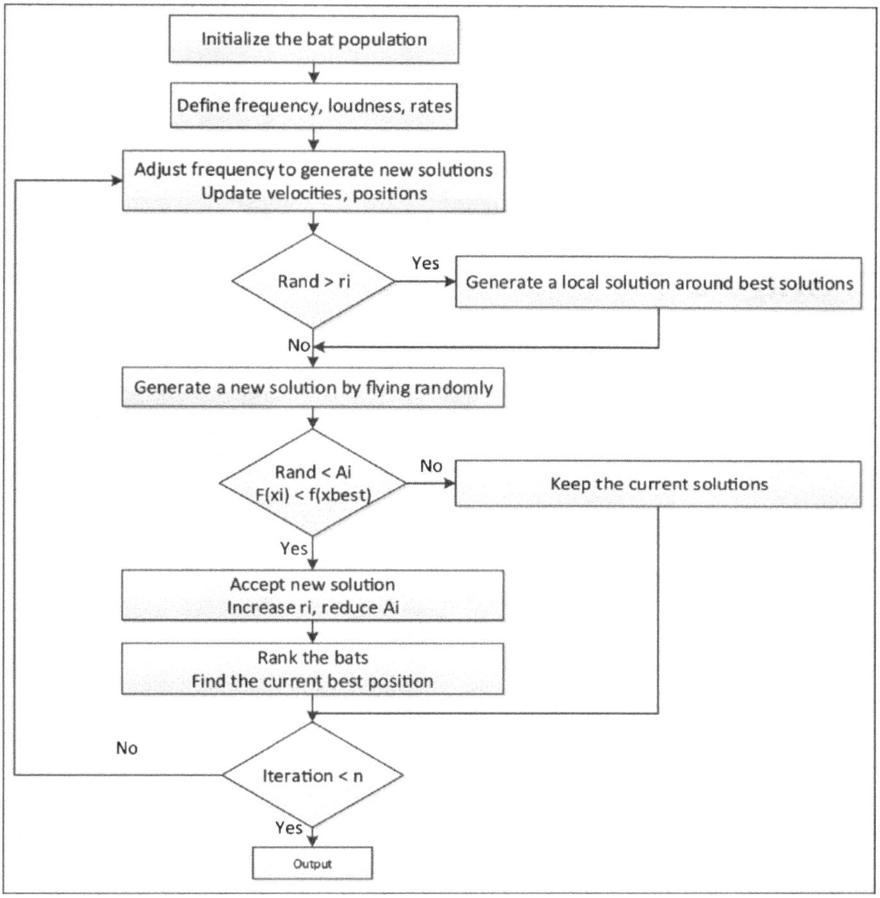

Fig. 2 Flowchart of BA

3 Performance Validation

The performance of the proposed OGAN model has been validated using a set of benchmark images. A sample set of images under diverse stages of severity are shown in Fig. 3. A set of three measures used to examine the results are sensitivity, specificity, and accuracy. Table 1 and Fig. 4 show the precision analysis of diverse models on the applied dataset. Under the classification of images under stage 1 grade, it is shown that the proposed OGAN model has achieved maximum classification with a precision of 96.56% whereas the AMD-ResNet has shown a poor performance by attaining a lower precision of 95.05%. Similarly, under the stage 2 grade, it is shown that the proposed OGAN model has achieved maximum

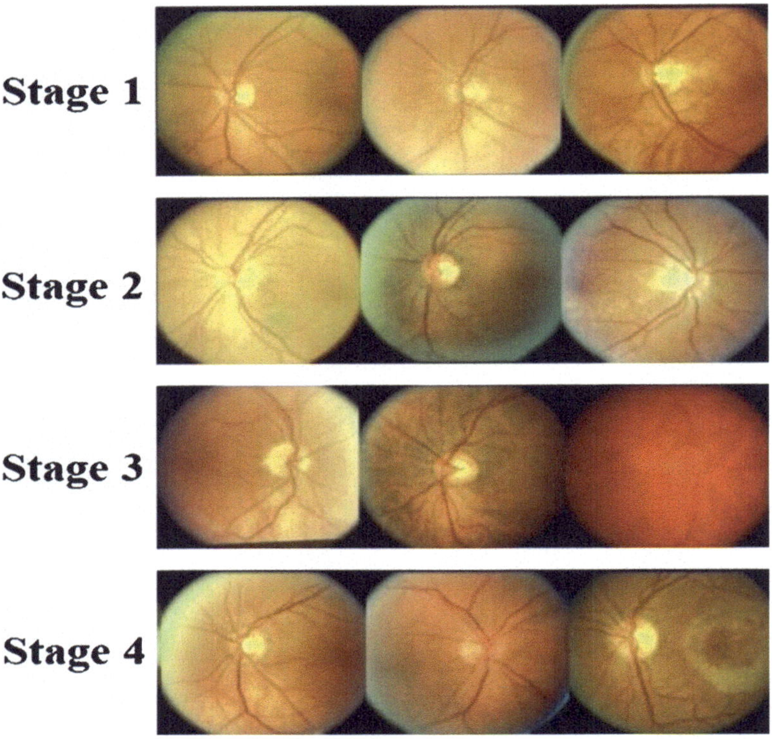

Fig. 3 Fundus images of various severity stages

Table 1 Precision of proposed model with existing methods for validation images

Input grades	OGAN	AMD-ResNet
Stage 1	96.56	95.04
Stage 2	97.37	95.06
Stage 3	97.54	96.22
Stage 4	93.87	92.46

classification with a precision of 97.37% whereas the AMD-ResNet has shown a poor performance by attaining a lower precision of 95.06%.

Likewise, under stage 3 grade, it is shown that the proposed OGAN model has achieved maximum classification with a precision of 97.54% whereas the AMD-ResNet has shown a poor performance by attaining a lower precision of 96.22%. Finally, under stage 4 grade, it is shown that the proposed OGAN model has achieved maximum classification with a precision of 93.87% whereas the AMD-ResNet has shown a poor performance by attaining a lower precision of 92.46%.

Table 2 and Fig. 5 show the recall analysis of diverse models on the applied dataset. Under the classification of images under stage 1 grade, it is shown that the

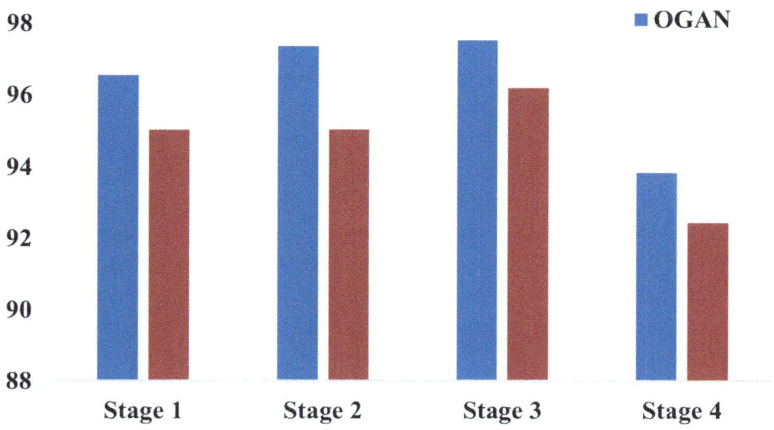

Fig. 4 Precision analysis of diverse models

Table 2 Recall of proposed model with existing methods for validation images

Input grades	OGAN	AMD-ResNet
Stage_1	95.89	94.95
Stage_2	97.91	97.03
Stage_3	94.87	93.45
Stage_4	95.09	94.96

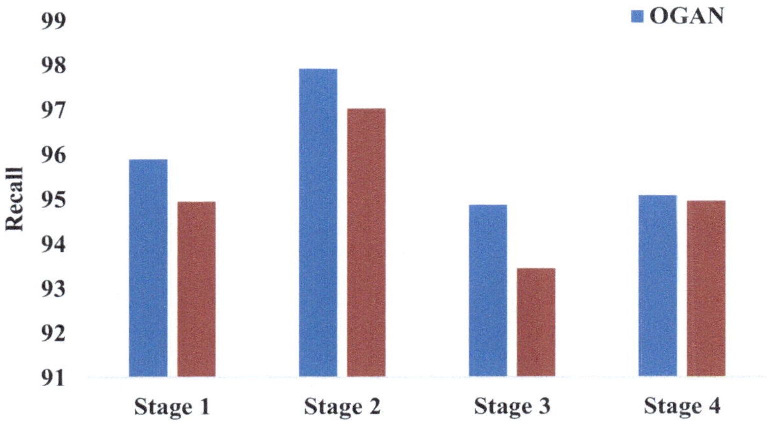

Fig. 5 Recall analysis of various models

proposed OGAN model has achieved maximum classification with a recall of 95.89% whereas the AMD-ResNet has shown a poor performance by attaining a lower recall of 94. 95%. Similarly, under the stage 2 grade, it is shown that the proposed OGAN model has achieved maximum classification with a recall of 97.91% whereas the AMD-ResNet has shown a poor performance by attaining a lower recall of 97.03%. Likewise, under stage 3 grade, it is shown that the proposed OGAN model has achieved maximum classification with a recall of 94.87% whereas the AMD-ResNet has shown a poor performance by attaining a lower recall of 93.45%. Finally, under stage 4 grade, it is shown that the proposed OGAN model has achieved maximum classification with a recall of 95.09% whereas the AMD-ResNet has shown a poor performance by attaining a lower recall of 94.96%.

An average accuracy analysis of the OGAN and existing models takes place as shown in Table 3 and Fig. 6. As shown, the manual diagnosis of the images resulted in a minimum accuracy of 90.60% which has been outperformed by the AMD-ResNet model with a slightly higher accuracy of 97.49%. However, the presented model has outperformed the previous model by attaining a maximum accuracy of 98.03%. The abovementioned tables and figures showed the superior performance of the proposed model over the compared methods by attaining a maximum accuracy of 98.03%.

Table 3 Accuracy of proposed model with existing methods for validation images

Methods	Accuracy (%)
OGAN	98.03
AMD-ResNet	97.49
Human	90.60

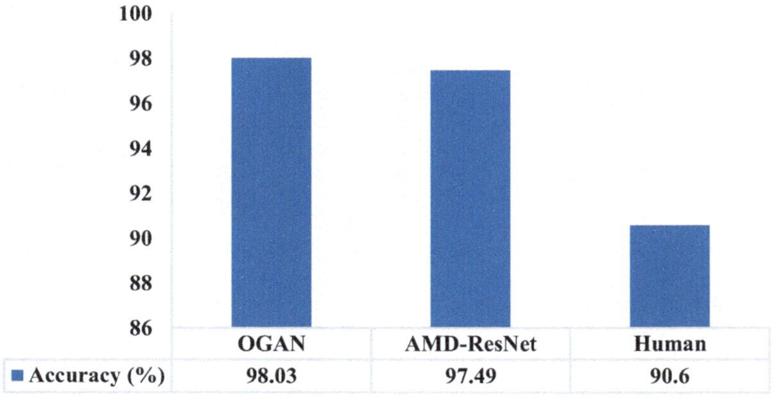

Fig. 6 Average accuracy analysis

4 Conclusion

IIoT comprises a collection of intelligent devices linked together to develop a system which observes, gathers, exchanges, and investigates the data. In this paper, a scalable cloud-oriented teleophthalmology structure by an IoMT to detect the AMD is projected. In the presented system, patients use a wearable camera for transmitting the retinal fundus photographs for a secured cloud drive storage for diagnosing the severity of disease as well as predictive progression examination. A projected OGAN helps to investigate the images to find as well as to compute AMD disease severity. The GAN would be optimized with the application of the bat method. The abovementioned tables and figures showed the superior performance of the proposed model over the compared methods by attaining a maximum accuracy of 98.03%.

References

1. National-Eye-Institute (NEI): Facts about age-related macular degeneration (Sept 2015). https://nei.nih.gov/health/maculardegen/armd_facts
2. Krishnaraj, N., Ezhilarasu, P., Gao, X.Z.: Hybrid soft computing approach for prediction of cancer in colon using microarray gene data. Curr. Signal Transduct. Ther. **11**(2), 71–75 (2016)
3. Mesko, B.: The role of artificial intelligence in precision medicine. Expert Rev. Precis. Med. Drug Dev. **2**(5), 239–241 (2017)
4. Dilsizian, S.E., Siegel, E.L.: Artificial intelligence in medicine and cardiac imaging: harnessing big data and advanced computing to provide personalized medical diagnosis and treatment. Curr. Cardiol. Rep. **16**(1), 441 (2014)
5. Gulshan, V., Peng, L., Coram, M., Stumpe, M.C., Wu, D., Narayanaswamy, A., Venugopalan, S., Widner, K., Madams, T., Cuadros, J.: Development and validation of a deep learning algorithm for detection of diabetic retinopathy in retinal fundus photographs. JAMA. **316**(22), 2402–2410 (2016)
6. Burlina, P., Freund, D.E., Joshi, N., Wolfson, Y., Bressler, N.M.: Detection of age-related macular degeneration via deep learning. In: 2016 IEEE 13th International Symposium on Biomedical Imaging (ISBI), pp. 184–188. IEEE, Washington, DC (2016)
7. Walton, O.B., Garoon, R.B., Weng, C.Y., Gross, J., Young, A.K., Camero, K.A., Jin, H., Carvounis, P.E., Coffee, R.E., Chu, Y.I.: Evaluation of automated teleretinal screening program for diabetic retinopathy. JAMA Ophthalmol. **134**(2), 204–209 (2016)
8. Hanen, J., Kechaou, Z., Ayed, M.B.: An enhanced healthcare system in mobile cloud computing environment. Vietnam J. Comput. Sci. **3**(4), 267–277 (2016)
9. Adam, M.K., Brady, C.J., Flowers, A.M., Juhn, A.T., Hsu, J., Garg, S.J., Murchison, A.P., Spirn, M.J.: Quality and diagnostic utility of mydriatic smartphone photography: the Smartphone Ophthalmoscopy Reliability Trial. Ophthalmic Surg. Lasers Imaging Retina. **46**(6), 631–637 (2015)
10. Bastawrous, A., Giardini, M.E., Bolster, N.M., Peto, T., Shah, N., Livingstone, I.A.T., Weiss, H.A., Hu, S., Rono, H., Kuper, H.: Clinical validation of a smartphone-based adapter for optic disc imaging in Kenya. JAMA Ophthalmol. **134**(2), 151–158 (2016)

11. Russo, A., Morescalchi, F., Costagliola, C., Delcassi, L., Semeraro, F.: Comparison of smartphone ophthalmoscopy with slit-lamp biomicroscopy for grading diabetic retinopathy. Am J. Ophthalmol. **159**(2), 360–364 (2015)
12. Ryan, M.E., Rajalakshmi, R., Prathiba, V., Anjana, R.M., Ranjani, H., Narayan, K.M.V., Olsen, T.W., Mohan, V., Ward, L.A., Lynn, M.J.: Comparison among methods of retinopathy assessment (CAMRA) study: smartphone, nonmydriatic, and mydriatic photography. Ophthalmology. **122**(10), 2038–2043 (2015)

Significance of IoT in the Agricultural Sector

Sushruta Mishra, Pradeep Kumar Mallick, and Debjit Koner

Abstract Agriculture has a crucial role to play in the overall growth of an agricultural nation. In our country, around 70% of the population are dependent on irrigation, while about one-third of the country's revenue is derived from agriculture. The situation pertaining to farming has always been an obstacle in a country's growth. The agricultural sector is boosting with the introduction of information and communication technology in this sector. Efforts are being made to enrich productivity and to decrement the cost of expenditure by utilizing the state-of-the-art technology and equipment. A feasible solution to this issue is smart agriculture by revolutionizing present-day conventional techniques of farming. To enhance effectiveness, productivity, and world market and to reduce manual labor, latency, and expenses, there is a necessity to divert toward the Internet of Things which will be gaining popularity in the coming years' time. IoT is the network of interconnected devices to transfer the sensed data without human involvement. Hence, to achieve high productivity, IoT in synergy with agriculture will provide smart farming. Our study in this chapter deals with various issues and scope of IoT technology in the agricultural domain. Various challenges concerning smart agriculture are presented in this chapter. Later a smart IoT-based architectural system model is illustrated and discussed in detail. Its implementation with its benefits is also presented with clarity.

Keywords Agriculture · Internet of Things (IoT) · Smart agriculture · Sensor network · Automation

1 Introduction

Internet of Things is a concept that makes the life of human beings easier. It helps networked equipment and sensors to transmit and receive information through the Internet in continuous streams of data. The IT systems can process this data and notify the user to take corrective measures. By 2015 the global population will touch 9.6 billion according to the report of the United Nations in 2013. So, to feed

S. Mishra (✉) · P. K. Mallick · D. Koner
KIIT Deemed to Be University, Bhubaneswar, Odisha, India

D. Gupta et al. (eds.), *Smart Sensors for Industrial Internet of Things*, Internet of Things, https://doi.org/10.1007/978-3-030-52624-5_12

this amount of population, the agriculture industry takes the help of IoT. Along with harsh weather changes and scarcity of rainfall, this industry must be prepared to feed the population. IoT can free humans by continuously monitoring the parameters. IoT can be used to attain the best produce by controlling the factors that can be controlled by humans, such as the soil moisture, soil pH, humidity, temperature, and rate of nutrient loss. Local farms can analyze this data with the help of another agency information, such as weather forecast, to improvise decision-making abilities. Proper irrigation techniques are very crucial for obtaining optimal crop yields. A very high volume of water will smother the roots of a plant and very less water will not help the plant to survive. Water requirements can be calculated with a high level of accuracy from the sensors placed in the fields. The combination of IoT and big data analysis is developing over the years and is being used as the key technologies. Also sensor technologies like gas sensors and environmental sensors are used in IoT applications which are widely helpful in building IoT models. Agriculture remains the fulcrum of the Indian economy. But usage of water in agriculture is more dependant on quantity of rainfall over a period. Also enhancing farm output is crucial to cope with the rise in demands of edible items for the growing population throughout the globe. Effective prediction of environmental conditions can result in an increase in agricultural output. The quality of crops depends on data gathered from field like temperature, moisture content in soil, and humidity conditions. Modern technologies can be precisely applied to improve agricultural productivity. IoT-enabled smart technology can be helpful in collecting massive amounts of data related to crops. At present, agriculture forms around 83% of the cumulative consumption of water in India [1]. Massive wastage of natural water occurs due to the ruthless usage of water resources. Hence it is an urgent necessity to develop advanced systems to prevent wastage of water with minimum intervention of manual labor. Over the last several years, farmers started to use software systems in organizing their data and tracking their transaction dealings with other consumers. They also utilized software systems to effectively monitor crops [2]. In an era of the Internet, smart agriculture is the need of the hour where farmers can collect and compute massive data from various heterogeneous devices. This would help the agricultural position in India to be more productive on transmitting accurate information [3].

2 Evolution of IoT in Agriculture

In India, agriculture is known to be a big source of income for a very large population and also contributes massively to the Indian economy. In 2013, the startup Avanijal (India) had devised an app-based solution to water fields, along with devising methodologies to save water. Around 17% of total India's agricultural sector contributes to the national GDP. More than 58% of the families rely upon it for their living. Adverse weather conditions and lack of skilled labor results in poor yield. One of the biggest challenges is irrigation for small farmers who use drip or sprinkler systems, as water is seen as a precious, finite resource. These can now be controlled remotely using various IoT sensor devices and activators using a mere mobile app. We require an active model that collects real-time data. All agricultural monitoring devices need to be strongly interconnected to have a proper decision-making system to maximize outputs.

3 Benefits of IoT in Agriculture

There are various reasons why IoT helps in smart agriculture. The olden methodologies of agriculture were very tiring and rigorous. Traditional farming methods consist of manual labor for tilling, sowing, and harvesting. And irrigation was completely interdependent on using non-hybrid seeds and weather conditions for rains. The only moment where conventional agricultural techniques are optimal compared to advanced farming techniques is an irony that better-quality food items were produced and that too in very small quantities. Even feeding the current growing population becomes very difficult. Conventional techniques use organic pesticides which make end product nutritious. Harvested agricultural items are clustered in the group of organic items that have a higher selling value which resulted in higher profits than the products of advanced methods. Modernized agricultural techniques include mechanized equipment and sensors for agriculture, sowing, and harvesting with hybrid seeds. Continuous monitoring of moisture levels in soil can guarantee accurate irrigation schedules with live weather forecasts from the Internet. Natural calamities and disasters can be predicted beforehand, and there would be less chance of the crops being affected by it. Therefore maximum profits can be obtained for every harvest season along with least crop damages by using this modern farming technology. IoT-enabled system can help in apprehending images of plants to detect for diseases, and then those images are then preprocessed and sent to remote laboratories. Also, the analysis power of computing can be so powerful that a correct pesticide or a fertilizer can be predicted. The utilization of fertilizer is an important part of a farming schedule that greatly affects farm productivity. Decisions on the application of chemicals and their crop-specific appropriate quantities need to be made by farmers.

4 Literature Survey

Lots of quality research work have been undertaken to enhance agricultural performance [4]. Proposed an Arduino-based technology in controlling soil moistening and roof covering of greenhouse. Here sensor-generated mathematical data are utilized such as moisture, pressure, temperature, etc., and the data are compared with the weather forecasting data samples, and then effective decision-making is done. Kalman filter helps in removing irrelevant noise from actuators. A research work conducted in [5] applied pH, temperature, humidity sensors, and fuzzy inference as input data from sensors. All sensor-related information is monitored on LCD and system nodes. Muhammad in [6] developed a system model approach to automate irrigation control issues with the application of an artificial neural network controller. This is successfully implemented and compared with the traditional ON/OFF device controller. It was observed that the neural network-based controller outperforms the traditional ON/OFF controller. In [7], a wireless network model using Zigbee technology was proposed to monitor and control humidity in the air, soil moisture content, and temperature value. The system implementation was done in a real-time environment with various components like sensor, humidity sensor, soil moisture content, water pump, relay, fan, and a buzzer. In [8], Zigbee integrates with a wireless network in transmitting the water value of soil and temperature measure of soil to the end user. This information is transmitted to the server through GPRS with the help of cellular nodes. Gautam and Reddy [9] discussed a remote coordination system model in which irrigation is undertaken with the help of an embedded hardware device. It used a creative Bluetooth and GSM oriented automated irrigation system. An innovative device called rain gun irrigation system was developed in [10] on the basis of an automated microcontroller where the soil is irrigated only when there is heavy scarcity of moisture level in the soil. It helped in conserving water to some extent in dry areas. Ref. [11] proposed an automatic distributed wireless network that helped inefficient usage of water for agriculture. Here various sensors like humidity and temperature sensors are placed at the root region of plants. In [12], an energy-constrained network for soil water optimization was developed to work well in a distributed environment. Here temperature sensors are arranged at the base of the saplings to detect the readings of plants with time. In [13], the uneven crop fertilization issue was handled where too little and too high fertilized soil was analyzed and concentrated accordingly. In [14], Mishra and his colleagues developed an adaptive feature optimization technique based on a genetic algorithm for efficient classification of dengue patients. In [15], the authors have implemented some vital biologically inspired computation techniques to classify various types of tumors using a multilayer perceptron as a classifier.

5 Use of IoT in the Agricultural Sector

The agricultural sector contributes to the highest GDP in India. But while overseeing the technological development deployed infield, it is found that the overall growth is not enormous. Recently it is seen that there is significant upgradation in methods having a relevant effect on several domains such as pollution control, agriculture, medical, congestion traffic, etc. In our country, farming is the primary occupation. The development of agriculture is very important that this occupation generates a huge income source. The Internet of Things can be successfully implemented in the agricultural domain. IoT helps to access sensed data and make accurate decisions from retrieved data like moisture of soil, sensors related to determining water level, water quality determination, etc. The use of IoT in the agricultural sector primarily helps in minimizing the wastage of water and the manual effort on the field for agriculture in order to save the time, money, and effort of the farmer. Here in this section, three different case studies based implementation are discussed where the efficient use of IoT technology in the agricultural field has been successfully shown.

5.1 Illustration of an IoT Framework for Automated Irrigation Based on Soil Moisture

The consumption of water is gradually increasing with every passing day leading to water scarcity. At present, not only saplings but also external plants at outdoor areas are getting very tough to deal with. Nowadays, there is an acute crisis of water everywhere where pacifying water availability has been a challenging task and become a critical job. The shortage of water resources is often seen in those countries where the country is economically weak. Hence it becomes a major issue in conventional agricultural systems. The challenges of the manual model comprise the physical effort put by farmers to restrict drip irrigation, water wastage, and wastage of time.

5.1.1 Illustration of the Irrigation Framework Model

To overcome the flaws, a smart irrigation system framework is proposed here which restricts the usage of water based on the soil moisture level. Three fundamental parts of the system are given: humidity sensing part, control section, and the output section. YL-69 soil sensor (a resistance-type sensor) can detect the soil humidity. ATmega328 microcontroller works on the Arduino platform which was the control unit. Depending upon the moisture level of soil, the control unit gives the output which can be controlled by switching it on and off (irrigation of fields) the system. The main working principle of the soil moisture sensor (in the plant) is based on the connection with the Arduino microcontroller, and other electronic components that are listed above are also connected as given in Fig. 1. The soil moisture is also checked by the sensor that forwards the information and parameters regarding the soil moisture to the microcontroller which is controlled by the pump. Whenever the moisture of the soil level drops below a threshold level, the pump is activated with the help of the microcontroller which sends a signal that delivered the required amount of water to the plant. The pump stops working, when plants get enough water. The complete system is powered by the help of power supply which again maintains the voltage drop range from 7 to 12 V. The relay module is a basic circuit comprising of a single transistor, a few resistors, and diodes, and the microcontroller digitally controls the relay model. Since the total framework is fixed in a little

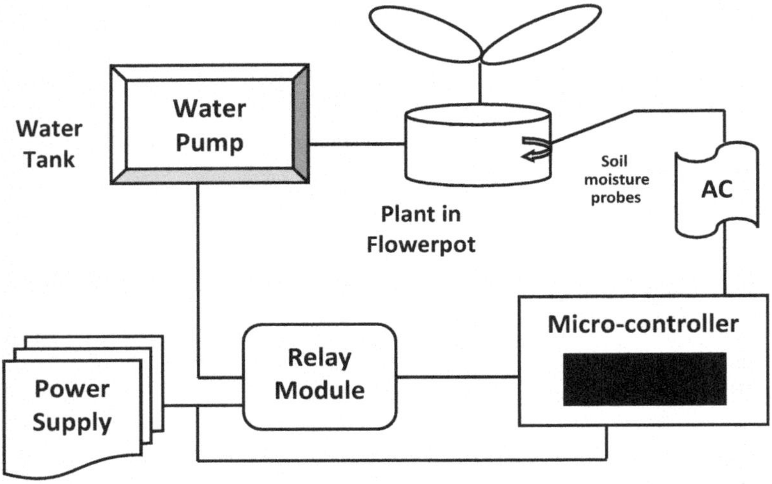

Fig. 1 System flow diagram

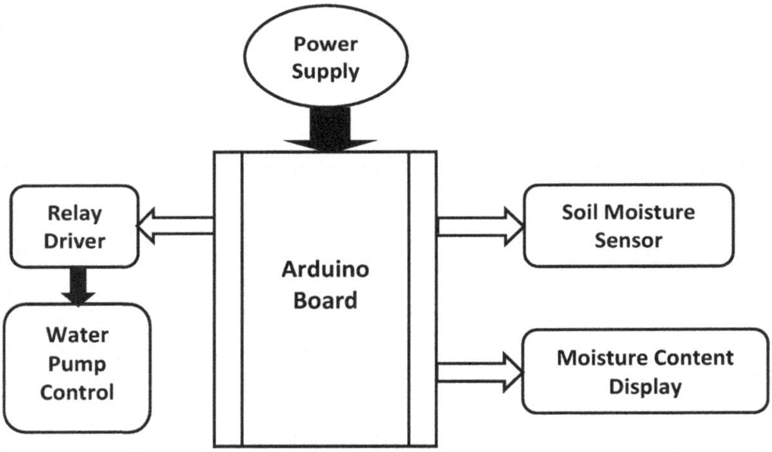

Fig. 2 Block diagram of smart irrigation system

box, for this purpose Arduino Uno is an ideal microcontroller in view of its measurements and the exhibition work. The amplifier circuit and probes are present in the soil moisture module. Potentiometer sets the threshold level. Real-time data about the moisture in the plant is an analog output which is utilized by the system. Whenever the relay module gets a command from the microcontroller, the model starts working as it is also connected by the relay model. The automated plant watering model and its circuit diagram of the smart irrigation system are shown in Fig. 2. The circuit incorporates an Arduino Uno board, a soil moisture sensor, an LCD

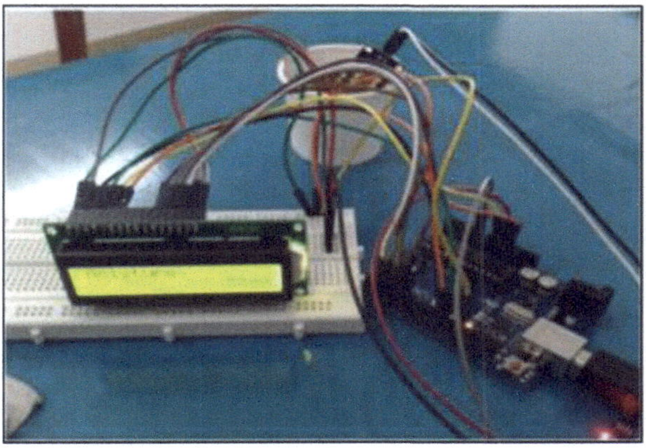

Fig. 3 Measuring the moisture level

display, a motor of 12 V water siphon, and a motor driver IC that will run the water pump. We can offer power to the Arduino board utilizing a 7–12 V divider mole or a connector plug or a solar panel.

The sensor detects the change in humidity, and when the soil moisture content values drop, it gives a signal to the microcontroller so it can reactivate the water pump (engine) as shown in Fig. 3. This thought can be executed by a programmed plant watering system for main roles. This circuit contains an Arduino Uno board, a soil moisture sensor, a 5 V motor pump, a motor driver L293D (IC1), and a motor driver IC to run the water pump. Irrigation activity and water sprinkler activity can be set using a threshold value. Whenever the sensor detects the moisture content before the threshold point, the sprinklers are turned on till the soil gets completely moist. The data from sensors is transmitted to an online database from where it is utilized and is made available to the website. The moisture content in the soil which is displayed on the website is divided into two types: low and high. When the moisture content is low, the pump is switched on. The threshold point varies on different types of soil used. Table 1 shows the test case analysis sample.

5.1.2 Results Analysis

The sprinkler in the system is made to switch on or off based on the recorded values of the sensors. This facilitates a farmer to oversee the entire irrigation method from his farm without too much manual intervention. Thus it makes the entire model smart and automated. Figure 4 denotes the value of sensor readings when soil is completely dry. Table 2 shows the readings records of two YL-69 soil moisture sensors over a period of 2 h. Among the two sensors deployed in the soil, one is placed over an irrigated soil, while another sensor is deployed in the soil where initial moisture availability is 72%. The main purpose of these two sensors is to record the readings taken at a time period of more than 2 h, and when the sprinklers are deactivated, we can notice the rate of reduction in the moisture content of the soil. The results of this

Table 1 Test case analysis

Soil status	Moisture content	Relay status	Water pump status	Test case status
Dry	700–1000	ON	ON	TRUE
Damp	300–700	OFF	ON	TRUE
Wet	<300	OFF	OFF	TRUE

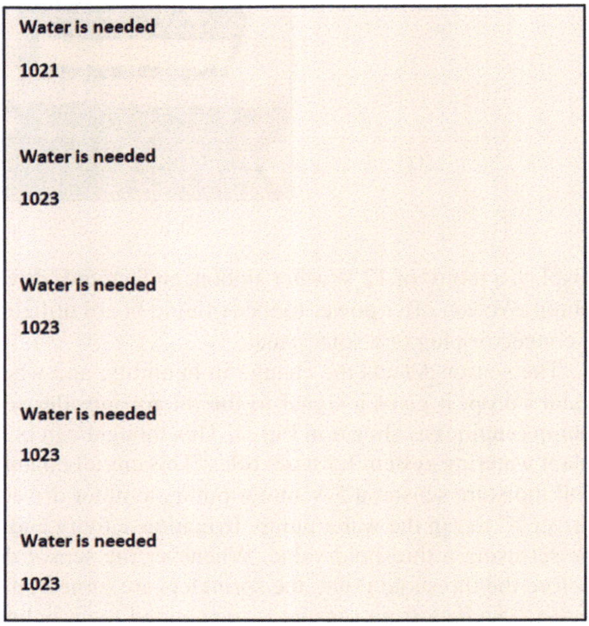

Fig. 4 Sensor value when soil is completely dry

Table 2 Readings of two YL-69 soil moisture sensors over a period of 2 h

Time (h)	Sensor S1 (%)	Sensor S2 (%)
2.00 PM	72.0	100.0
2.10 PM	70.0	100.0
2.20 PM	69.0	100.0
2.30 PM	64.5	99.5
2.40 PM	61.0	98.0
2.50 PM	58.0	96.0
3.00 PM	56.7	95.5
3.10 PM	54.2	95.0
3.20 PM	50.9	94.5
3.30 PM	48.1	94.0
3.40 PM	45.0	93.0
3.50 PM	44.5	91.2
4.00 PM	40.0	89.0

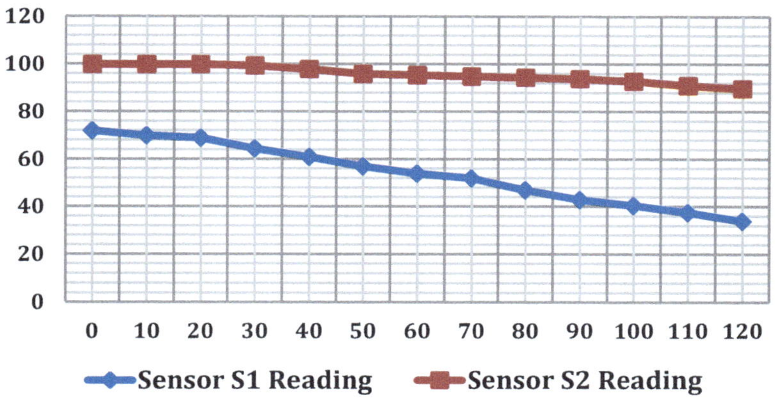

Fig. 5 Comparison analysis of moisture content sensor readings

observation are shown in a graphical manner in Fig. 5. It can be clearly seen that the rate of reduction of moisture level in over-irrigated soil is much less when compared to that of the soil with an initial moisture level of 72%. Hence the soil with sensor S1 has a constantly decreasing curve with respect to the time period. The curve for sensor S2 decreases at a slow pace and even after 2 h its moisture content is 89%. 800 is the threshold moisture value. When the sensor detects that the soil moisture is greater than 800, that means soil moisture is low and can be irrigated. When sensor detects that the soil moisture is less than 800, that means the soil has enough water.

5.1.3 Summary

With the advancement of technology, smart irrigation comes into the forefront. In our research work, an IoT-based smart irrigation framework has been developed which is dependent on the moisture level of the soil. Various hardware components were integrated into the development of this automated irrigation framework. With the help of moisture sensors, the water content in soil can be computed successfully. Water is provided to the soil based on the scarcity of soil moisture through an Arduino triggered model. It was successfully implemented in an irrigated region and was observed to be functioning efficiently. Thus this smart irrigation model restricts the manual intervention to a great limit. Further, this proposed model can be upgraded into remote areas with more functionality.

5.2 Illustration of a Multidisciplinary Approach for Smart Agriculture

In another discussed work concerning use of smart agriculture, a multidisciplinary framework is presented. Various modules associated with it are discussed in the following section.

5.2.1 Discussed Multidisciplinary Approach for Smart Agriculture

In this presented architecture, we use a precision agricultural approach. The approach is all about maintaining the optimum quality of crops and their productivity. Real-time data analysis of the crops, soil nutrient levels, and the environment parameters are taken into account. The multidisciplinary model of the presented architecture as shown in Fig. 6 consists of five modules. They are:

- Sensor Kit Module
- MobileApp Module
- AgroCloud Module
- Big Data Mining, Analysis and Knowledge Building Engine Module
- Government and Agro-Banks UI

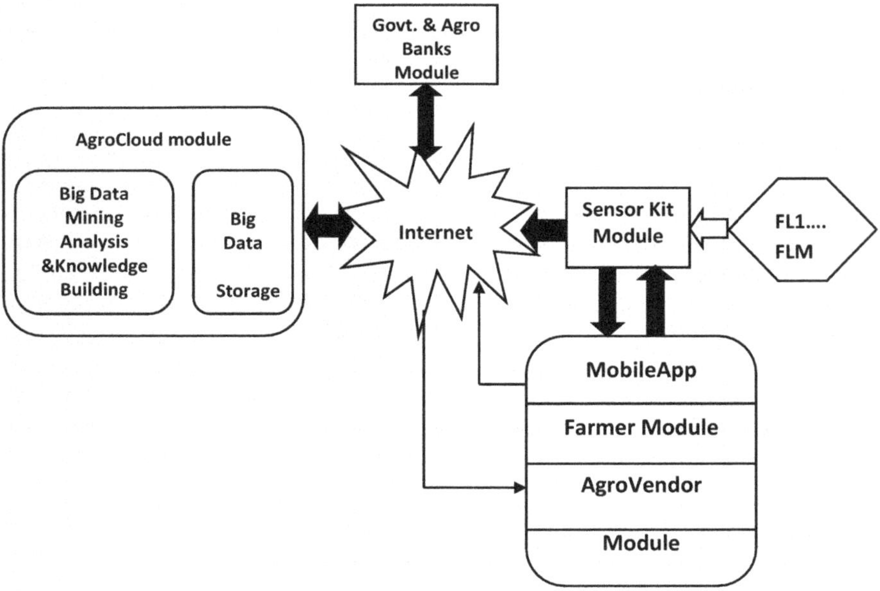

Fig. 6 Multidisciplinary framework for smart agriculture

The Sensor Kit unit is the robust integrated IoT component embedded inside soil as well as surrounding sensors. MobileApp unit functions to act as a front-end and helps users to communicate with the system. AgroCloud Module provides vast memory. Knowledge Building Engine Module and Big Data Mining and Analysis help in exchanging data among users. Information related to agricultural loans and schemes is provided by the Government and Agro-Banks user interface which is a web interface.

Sensor Kit Module

It forms an integral component of the framework which performs sampling of soil at regular intervals to extract soil property values. Figure 7 shows the Sensor Kit module which is quite cheap and a robust kit that forms the BeagleBone Black which is equipped with IoT and storage integrated with faster processing power. The BeagleBone Black is an economical IoT integrated with environmental and soil sensors for the collection of soil parameters as well as environmental situations. Geographical data is detected by the GPS sensors. Soil parameter sensors are the soil sensors, soil water level sensor, and potassium (K), nitrate (N), and phosphorus (P) sensors being embedded with the IoT component.

ModuleApp Module

It has three components. They are:

- UI for a farmer

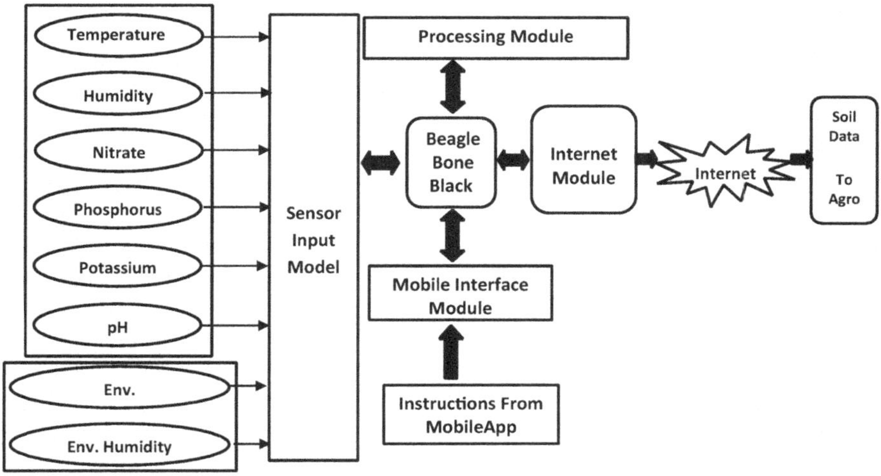

Fig. 7 The Sensor Kit Module

- UI for an agro-marketing agency
- UI for agro-vendors which include pesticide suppliers, fertilizers, and seed suppliers

In the beginning, the end user registers with a few personal details like identity information, user category, location, and other vital data. If the end user is a farmer, then he/she is to fill in the data about the farmland regarding the location of the farm and the total area for each farmland. Sensor Kit obtains the necessary instructions from the MobileApp module. This information is sent and stored on AgroCloud. Sensor Kit retrieves and transmits the soil-related data to the cloud when the cultivation of seeds is in continuous development. Through this app, the farmers receive notifications related to fertilizers and their market value for better quality crop results and spending economically for the harvest. The app sends notifications at regular intervals. Total production data per crop and current soil features are taken from the farmer and stored in cloud storage after the crop is harvested. Agro-marketing agencies purchase harvested crops from farmers and transmit regular notifications about the variations in prices of particular crops the farmer wants to sell. Agro-vendors are assigned responsibility to sell fertilizer, pesticides, and agricultural-based entities. It gives an interface for the farmers to gain knowledge and insights about the cost of equipment and to analyze better cost-effective mechanisms. The mobile application module is depicted in Fig. 8.

AgroCloud Module

To use this architecture, the user must register with AgroCloud with the help of the MobileApp. AgroCloud, which is a big data storage, is responsible for storing farmer-related details, agro-marketing agent data, agro-vendors and service

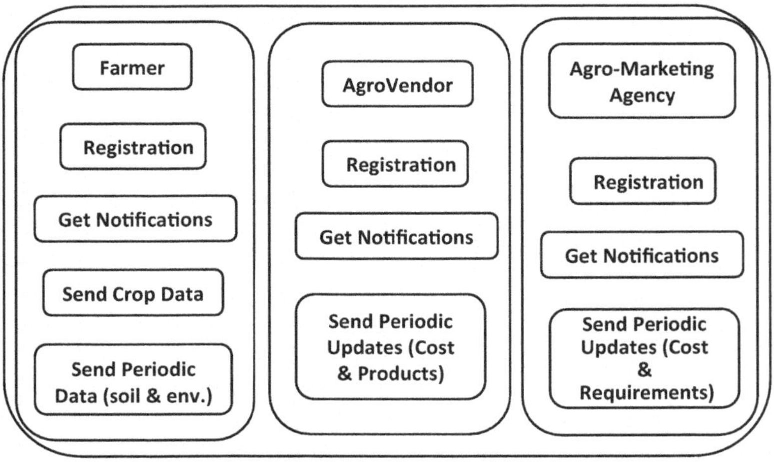

Fig. 8 MobileApp Module

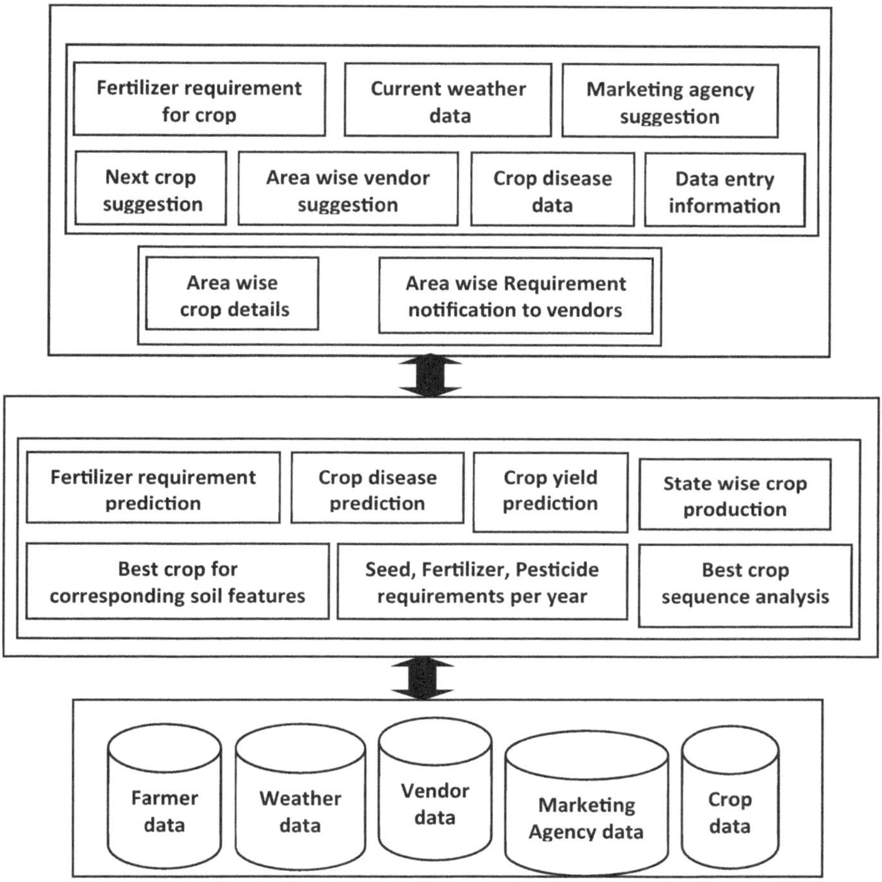

Fig. 9 AgroCloud module

providers (fertilizer, seed, agro devices, and pesticide providers), government schemes, and details about bank loans for farmers. This module saves the periodic data retrieved through a sampling of soil and environment data. The more the number of end users, the more the data generated, eventually resulting in a "big data." The architecture of the AgroCloud is depicted in Fig. 9. The AgroCloud comprises the big data mining, big data storage, analysis, knowledge building engine, and application module.

Big Data Mining, Analysis, and Knowledge Building Engine

It exists in AgroCloud, which plays a decision-making for the requirements of fertilizer of the present harvested crop. Decisions are implemented on the basis of present soil parameters for optimum productivity, crop-related disorder prevention, best crop sequence analysis, and accurate irrigation of a particular crop. The database also keeps track of the type of crop grown for a given region and the total crop cultivated in that region. On the basis of this, the present requirements of customers can be analyzed, and this will in turn help in the reduction of cost of each agro-product.

This data repository keeps collecting data over decades in a continuous manner. So for every agriculture land, inference output along with data analysis could be used to calculate the best crop sequence for better production and good soil health, thereby helping farmers to take corrective measures and help farmers to increase their production. Big data analytics may be undertaken to predict the outcome of each entity on the basis of the previous history of the database.

Government and AgroBanks UI

With the help of this unit, the agriculture department is capable of acquiring information about incentives, subsidies, and the latest schemes for the agricultural sectors and farmers. AgroBanks has been an integral part of a developing nation by providing financial solutions to the farmers. Also, agricultural banks find a medium to give loans through this interface and reach out to the farmers. Farmers and other beneficiaries can obtain this information as it is saved on the cloud, and they would receive periodic notifications to make their life simpler, rather than going physically to inquire about the schemes to the government offices.

5.2.2 Discussion

The farming industry would gradually be more important in the coming years. As per UN Food and Agriculture Organization, around 70% of surplus food is to be generated by 2050 to meet the needs of the developing population around the world. Figure 10 shows the trends of IoT over the years and it is gaining popularity in the world market.

Big data is to provide predictive views to future outcomes of farming, can follow real-time functional decisions, and rediscover business functionalities for easier, creative theories and game-changing strategic methodologies. The process of decision-making will be a hybrid of human and computational elements. The vast approach of big data is likely to bring changes to both farm infrastructures and wider food chain in new unexplored ways. Figure 11 gives the push factors and pull factors of big data in agriculture.

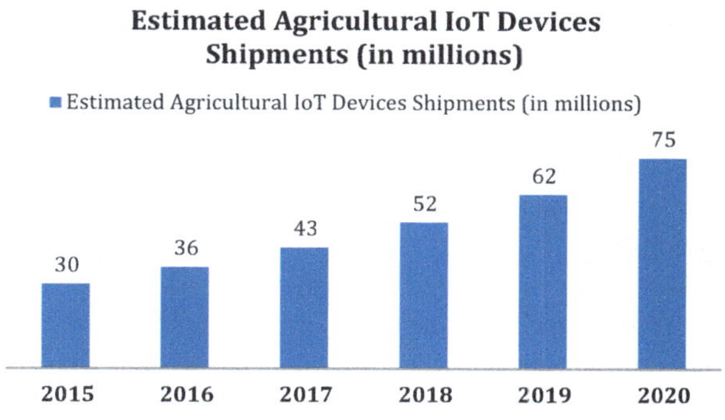

Fig. 10 Future trend of IoT devices over time

Push factors

- **General technological developments**
  - Internet of Things and data-driven technologies
  - Precision Agriculture
  - Rise of ag-tech companies
- **Sophisticated technology**
  - Global Navigation Satellite Systems
  - Satellite imaging
  - Advanced (remote) sensing
  - Robots
  - Unmanned Aerial Vehicles (UAVs)
- **Data generation and storage**
  - Process-, machine- and human--generated
  - Interpretation of unstructured data
  - Advanced data analytics
- **Digital connectivity**
  - Increased availability to ag practitioners
  - Computational power increase
- **Innovation possibilities**
  - Open farm management systems with specific apps
  - Remote/computer-aided advise and decisions
  - Regionally pooled data for scientific research and advise
  - On-line farmer shops

Pull factors

- **Business drivers**
  - Efficiency increase by lower cost price or better market price
  - Improved management control and decision-making
  - Better local-specific management support
  - Better cope with legislation and paper work
  - Deal with volatility in weather conditions
- **Public drivers**
  - Food and nutrition security
  - Food safety
  - Sustainability
- **General need for more and better information**

Fig. 11 Push and pull factors in agriculture

In this presented model, there was an evolution of a multidisciplinary framework related to IoT-based agriculture with the help of five vital techniques which include Sensors, Internet of Things, Cloud Computing, Big Data Analysis, and Mobile Computing. By soil sampling in real-time, farmers are capable of getting present requirements of fertilizers related to their crops. It is a vital necessity for Indian farmers to combat and reduce their expenditure for their crop fertilizers, keeping the land fertile.

Since data is gathered over several years for the soil and seed growth conditions, this framework facilitates Big Data analytics for the optimal sequence of the crop, total crop production, and other data of interests that can be analyzed. The interconnection of different technologies will simplify the circulation of harvested crops in the agriculture market, henceforth expanding the profits of farmers. This will benefit the farmers through notifications and keep farmers updated about the suitable schemes available by the Ministry of Agriculture. The cost farmers incurred will be thereby reduced.

5.3 IoT-Based Analytics Model for Crop Water Requirement

In this framework, sensors are deployed to determine several agriculture field metrics like temperature, humidity, and moisture level of soil. This information from the field is gathered and stored in the server. Therefore, proper analysis is carried out to compute the water requirement of a crop. This model is usually applied on a local level where this IoT-oriented analytics may be useful for farmers in providing new insights and enhancing decision-making process.

5.3.1 Overview of Model Discussed

The range of the presented model is limited to only five major crops, i.e., wheat, cotton, groundnut, rice, and sorghum. This section explains the working and the design of the model, with a flow diagram which is explained. The system contains three main modules:

1. Water requirement measurement and calculation.
2. Using soil NPK values, the fertilizer requirement is generated.
3. Generation of irrigation schedule.

The first module comprises the following stages:

- The seed is selected along with its phase.
- On the basis of location, climate information is computed and shown.
- Cultivation land is inputted in m^2 for the specific crop.
- On the basis of the above input metrics, the moisture requirements for that crop are determined using a predefined method.

Moisture for crop requirements is computed through a three-phase method:

1. Climatic effects: temperature, humidity, wind, sunshine hours, radiation, wind speed, sunshine hours, and radiation values for the desired location are obtained.
2. The crop characteristics are determined.
3. Local condition's effects are taken into account.

The water demands may be estimated for various phases of the crop which include initial phase, development phase, mid-late phase, and harvest phase.

The second module comprises the following steps:

- After the testing of the soil, NPK soil is the input value from the test report of the soil.
- The database stores the advised dose of NPK for the crops. On the basis of input, the model does recommendation of fertilizers so that the recommended section may be taken.
- The fertilizer name and the rate of quantity to be applied per acre should be displayed.

The third module includes the following steps:

- The crop is selected.
- Time period for day of sowing is entered.
- The crop water (L) value is entered (computed in module 1).
- The irrigation counts of the crops are selected.

The overall flowchart of the discussed model is shown in Fig. 12.

5.3.2 Result and Analysis

This model has combined three different modules to create the model implementable as well as the agricultural process effective and simpler. The following figures take input about the details of the crop from a web page and provide the output in the form of tables and irrigation charts. Figure 13 shows the snap of the web page where user is to select the crop and its corresponding stage. Land available for cultivation is entered by user in Fig. 14, crop water requirements are depicted in Fig. 15, while the soil test reports are entered by user in Fig. 16.

Here, we use the Penman-Monteith algorithm. The Penman equation describes evaporation (E) from an exposed surface of the water. This algorithm provides the ability to determine the impact of variations of cultural guidelines and crop variety enhancements by changing metrics in the computation techniques. In addition to weather catastrophe measurements, the Penman equation is sensitive to vegetation-specific parameters, i.e., for different varieties of crops. The information for the particular crop is generated and retrieved from the database. This model promotes the usage of water efficiently by determining the evapotranspiration rates accurately. Fertilizer requirement is shown in Fig. 17.

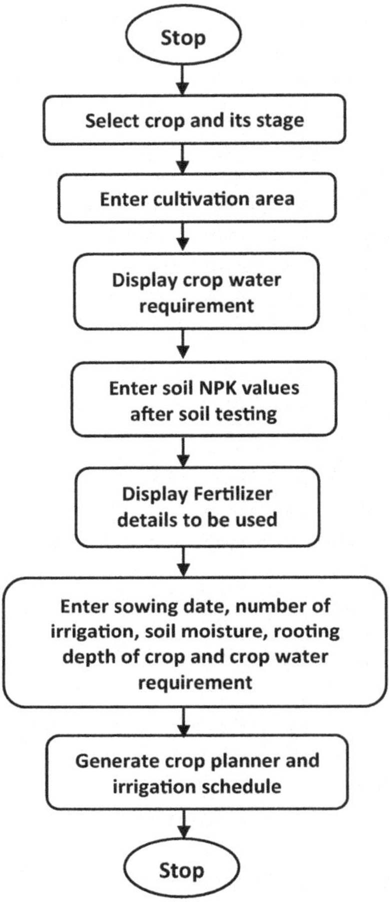

Fig. 12 Flowchart for the presented model

The irrigation schedule is depicted in Fig. 18, after careful estimation of the input values with sensor detection as well as the parameters fetched from the database along with the weather conditions that were determined. This project is implemented locally for the benefit of farmers and to support efficient decision-making strategies.

5.3.3 Advantages of the Discussed Model

Comparing the advantages over conventional methods, the advantages are:

- The efficiency of farming with IoT is high.
- Meets the needs of experienced and inexperienced farmers.

Fig. 13 Snap of user select the crop and its stage

Fig. 14 User enters the land available for cultivation

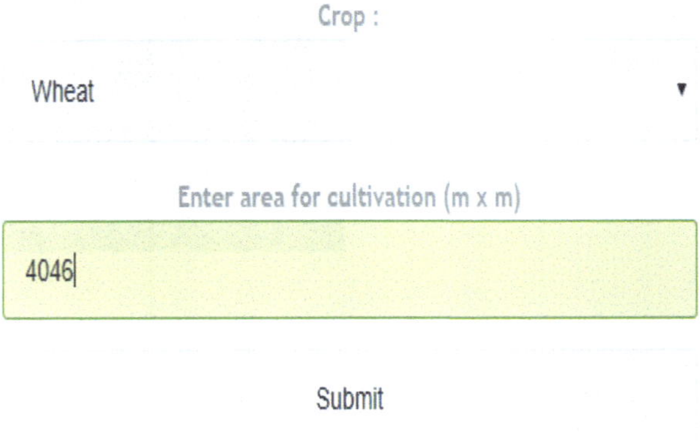

Fig. 15 System calculates the crop water requirements

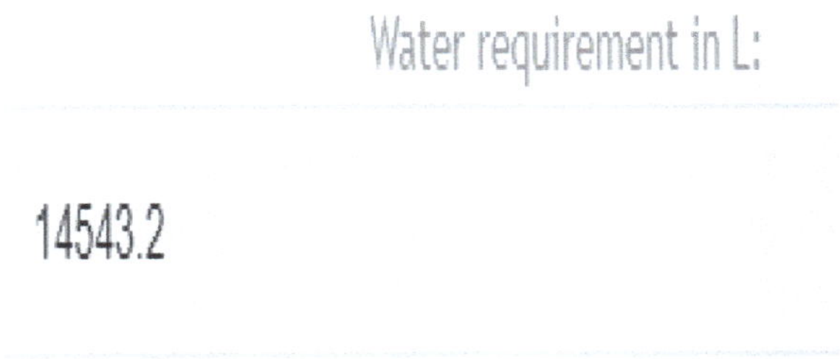

Fig. 16 User enters the reports of the soil test

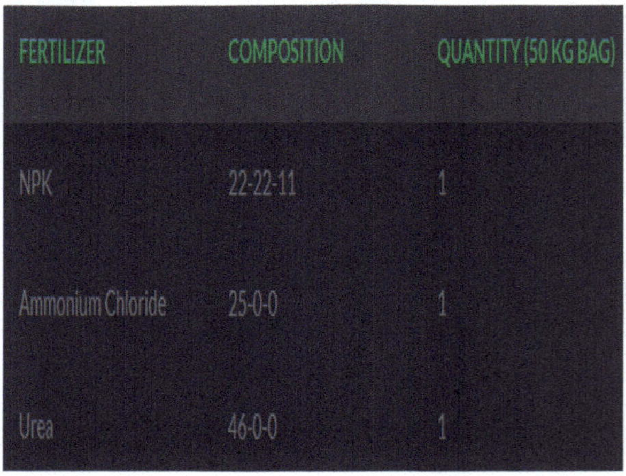

FERTILIZER	COMPOSITION	QUANTITY (50 KG BAG)
NPK	22-22-11	1
Ammonium Chloride	25-0-0	1
Urea	46-0-0	1

Fig. 17 System shows the fertilizer required by the system

- Problems of excessive irrigation and improper amounts of fertilizers can be solved.
- The irrigation schedule generated is flexible and dynamic. It can be rearranged to farmer's needs.

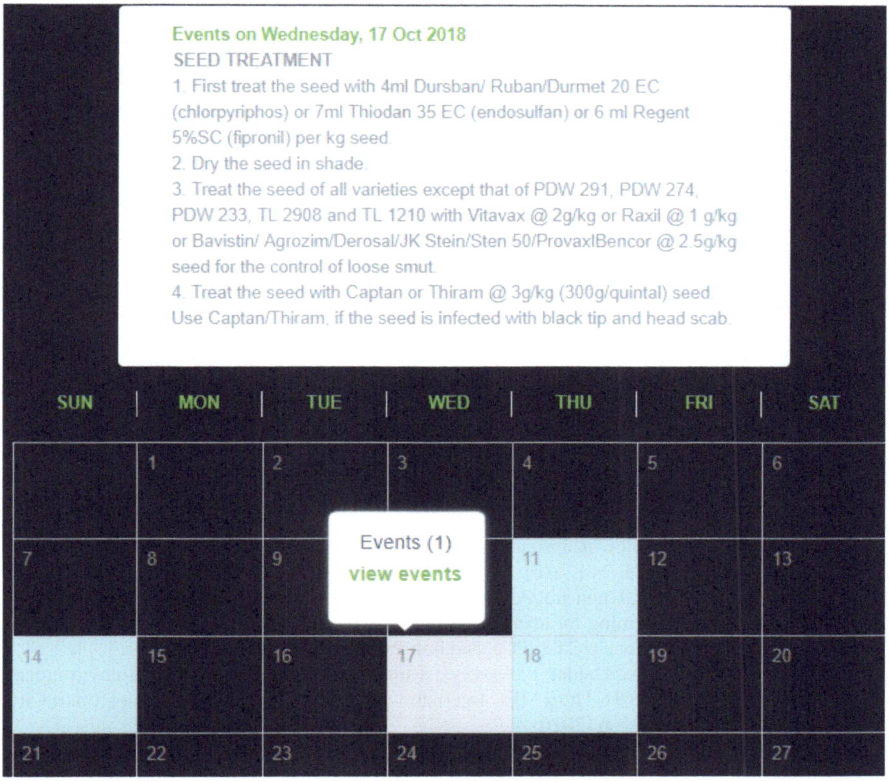

Fig. 18 System shows the irrigation schedule

5.3.4 Summary

The above smart agriculture model is designed mainly for specific crops and has deployed farmer-centric data which can meet the requirements of local farmers at rural sectors. The challenging issues of excess water application and poor use of fertilizers may be prevented to restrict usage of water and obtain optimal yield.

6 Conclusion

Agriculture plays an important role in determining the overall progress of our country. With the advancement of technology, smart irrigation comes into the forefront. In our study, significance and evolution of IoT in agricultural sector have been extensively discussed. Three different smart agriculture-based frameworks are discussed in detail. At first an IoT-based smart irrigation framework has been developed which is dependent on moisture level of soil. Various hardware components were integrated in the development of this automated irrigation framework. It was successfully implemented in an irrigated region and was observed to be functioning efficiently. Similarly in another discussed case study, a multidisciplinary framework for smart agriculture was presented along with its modules. The third case study presents a smart agriculture model for better crop management. The overall GUI and its application and its implementation at user end were also discussed. Hence we can infer that with the constant rise in technology, IoT is a latest revolutionary concept which can be successfully used in agriculture domain, and it thereby helps in restricting the manual work of farmers and end users.

References

1. Hegde, N.G.: Water scarcity and security in India. BAIF Development Research Foundation, Pune (2014)
2. Batte, M.T.: Changing computer use in agriculture: evidence from Ohio. Comput. Electron. Agr. **47**, 1–13 (2005)
3. Csótó, M.: Information flow in agriculture – through new channels for improved effectiveness. J. Agric. Inform. **1**(2), 25–34 (2010)
4. Putjaika, N., Phusae, S., Chen-Im, A., Phunchongharn, P., Akkarajitsakul, K.: A control system in an intelligent farming by using Arduino technology. In: 2016 Fifth ICT International Student Project Conference (ICT-ISPC), Nakhon Pathom, pp. 53–56 (2016)
5. Abdullah, A., Enazi, S.A., Damaj, I.: AgriSys: a smart and ubiquitous controlled-environment agriculture system. In: 2016 Third MEC International Conference on Big Data and Smart City (ICBDSC), Muscat, pp. 1–6 (2016)
6. Muhammad Umair, S.: Automation of irrigation system using ANN based controller. Int. J. Electr. Comput. Sci. **10**(2), (2010)
7. Chikankar, P.B., Mehetre, D., Das, S.: An automatic irrigation system using ZigBee in wireless sensor network. In: 2015 International Conference on Pervasive Computing (ICPC), Pune, pp. 1–5 (2015)
8. Gutiérrez, J., Villa-Medina, J.F., Nieto-Garibay, A., Porta-Gándara, M.Á.: Automated irrigation system using a wireless sensor network and GPRS module. IEEE Trans. Instrum. Meas. **63**(1), 166–176 (2014)
9. Gautam, I., Reddy, S.R.N.: Innovative GSM based remote controlled embedded system for irrigation. Int. J. Comput. Appl. **47**(13), 1–7 (2012)
10. Suresh, R., Gopinath, S., Govindaraju, K., Devika, T., SuthanthiraVanitha, N.: GSM based automated irrigation control using raingun irrigation system. Int. J. Adv. Res. Comput. Commun. Eng. **3**(2), (2014)
11. Gutiérrez, J., et al.: Automated irrigation system using a wireless sensor network and GPRS module. IEEE Trans. Instrum. Meas. **63**(1), 166–176 (2014)
12. Nisha, G., Megala, J.: Wireless sensor network based automated irrigation and crop field monitoring system. In: Sixth international conference on advanced computing (ICOAC) (2014)
13. Ruan, J., Liao, P., Dong, C.: The design and research on intelligent fertigation system. In: 2015 Seventh International Conference on Intelligent Human-Machine Systems and Cybernetics (2015)
14. Mishra, S., Tripathy, H. K., Panda, A. R.: An improved and adaptive attribute selection technique to optimize dengue fever prediction. International Journal of Engineering & Technology. **7**(3.34), 480–486 (2018)
15. Mishra, S., Tripathy, H. K., & Mishra, B. K.: Implementation of biologically motivated optimisation approach for tumour categorisation. International Journal of Computer Aided Engineering and Technology, **10**(3), 244–256 (2018)

Soil Moisture Sensor Nodes in IoT-Based Drip Irrigation System for Water Conservation

K. Muruganandam and Usha Chauhan

Abstract Agricultural data analysis is an important challenge for the data mining community present not only in India but also in the world. This is due to the fact that majority of the cultivation lands in the globe suffer highly from shortage of irrigation water. In such a scenario, drip irrigation techniques have been developed by researchers to reduce the water consumption in dry areas. However, due to the lack of guidelines and systematic techniques to utilize the water and electricity in an optimistic manner, the agricultural sector faces water problem in many places. Therefore, the overheads of farmers using the conventional drip irrigation have become high, and they have to manually visit and monitor the lands frequently. Agricultural data can be collected more efficiently by the use of sensors for data collection and also for effective analysis. A smart irrigation system can be built with smart sensor networks for collecting field values and can be analyzed using rules for effectively watering the plants. Hence, a new sensor network-assisted irrigation system and rule-based analysis model have been developed in this research work to enhance the efficiency of water usage.

Keywords Soil moisture sensor nodes · Irrigation system · Water conservation · Wireless sensor networks

1 Introduction

The wireless sensor networks (WSNs) are extensively used in many applications like precision agriculture, environment monitoring, habitat monitoring, battle-field surveillance, disaster rescue operations, etc. The WSN nodes solely depend on battery for power, and they are deployed mostly in open and harsh environmental conditions and are expected to work unattended for a long duration.

K. Muruganandam (✉) · U. Chauhan
SEECE, Galgotias University, Greater Noida, India
e-mail: muruganandam@galgotiasuniversity.edu.in; usha.chauhan@galgotiasuniversity.edu.in

D. Gupta et al. (eds.), *Smart Sensors for Industrial Internet of Things*, Internet of Things, https://doi.org/10.1007/978-3-030-52624-5_13

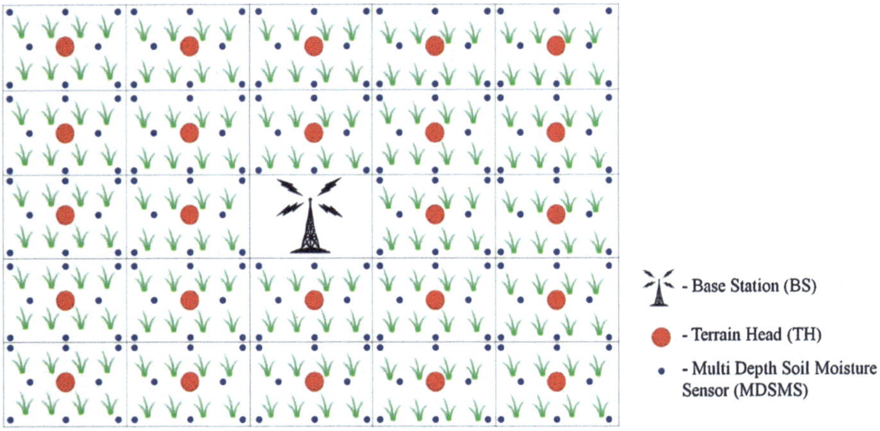

Fig. 1 Farmland setup of sensor nodes

Akbas et al. [1] wireless sensor network (WSN) subsist huge values of sensor hubs expand over a geological place for monitoring substantial components like temperature, vibrations, humidity, and seismic events. Every node hub in the sensor of WSN is a small device that includes three basic components: a sensing hub [2] for data procurement of the substantial areas, a transform mutual system for base data conversion and storage, and without wire transmission subsist for data transform to a paramount compilation point (sink hub node or base hub station) shown in Fig. 1. The main objective of the proposed system is to conserve the water in dry areas by using drip irrigation system with the help of intelligent wireless sensor networks.

Irrigation is a method of transporting water to crops to maximize the amount of crops produced. Various types of irrigation systems in present use not handed the water in a competent way. To improve the accuracy and precision of sensors for sensors data collection and investigation [3]. Moreover, wireless sensor networks are used in many applications like agricultural activities, natural disaster management systems, military applications, forest fire monitor, underwater processing, industrial automations, hospitality systems for elder people, smart home applications, etc. [4, 5]. A smart irrigation system can be built with smart sensor networks for collecting field values and effectively watering the plants [6]. Hence, a new sensor network-assisted irrigation system has been developed in this research work to enhance the efficiency of water usage. The main objective of this chapter is to conserve the water in dry areas by using drip irrigation system with the help of a multi-depth soil moisture sensor using wireless sensor networks.

The organization of the paper is described as follows. Section 2 analyzes various literatures and considered their future suggestions for the advancement of such segmentation methodologies. Some traditional concepts like IDIS models have been proposed and describe the algorithm flow pseudocode and architecture explanation in Sect. 3. The proposed IDIS method results are discussed and compared to the conventional methods tabulated under Sect. 4. Finally, the outcome is summarized in Sect. 5.

2 Related Works

Anand et al. [7] proposed an in-network data aggregation approach for aggregating spatial data at multiple resolution points, but did not deal with feature region extractions of construction and maintenance. Singh et al. [8] and Kumar et al. [9] addressed the problem of building topographic maps, and Viji and Prasanna [10] demonstrated how the information stored in topographic maps can be used for efficiently routing and resolving topographic queries.

The energy efficient communication module proposed in this thesis is the extension of the work done by Zhang et al. [11] and Ahmad and Khan [12]. The former proposed the idea of spatially clustering regions of the network with similar trends and patterns. The latter proposed a technique of using the AR (autoregression) models to approximately answer queries based on time series forecasting; this idea is used in this thesis to compute the AR model coefficients which serve as the local drafts of the observed data at each node. The cluster head then uses these AR model coefficients to predict the sensor readings without directly communicating to the members. In this research paper, the minimum spanning tree clustering technique is adopted for the formation of feature regions.

3 System Design

In this research work, an Intelligent Drip Irrigation System (IDIS) has been designed and implemented. This system consists of two important components, namely, an efficient soil moisture sensor unit (ESMSU) and an agriculture control station (ACS). The ESMSU is used to measure the soil moisture accurately using a newly proposed multi-depth soil moisture measurement algorithm. These soil moisture values help the farmers to conserve water and electricity and to avoid both the excess and shortage of water supply in irrigation. Moreover, the new protocol called ESMSU communication protocol has been proposed in this research work to collect the soil moisture data from the network of ESMSUs with less battery power consumption. This power conservation in ESMSUs is useful to prolong the lifetime of the sensors present in the network of ESMSUs. The ACS is responsible for collecting the data from sensor nodes and processing them accurately by using the data collection module for an effective data collection. This data is useful for agriculture scientists to guide the farmers through the farmer web portal and short message service (SMS) and also to find the rules (Fig. 2).

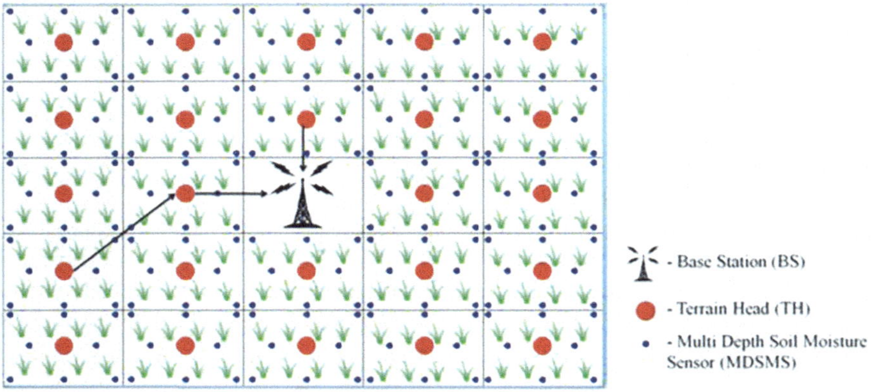

Fig. 2 Soil sensor nodes for communication

Fig. 3 Architecture of the ESMSU

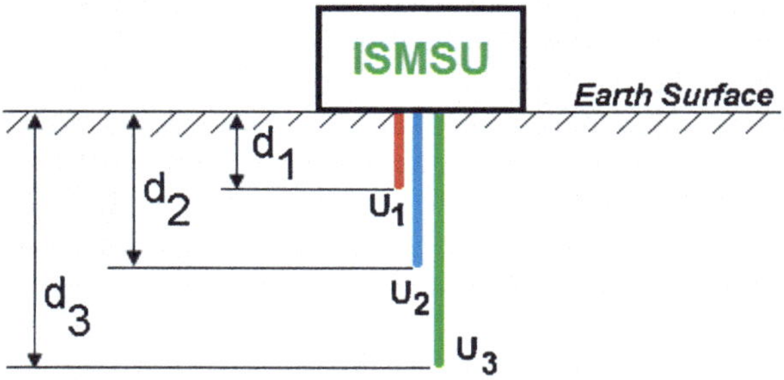

In the proposed work, soil moisture of the cultivation land is measured at different depths based on the root length of the plants over different time periods. The proposed soil moisture sensor unit (ESMSU) is used in the proposed IDIS (Fig. 3).

This intelligent sensor unit has multiple soil moisture units, namely, U1, U2, and U3, with multiple depths d1, d2, and d3, respectively. The proposed design can be extended to support many levels of depths based on the requirements. This intelligent sensor unit is capable of activating individual sensor unit or combination of sensor units based on the commands received from the control station (Fig. 4).

This configurable nature of the sensor unit adds higher flexibility in soil measurement of the land based on agricultural scientist's recommendation or farmer's choice. The procedure to configure the sensor unit is given in the algorithm for ESMSU configuration. For power saving, generally the transmission unit of all ISMSUs is turned off. Timers are used to wake up the transmission units of ESMSU for data transmission in a well-planned timing set by the ACS (Fig. 5).

Similarly, each ESMSU measures the soil moisture in defined intervals with timers. Moreover, each ESMSU is equipped with memory for storing the soil moisture records (SMR). The sensor unit's depth and number of records to be collected at each session are dynamically configurable by the ACS.

Fig. 4 Algorithm for ESMSU configuration

Input: Data received from DBC

Output: Selection of sensor unit(s)

Step 1: Begin

Step 2: receive data from DBC

Step 3: extract ID_DBC from CONFIG_DBC

Step 4: if ID_DBC DBID_DBC and ID_ISMSU != ID_SELF then

Step 5: ignore CONFIG_DBC

Step 6: else if DEPTH=di (i = 1,2,3) then

Step 7: select Ui

Step 8: else if DEPTH=di && dj (i,j = 1,2,3; i j) then

Step 9: select Ui , Uj

Step 10: else if DEPTH= di && dj && dk (i,j = 1,2,3; i j k) then

Step 11: select Ui , Uj , Uk

Step 12: end if

Step 13: End!

3.1 Agriculture Control Station

The information is sent from sensor nodes to the agriculture control station in specific intervals about the soil moisture data from the network of ISMSUs deployed in the field. The ACS retrieves the data from the soil moisture database (DBSM). The ACS consists of farmer database (DBF) which is used to store the details about farmers, their lands, and the unique identification number of the ESMSUs deployed in their lands. The vegetation preprocessing unit (VP-Unit) present in the ACS preprocesses the raw data retrieved from the DBSM. This unit extracts the date, time, identification number of ESMSUs, soil moisture value at different depths, and the battery status of ISMSUs and directs these data to the classification unit. The preprocessed data is further classified using the rule-based classification algorithm by the classification unit, and the classified results are stored in the classified database (DBCLASS). The information about the battery condition helps the farmers to recharge or replace the battery of ESMSUs if necessary.

The soil moisture at a particular depth alone is not a deciding factor for watering the plants, since the water holding capacity of each layer in the field is highly dependent on the adjacent layers. If the bottom soil layer is too dry, then the water supplied at the upper layer will be drained soon. Hence in the real-time environment, the soil moisture content to be maintained in a particular depth is highly dependent on the soil moisture content of the upper and bottom layers. The proposed scientific processing unit accesses the soil moisture data from the classified database and analyzes it for calculating the actual quantity of water to be supplied in the plant's root in different crop periods (CP).

Input : TRIGGER_TIMER

Output : FILE_SMR

Step 1: Begin

Step 2: initialize NO_OF_RECORDS

Step 3: open FILE_SMR

Step 4: if SUi (i = 1,2,3) enabled then

Step 5: while (NO_OF_RECORDS !=0)

Step 6: read TIME_ISMSU

Step 7: read PIN_VALUE_SUi

Step 8: write PIN_VALUE_SUi,TIME_ISMSU, ID_ ISMSU, di to FILE

Step 9: decrement NO_OF_RECORDS

Step 10: end while

Step 11: else if SUi && SUj (i,j = 1,2,3; i j) enabled then

Step 12: while (NO_OF_RECORDS !=0)

Step 13: read TIME_ISMSU

Step 14: read PIN_VALUE_SUi , PIN_VALUE_SUj

Step 15: write PIN_VALUE_SUi, PIN_VALUE_SUj, TIME_ISMSU, ID_ ISMSU, di to FILE

Step 16: decrement NO_OF_RECORDS

Step 17: end while

Step 18: else if SUi && SUj && SUk (i,j,k = 1,2,3; i j k) enabled then

Step 19: while (NO_OF_RECORDS !=0)

Step 20: read TIME_ISMSU

Step 21: read PIN_VALUE_SUi, PIN_VALUE _SUj, PIN_VALUE_SUk

Step 22: write PIN_VALUE_SUi, PIN_VALUE_SUj, &&

PIN_VALUE_SUk, TIME_ISMSU, ID_ISMSU, di to FILE

Step 23: decrement NO_OF_RECORDS

Step 24: end while

Step 25: end if

Step 26: close FILE_SMR

Step 27: End!

Fig. 5 Algorithm for a multi-depth soil moisture measurement

As a whole, the proposed system helps both the scientists and farmers to frequently examine the dynamic nature of soil moistures in the cultivation fields without frequently visiting the lands with manual interventions. Based on the dynamic nature of the lands, the scientists can guide the farmers about the duration and quantity of water supply via farmer web portal and SMS. The proposed approach significantly increases the accuracy of soil moisture measurement with the help of multi-depth sensors and hence conserves the water in dry areas. This water conservation extends the water availability, where water is a scarce resource. The proposed system also conserves the battery energy in ESMSUs and saves electricity.

3.2 *Experimental Setup*

In this proposed experiment, the ESMSU is built on RPi board, where soil moisture sensors (FC-28) are interfaced with the analog-to-digital conversion (ADC) connection board to count the soil moisture value. Three FC-28 sensors are placed in the field at three different depths (10, 20, and 30 cm) and interfaced with each ESMSU. Also a Wi-Fi module is interfaced with each ISMSU. In the proposed experiment, sensor hubs transform the data ESMSU using a Wi-Fi protocol.

In a well-planned scheduling, the ESMSU measures the soil moisture values at 10, 20, and 30 cm depths, date and time of measurement, and battery status and writes to the Secure Digital (SD) card. The sensor nodes collect data from the field in well-defined schedules and collect the values stored in the SD card of ESMSU. The area has a tropical climate with relatively high rainfall of about 15–41 mm with a peak in summer (March to October). The temperature ranges between 20 and 21 °C in winter and between 38 and 42 °C in summer. The brinjal crop with drip irrigation is considered for this experiment as shown in Fig. 4.5. The distance between adjacent plants is 40 cm, and the drip irrigation pipe is designed to watering the plants near the root region.

4 Result and Discussion

This section gives the result and analysis of the proposed system. The performance of the proposed system has been evaluated and compared with the existing static depth soil moisture measurement approach [13]. The experiments were conducted for 15 days continuously in the depths of 10, 20, and 30 cm, respectively. The expected soil moisture (ESM) value is a reference value that can be configured with the user's choice as it highly depends on the nature of soil, crop type, and crop period. From the experiments, it is found that the proposed system provides a high level of flexibility to the agriculture scientists to set these values.

We can find the similar straight line fitting to the data when the land depths are 10, 20, and 30 cm. The ESM values and the ASMR calculated by both the proposed and existing static measurement approaches proposed by Gutierrez et al. [14] in different days are tabulated in Tables 1, 2, and 3, and comparison graphs are shown in Figs. 6, 7, and 8.

From Table 1, it is observed that the proposed model predicts the soil moisture more accurately than the static measurement approach at 10 cm depth.

Table 1 Volumetric water content of the soil in 10 cm depth

Expected soil moisture (ESM) (%)	ASMR calculated by the proposed approach (%)	ASMR calculated by static measurement approach (%) [14]
0.06	0.056	0.012
0.065	0.06	0.019
0.07	0.065	0.019
0.075	0.07	0.026
0.08	0.074	0.029
0.085	0.081	0.03

Table 2 Volumetric water data of the soil in 20 cm depth

Expected soil moisture (ESM) (%)	ASMR calculated by the proposed approach (%)	ASMR calculated by static measurement approach (%) [14]
0.08	0.076	0.04
0.082	0.077	0.041
0.084	0.078	0.043
0.086	0.0799	0.044
0.088	0.082	0.045
0.0899	0.084	0.046

Table 3 Volumetric water data of the soil in 30 cm depth

Expected soil moisture (ESM) (%)	ASMR calculated by the proposed approach (%)	ASMR calculated by static Measurement approach (%) [14]
0.09	0.083	0.06
0.091	0.084	0.061
0.092	0.085	0.062
0.093	0.086	0.064
0.094	0.087	0.066
0.095	0.088	0.067

From Table 2, it is observed that on all the 6 days of observation, the proposed approach provides more accurate results than the static measurement approach.

From Table 3, it is seen that the proposed approach provides matching results with the estimated moisture. This is due to fact that the proposed model uses intelligent rules for making an effective decision on the water usage based on the management of soil moisture.

From Fig. 7, it is observed that the volumetric content at 20 cm depth is measured more accurately by the proposed methodology compared with conventional static measurement approach.

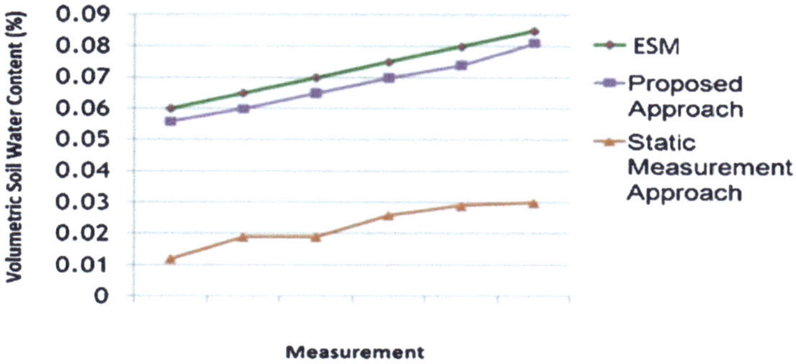

Fig. 6 Soil moisture data at 10 cm depth

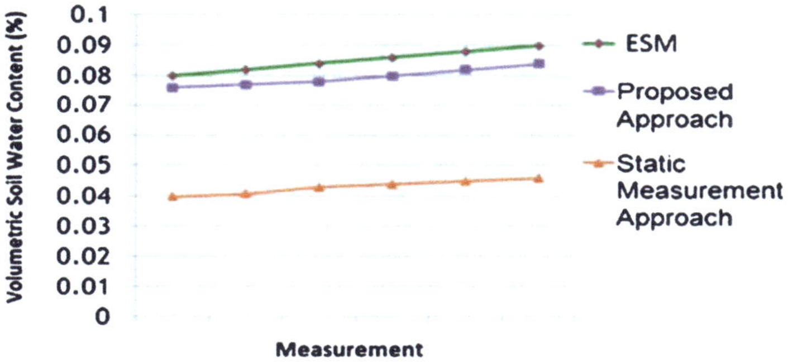

Fig. 7 Soil moisture data at 20 cm depth

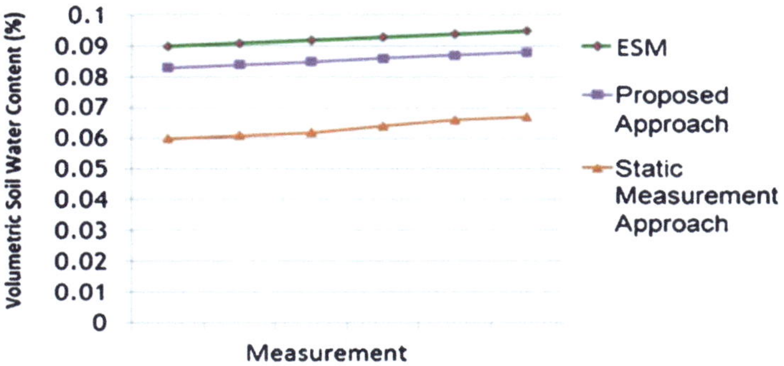

Fig. 8 Soil moisture data at 30 cm depth

Table 4 Volumetric water content comparison

Depth of 10 cm		Depth of 20 cm		Depth of 30 cm	
Deviation in the proposed method (%)	Deviation in static measurement approach (%) [14]	Deviation in the proposed method (%)	Deviation in static measurement approach (%) [14]	Deviation in the proposed method (%)	Deviation in static measurement approach (%) [14]
0.004	0.048	0.004	0.04	0.007	0.03
0.005	0.046	0.005	0.041	0.007	0.03
0.005	0.051	0.006	0.041	0.007	0.03
0.005	0.049	0.0061	0.042	0.007	0.029
0.006	0.051	0.006	0.043	0.007	0.028
0.004	0.055	0.0059	0.0439	0.007	0.028

From Fig. 7, it is observed that the volumetric content at 20 cm depth is measured more accurately by the proposed methodology when compared with conventional static measurement approach.

From Fig. 8, it is observed that the volumetric content at 30 cm depth is predicted more accurately by the proposed methodology when it is compared with conventional static measurement approach.

From Table 4, it is observed that the proposed multi-depth soil moisture measurement approach outperforms the existing static measurement approach in terms of accuracy in the calculation of soil moisture to be maintained in different depths at different periods. The efficiency and precision lead to exact watering of agriland and increase the product yield. Also, this precise measurement conserves water and electricity by avoiding unnecessary operation of drip irrigation pump.

5 Conclusion

The proposed Intelligent Drip Irrigation System (IDIS) is formed with wireless sensor networks, where each ESMSU is a sensor node. The networks of ESMSUs are deployed in the agriculture field with a fixed topology based on the requirements. The data collected by the sensor network using the sensor nodes are processed by the ACS. In this proposed model, generally the transmission units of all ESMSUs are turned off for effective power saving. Timers are used in this model to wake up the transmission units of ESMSU which are used for data transmission in a well-planned timing which is set by the ACS. Similarly, each ESMSU measures the soil moisture in defined intervals with timers. Here, each ISMSU is equipped with memory for storing the soil moisture records (SMR). The sensor unit's depth and number of records to be collected at each session are dynamically configured by the ACS. The ACS consists of a local coordinator and region heads. The raw data obtained from the field is sent to preprocessing and also for dynamic terrain management and to the decision manager to provide recommendation to the farmers.

References

1. Akbas, A., Yildiz, H.U., Tavli, B., Uludag, S.: Joint optimization of transmission power level and packet size for WSN lifetime maximization. IEEE Sensors J. **16**, 5084–5094 (2016)
2. Li, J., Shen, C.: An energy conservative wireless sensor networks approach for precision agriculture. Electronics. **2**, 387–399 (2013)
3. Figueroa, M., Pope, C.: Root system water consumption pattern identification on time series data. Sensors. **17**, 1–21 (2017)
4. Gajendran, E., Prabhu, S.B., Pradeep, M.: An analysis of smart irrigation system using wireless sensor. Multidiscip. J. Sci. Res. Educ. **3**(3), 230–234 (2017)
5. Grichi, H., Mosbahi, O., Khalgui, M., Li, Z.: RWiN: new methodology for the development of reconfigurable WSN. IEEE Trans. Autom. Sci. Eng. **14**, 109–125 (2017)
6. Kuppannan, P., Ramarao, R., Samiappan, S., Malik, R.P.S.: Estimating technical and irrigation water productivities in rice production in Tamil Nadu, India. Irrig. Drain. **66**, 163–172 (2017)
7. Anand, C., Sadistap, S., Bindal, S., Botre, B.A., Rao, K.S.N.: Wireless multi-sensor embedded system for agro-industrial monitoring and control. Int. J. Adv. Netw. Serv. **3**, 1–10 (2010)
8. Singh, P., Manna, S., Singhb, S.: Energy efficient clustering protocol based on improved metaheuristic in wireless sensor network. J. Netw. Comput. Appl. **83**, 40–52 (2017)
9. Kumar, M., Kumar, R., Rajput, T.B.S., Patel, N.: Efficient design of drip irrigation system using water and fertilizer application uniformity at different operating pressures in a semi arid region of India. Irrig. Drain. **66**, 316–326 (2017)
10. Viji, R., Prasanna, P.R.: Assessment of water holding capacity of major soil series of Lalgudi, Trichy, India. J. Environ. Res. Dev. **7**, 230–234 (2012)
11. Zhang, W., Han, G., Feng, Y., Lloret, J.: IRPL: an energy efficient routing protocol for wireless sensor networks. J. Syst. Archit. **75**, 35–49 (2017)
12. Ahmad, A., Khan, S.: Water and energy scarcity for agriculture: is irrigation modernization the answer? Irrig. Drain. **66**, 34–44 (2017)
13. Prijono, S., Bana, S.: Study of soil moisture on coffee plantation in dry land using neutron probe in Malang, East Java. Bull. Env. Pharmacol. Life Sci. **4**(2), 135–143 (2015)
14. Gutierrez, J., Villa-Medina, J.F., Nieto-Garibay, A., Porta-Gandara, M.A.: Automated irrigation system using a wireless sensor network and GPRS module. IEEE Trans. Instrum. Meas. **63**(1), 166–176 (2014)
15. Li, Y., Gao, W., Wu, C., Wang, Y.: Deployment of sensors in WSN: an efficient approach based on dynamic programming. Chin. J. Electron. **24**, 33–36 (2015)

Precision Agriculture Using Advanced Technology of IoT, Unmanned Aerial Vehicle, Augmented Reality, and Machine Learning

Vijayakumar Ponnusamy and Sowmya Natarajan

Abstract Agriculture is one of the primary processes for quality food production in the globe. Unfortunately, the productivity of agriculture is very low, and many factors affect the yield level of it. Precision agriculture (PA) is one of the solutions for the above problem. PA uses site-specific crop management concept based on measured data using sensors and data analytics to find the root cause of yield reduction. Precision agriculture automates farming which involves the collection of data and analysis of them for better decision-making to gain high yield and quality of the agricultural product. The agriculture system integrated with data analytics and machine learning is called as smart farming or smart agriculture The goal of smart agriculture is to develop a decision-making support system for farming management. The precision smart agriculture can be enhanced with the help of latest technologies of Internet of Technology (IoT), unmanned aerial vehicle (UAV), augmented reality (AR) system, and machine learning (ML) algorithms. This chapter focuses on the illustration and utilization of those advanced technologies for smart farming.

Keywords Augmented reality · Data analytics · Drone · Machine learning · Precision agriculture · Smart farming · IoT · Unmanned aerial vehicle · Virtual reality

1 Introduction

Besides feeding the population of a nation, agriculture plays a vital role in the contribution of exports and also helps to boost the GDP of our country [1]. In recent years precision agriculture has become one of the most famous applications of IoT. IoT technologies will help the farmers to enhance productivity at less expense [1]. Precision agriculture is a process that makes the farming to be more accurate

V. Ponnusamy (✉) · S. Natarajan
ECE Department, SRM Institute of Science & Technology, Kattankulathur, Tamil Nadu, India
e-mail: vijayakp@srmist.edu.in

D. Gupta et al. (eds.), *Smart Sensors for Industrial Internet of Things*, Internet of Things, https://doi.org/10.1007/978-3-030-52624-5_14

and controlled for raising livestock, growing of plants, and yield of crops [2]. Smart farming says thriving agriculture with technology. The adaptation of latest technologies like IoT, unmanned aerial vehicles or drones, augmented reality, virtual reality, and machine learning-based data analytics is happening in agriculture. This new technology makes a revolutionary change in the field of agriculture process and makes more economical efficient farming. This chapter aims to review how these latest technologies can be applied in precision agriculture or smart agriculture. The forthcoming part of the chapter is organized as follows. In Sect. 2, uses of IoT for the precision agriculture are discussed. In Sect. 3, the use of UAV or drone technology is presented, and Sect. 4 covers the virtual reality or augmented reality system application for smart farming. Section 5 analyzes the use case of machine learning in data analytics of agriculture data. Then, finally, this chapter closes with some general conclusions.

2 Internet of Things (IoT) and Precision Agriculture

In India, agriculture is the primary resource of livelihood for two-thirds of the entire population. Agricultural land occupies approximately 43% of India's geographical land cover [3]. Currently, traditional farming practices are used by the farmer, which are inefficient and not providing good yield or good food. Even though the farmers are working hard to attain profitability, it fails in many instances. Precision agriculture is one of the solutions to improve the agricultural system using Information Communication Technology (ICT) [2]. IoT is currently molding different aspects of human life. It is predicted that by the year 2050, agriculture using IoT will increase nourishment productivity by 70% and furnish foods up to 9.6 billion human beings [4]. IoT is coming together with electronics engineering and computer science to solve the traditional farming problems. This relies on specialized equipment and software/IT services. IoT facilitates to access real-time data about the conditions of crop growth, soil nutrient content, and stress caused due to lack of water. The data analytics software in IoT uses the data to provide precise information to the farmers. IoT makes it possible for farmers to receive highly accurate real-time information about agricultural processes like irrigation, planting seeds, harvesting, and management of farming tasks without investing in labor [5].

The pictorial representation of IoT agriculture system in Fig. 1 shows how IoT can interact with the real environment freely in noninvasive way. The smartphone or tablet can be used to capture image data from the field, and cloud computing can be used to analyze them. Through the data analytics and tracking of the image of particular farmland, it can be possible to know its soil erosion, variable rate of fertility, water stress, crop yield, diseases in plant, etc. [2].

In the following sections, the benefits of acquiring IoT in precision farming and a cloud-based IoT technology for agriculture are going to be discussed. Then an efficient IoT platform called AgriTalk for agricultural application is presented with an application of turmeric cultivation.

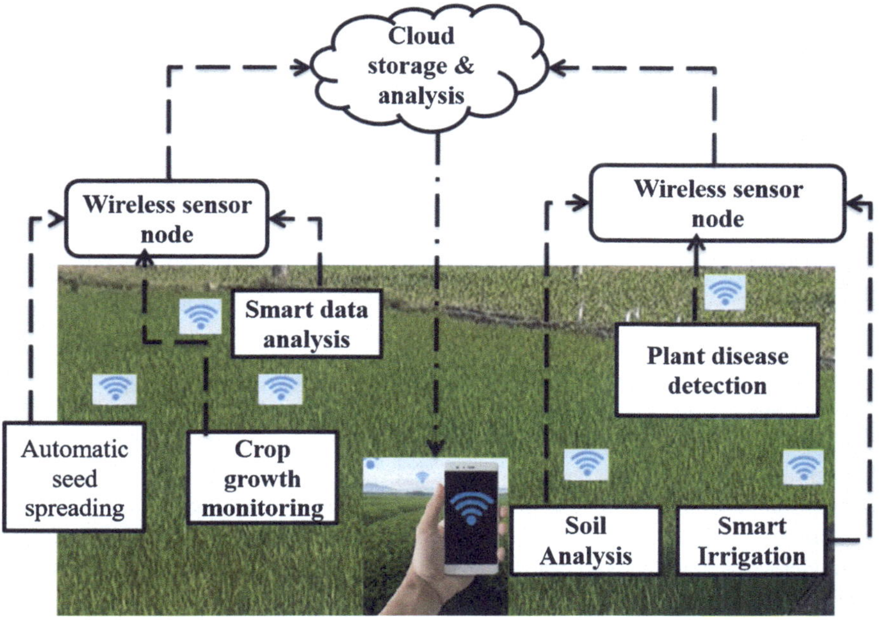

Fig. 1 Pictorial representation of IoT in agriculture and its footprints

2.1 The Benefits of Adopting IoT in Agriculture

2.1.1 Sustainable Farming Based on Climate Conditions

It enables the farmers to access the weather conditions. The sensors which are deployed in and around the agriculture field continuously monitoring the climate condition thereby allow the farmers to select the crop type to cultivate for better yield [6].

2.1.2 Decision Support for Improving Productivity

The purpose of precision farming is to examine generated data via sensors and make information extraction and decisions to react. IoT makes the information extraction and decision-making very easy; thereby the profit of the farmers can be increased [6].

2.1.3 Automated Smart Greenhouse

IoT, with a certain set of instructions, provides the facility to the farmer in automatically tuning the weather conditions in the greenhouse. The sensors deployed on the greenhouse help to know the information about the pressure, humidity, temperature, and light levels. This IoT technology has eliminated the use of human being involvement, thus making the whole process to be cost-effective.

2.1.4 Quality Assessment on the Field by Data Analytics

IoT systems also provide applications in the analysis of livestock and crop quality. Thereby the farmers can know about the harvest timing of crops based on the quality. It also helps the farmer in getting the right price for their product based on quality data [6].

2.1.5 Spatial Variability Analysis and Large-Area Coverage Using Agricultural Drones

Agriculture land has a high degree of spatial variability, which needs to be analyzed to label the particular issue in a specific location. The ground and aerial drones are used for location-wise assessment of crop health monitoring, planting, spraying of pesticides, and field analysis for crop yield monitoring. The IoT drones allow for wrapping a huge area of field by taking less time and cost-effectively [6].

2.2 Cloud-Based IoT Architecture for Precision Agriculture

Cloud computing is a mechanism of delivery of on-demand computing resources like storage or computing processor from the applications through the data centers over the Internet on a *pay-for-use basis*. IoT-based precision agriculture system deployment requires a high level of scalability since it has to collect data from a wide range of locations. It also demands high storage and high computing power to analyze large volumes of data in real time, which doubts the affordability of it by the farmers. Those challenges can be addressed economically by the cloud system because of the following characteristics: (1) Cloud allows elastic resource creation, which enables users to scale up or down quickly based on the application demand. (2) It provides metered services, i.e., pays only for what you use. (3) Cloud is a purely self-service system which makes the application to create the needed resources without depending on any other entity [7].

Wireless sensor networks (WSNs) are vital components of the IoT. It enables the collection of data from the sensing field through wireless communication. A cloud-based IoT system consists of WSN sensor devices which collect data and communicate to the cloud through wireless Internet connectivity [7]. Once the data gets into the cloud, data analytic software processes it and then decides to act, such as sending an alert or automatically adjusting the actuator/sensor without user involvement.

Figure 2 shows the cloud-based IoT system. It has mainly three layers. The first layer is the front end layer with a collection of sensor nodes which has sensors/

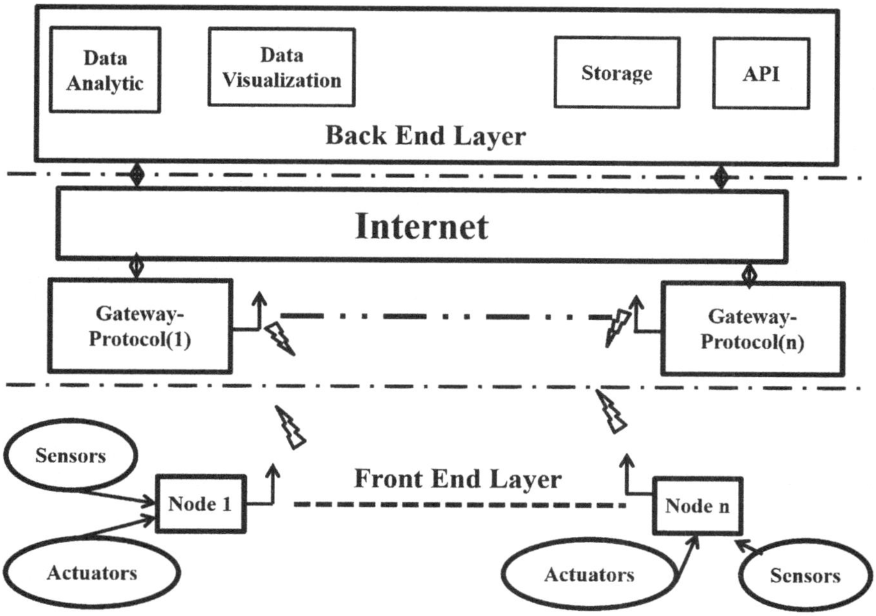

Fig. 2 Cloud-based IoT system

actuators/mechanical devices. Those sensor nodes are capable of collecting or interacting with the physical features from the real environment and have microcontroller to process/control them. The wireless module in this layer communicates the raw or preprocessed data to the next layer called the gateway layer. The gateway layer relays the collected data to the cloud servers in the back end for storage and data analytics. The third layer is the back-end layer, which is responsible for facilitating the end users by providing data storage, data analytics, data visualization, and software tools for accessing the data [7]. The gateway layer also forwards requests/commands from the back-end layer to the actuators/sensors in the nodes. The command from the back-end layer enables taking appropriate action such as spraying chemicals or fertilizers, watering the plants, etc.

2.3 An Efficient IoT Platform for Precision Farming

A low-cost scalable IoT platform called AgriTalk is proposed [4] for outside cultivation. It provides a user-friendly GUI called AgriGUI that allows to easily configure the sensor and actuators through graphical connections on a mobile dashboard. Calibration of the sensor on the field is another significant challenge that is addressed through AgriGUI functionality easily. The farmers can make use of these AgriTalk resources through the web browser by uploading any specific application. The AgriTalk performance is evaluated on turmeric cultivation. The result of cultivation shows turmeric quality has been enhanced around five times more compared with the conventional IoT system. Failure detection of the sensor is another important aspect covered in work, which may frequently happen in the agricultural field [4]. AgriTalk uses two simple ways to detect the potential abnormality of IoT devices automatically. In the first method, AgriTalk automatically compares the data of nearby sensors of the same type. If the values deviate more from a neighbor, then it is classified as a faulty one. In the second method, AgriTalk monitors the power consumption of the sensors. If it is abnormal, then it is classified as defective. If any sensor is detected as faulty, the AgriTalk alerts the farmer to replace the sensor. The response time of IoT is another crucial factor for real-time monitoring. The response time of the AgriTalk is reported as it can respond quickly to the dynamic change of the field environment condition by taking only less than 0.2 s [4]. Pest-based diseases in the plant are unavoidable. The pesticides are used to control the diseases which produce toxic content in the agricultural product. Disease prediction is another important factor in agriculture, which is carried out by the regression mechanism in the AgriTalk platform. A self-developed biopesticide is used before the diseases occur based on the prediction result to avoid toxic generation. Maintaining the acidification level of the soil is another critical issue, especially for turmeric cultivation. Due to acidification of soil after turmeric harvest, the land typically lies fallow for about 2 years [4]. AgriTalk uses the pH and the EC sensors to monitor and prevent acidification, which allows the land to continuously grow turmeric without land fallowing.

3 Unmanned Aerial Vehicle (UAV) and Precision Agriculture

Agriculture drones/UAVs are a perfect example of IoT applications in agriculture. The UAV is also called drone in general. There are two types of drone available: (1) ground-based and (2) aerial-based. Many countries are starting to use UAVs in their farmlands. A UAV is an aircraft without being operated by the human pilot but controlled by the radio channel. There are different types of UAV structures available for the field monitoring: (1) single rotor, (2) quadcopter, (3) hexacopter, and (4) octocopter [1]. UAVs can perform real-time monitoring at a more accurate and low cost. Agricultural drones let the farmers see their fields from the sky. Sky monitoring enables them to be supportive in agriculture and proving astounding performance in various assessments like producing warning signals to measure the field data, detect a specific event in the field, real-time crop monitoring, crop growth reporting, soil nutrient variation monitoring, irrigation problems, moisture content assessment, and yield prediction [8]. It also employed in spraying precious amounts of fertilizers on the infected areas. In the case of spraying fertilizers, it reduces chemical cost, time consumption, and labor cost. The drone technology will give a high-tech makeover to the agriculture industry by making use of real-time data collection and processing [1, 9].

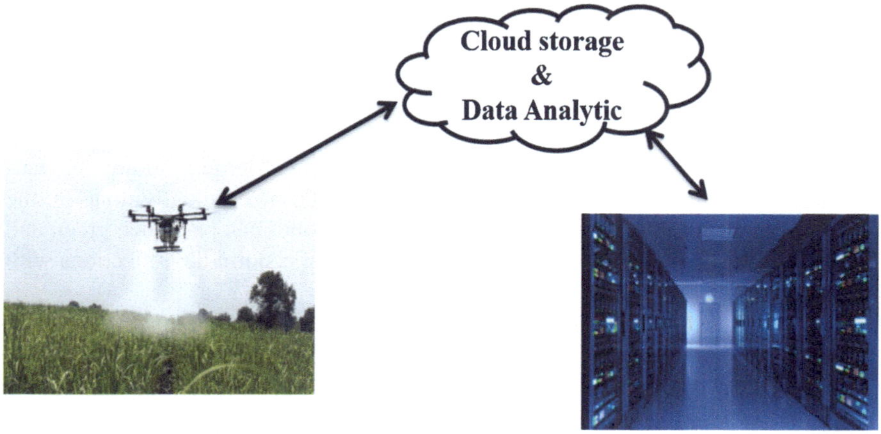

Fig. 3 Drone-based smart agriculture

The limitations of drones are the limited onboard computational power and storage. In real-time image analysis of agriculture multispectral data, it may not be able to cope up. This kind of high computational intensive process can be carried using cloud resources. Figure 3 shows a typical cloud drone deployment for the agricultural application. The drone will collect the on-field image or data and transmit it to the cloud. The cloud server will take the data from the cloud and process it. Then, the cloud server communicates the result or control information to the drone to initiate appropriate control action.

The benefits and limitations of using the drone for precision agriculture are discussed below, followed by existing applications of drones with precision agriculture, and their implementations were discussed.

3.1 Benefits of Using Drone in Agricultural Field

There are lots of benefits to using drones in the agriculture field because of its high degree of spatial and temporal resolution and capability of covering a large area. Some of the benefits are reviewed below:

1. *3D Mapping*: Drone could be used for the development of accurate 3D mapping easily. This 3D mapping can be used to conduct soil analysis. 3D mapping plays a vital role in seed planting patterns. Even after planting, the map is useful for getting soil property information like its moisture, temperature, and nutrient content, which are helpful for irrigation and nitrogen level management.
2. *Low-cost planting*: Low-cost precious planting with the large covered area is possible using drones. The drone is also employed to shoot pods containing seeds and plant nutrients into the earlier prepared field.

3. *Time-series animations*: Time-series animated data generation enables us to visualize the crop development effectively. The flying drones can efficiently be used to develop such kind of time-series animations [10].
4. *Ease of deployment*: Unlike traditional aircraft, it is much easier and cheaper to deploy the drones on the agricultural field monitoring.
5. *Huge area stress assessment*: Stress assessment on the large field area is challenging and time-consuming. Stress assessment can be efficiently done by the green and near-infrared reflection of the plants. Drones can pass such green and near-infrared rays on a large area on the fly and can acquire the reflections with less time.
6. *Location-based efficient irrigation*: Drones equipped with thermal, hyperspectral sensors and GPS can give sky view and make easier identification of parts of the field with the exact location that has become dry and wet. This way, the designated areas can be attended to promptly, making irrigation precise and timely.
7. *Efficient spraying*: By dynamically adjusting the altitude of flying, a drone can spray the exact amount of the fertilizer evenly on the required location in real time [10].

3.2 Challenges of Using Drone in Agricultural Field

1. *Limited flight time and range*: Most of the drones have a short flight time of between 20 min and an hour. The flight range also limits the radius that can be covered during every flight time. Drones with longer flight time and a longer span of coverage are relatively costlier [10].
2. *Cost*: Drones with some significant features for use in the agricultural field are quite expensive. For some drones, the high price is inclusive of hardware, software, tools and imaging sensors, desired cameras, and processing software.
3. *Laws*: Since the usage of drones is commercial, the farmer needs to undergo FAA (Federal Aviation Administration) operator training to acquire a remote pilot certificate or hire an operator with such qualifications. FAA demands that drones be flown at an altitude of not more than 400 feet [10].
4. *The airspace interference*: Drones share the same airspace with manually human-crewed aircraft. Hence it is prone to interference.
5. *Investment on connectivity*: Any farmers intending to use drones have to invest in connectivity.
6. *Climate dependent*: Drones flying are affected by climatic conditions. If it is more windy or rainy outside, it may not be able to fly [10].

3.3 UAV for Crop Monitoring

The assessment of crop condition information at early stages in the crop growing season is one of the significant processes to estimate yield or prepare proper management techniques. Mostly, the remote sensing applications using multispectral satellite data have been used to monitor large-scale crop field, which is a costly one. The drones with a high degree of spatial and temporal resolution are the better alternate choice for crop monitoring.

In drone-based monitoring of the agriculture field, the vegetation index parameter is used to quantify the crop monitoring application. Among many vegetation index, Normalized Difference Vegetation Index (NDVI) is mostly used [11]. NDVI is a graphical indicator mainly used in remote monitoring measurements. It is a parameter that enables us to show the differences in the performance of the same seed within a specific row.

NDVI quantifies vegetation by calculating the difference between near-infrared (which vegetation strongly reflects) and red light (which vegetation absorbs) channel.

$$\text{NDVI} = (\text{NIR channel value} - \text{RED channel value}) / (\text{NIR channel value} + \text{RED channel value})$$

Near-infrared channel data are collected by capturing optical light in the ranges between 750 and 2500 nm. The red spectrum data are collected by scanning the optical spectrum ranges between 620 and 750 nm [12]. The drone can be used for the specific crop monitoring with impact analysis of various factors on the growth of the crops.

In this direction, the monitoring of sugarcane growth for varying nitrogen application rates using UAV is presented [13]. A 3D LiDAR mapping and multispectral imaging data are used to generate time-series data over the growing cycle of sugarcane to accurately measure crop height. Correlation analysis is also carried out between measured growth height and the sugarcane biophysical parameters of the stalk population, total fresh biomass, and cane yield to assess the quality of sugarcane. The work only analyzed the impact of nitrogen nutrient on the growth of sugarcane, but another critical nutrient like potassium is not considered. The growth analysis needs to be carried out concerning multiple nutrient and environmental conditions during the growth cycle.

3.4 Soil Analysis Using Drone

The drone can able to produce precise 3D maps for soil analysis. It helps the farmers to find direct seed planting patterns, the irrigation level, nutrient consistency, moisture content, and soil erosion before the beginning of the crop cycle. The 3D imagery NDVI map makes it possible to identify the photosynthetic activity of the plants, which shows which region has high cation exchange capacity. Cation exchange capacity (CEC) is a measure of cation retaining capability of soil. CEC is used as a measure of soil fertility and indicates the ability of the soil to retain the nutrients. Figure 4 shows a sample map of NDVI and cation exchange capacity. The region-wise soil nutrient information can be extracted from photosynthetic activity.

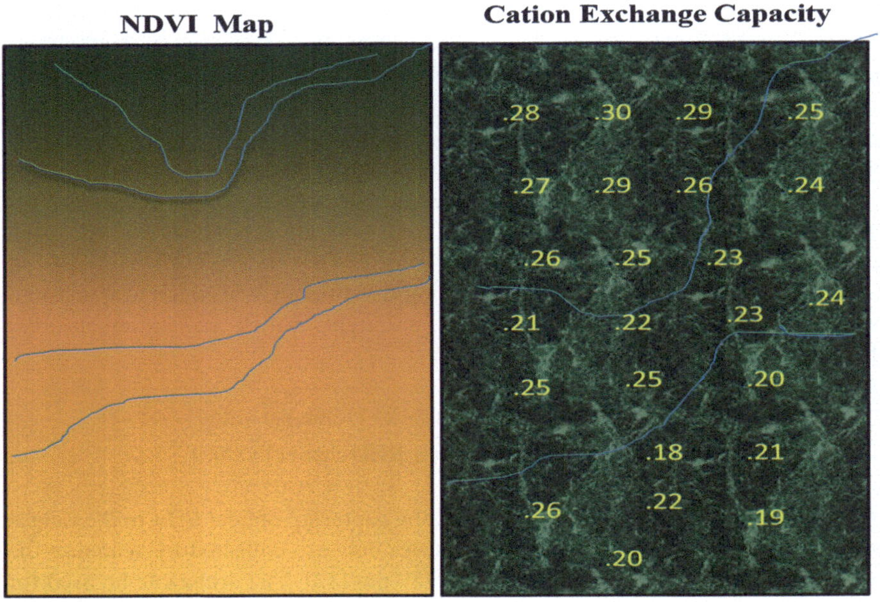

Fig. 4 NDVI map and CEC map

Soil residual analysis is another application of drone, which is called crop residual cover (CRC) analysis. CRC is the critical parameter for sustainable agriculture. A drone-based soil residual cover estimation using RGB image processing is presented with a comparison of the performance with Landsat 8 OLI data satellite image processing [14]. The optimal distance to capture the image by the drone is an essential parameter in soil content analysis. The study reported that at 10 m distant, an accepted correlation of 0.84 is obtained on the ANOVA test. Even though the result obtained by drone is slightly inferior compared to that of satellite images, it can be employed for CRC analysis because of other benefits like the lowest cost, accessibility, and the highest spatial and temporal resolution. The drone can be employed for the soil sampling process and also for fixing the sampling point from aerial view. One of such methods is discussed in the augmented reality section.

3.5 Health Assessment Using Drone

Drones with multispectral or thermal sensors can be able to detect and identify the dry part of the field or the regions which need further improvement. Drones with those sensors are also able to measure the vegetation index and provide a health indication of the crop. The sensors use five spectral bands to capture the images to provide more reliability.

With the infrared and visible light sensors, it is possible to identify the area which was infected by fungus or bacteria, and this analysis will help the farmers in preventing them from spreading to another region of the crop field. Multispectral sensors can detect the diseased area even before human eyes capture it [15]. Drones can help to assess crop health and spot fungal infections with visible and NIR light. Once the sickness is identified, farmers can monitor and apply remedies more precisely.

3.6 *Aerial Spraying Using Drone*

Aerial spraying of water, pesticides, and fertilizer is one of the leading applications of the drone in precision agriculture. Making a proper amount of droplet deposition density during the spraying operation using the drone is a challenging one. An optimal drone parameter selection method is proposed to solve this issue [16]. The UAV flying speed and spraying rate are selected for proper deposition density. The experiment is conducted for the flying speed of drone at 2, 4, and 6 m/s at the constant height of 2 m above the rice field. The spraying of the organic liquid fertilizer is carried out with the spraying rate of 0.75, 1.5, 2.25, and 3.00 L/min.

The droplet deposition density is calculated using statistical analysis and water sensitivity paper. The research concluded that for good deposition density, the drone is to be operated with a speed of 2 m/s at the spraying rate of 3.00 L/min.

4 Augmented Reality in Precision Agriculture

Virtual reality (VR) is a technology that superimposes a computer-generated image on a user's view of the real world, thus providing a composite view called virtual reality. When a virtual reality system is designed to help to execute a specific task through augmenting needed information,it is called augmented reality (AR) [17]. In other words, AR input from the real environment to the application must be provided automatically through the eyeglass. Under AR, the guidance information will come to the user automatically without any need to search for it. AR system consists of three primary features: (1) a combination of virtual reality worlds, (2) real-time interaction, and (3) an accurate 3D registration of virtual and real objects [18]. Wearable augmented reality glasses are the obvious choice of hardware for farmers, as the virtual content will be in the user's view and farmers are free to operate their hands [19]. Combining IoT- and AR-based environment makes to superimpose IoT information into physical objects and interpretation of such information [20].

The benefit of augmented reality to the farmer can give useful information that is unavailable to their senses and help them to perform real-world agricultural tasks [19]. It can aid the farmers in different ways, like inspecting the field, identifying the species of insects, and also determining the ways to handle it. Since AR set hardware has limited computational capability, real-time image data processing may not be done within the AR set. So, we can use a cloud computing platform for real-time data analysis. In the below section, use cases of AR for smart agriculture and an AR system design approach using cloud computing and machine learning are presented for the application of field plant analysis. This study helps to know about the detailed architecture of AR in smart farming, which helps in the identification of plant diseases and toxic contents. The subsequent section describes how the soil analysis could be made using AR combined with the drone device, integrating augmented reality (AR) and IoT for better visualization of crop growth and tractor navigation system using AR.

4.1 Use Cases of AR for Smart Agriculture

1. *On-field plant analysis*: AR system helps the farmers to detect diseases and toxic content in the plant in real time with visual annotated information about the disease and toxic items. In the below section, one of the system models for plant analysis is presented.
2. *Land examination/soil sampling*: With AR, it is easier for farmers to analyze their land and determine which crop would flourish under the given conditions. This kind of land examination will minimize the loss and help the farmers to enhance the quality of the land. A case study of soil sampling with AR and UAV is presented in the forthcoming section.
3. *Better data visualization*: AR system can be integrated with the IoT system to support better data visualization. One of the integrated systems is discussed in the below section.
4. *Pest control management system*: Using AR farmers can have a closer look at the pests infesting their crops. Some types of pests help the farmers and few harmful pests, which damage the crop production, and it needs to be eliminated.
5. *Access to weather-related information*: With AR weather information can be used to choose which crop to cultivate under given weather conditions. This type of weather condition-based decision-making will minimize crop losses and maximize yield [21].

4.2 Cloud-Based AR System for On-Field Plant Analysis

Since the computational power and memory of the AR set is limited, the AR set is used for image data collection and visualization only. The power of the cloud server is utilized to carry out the computationally intensive data analytic task.

Figure 5 shows the representation of such cloud-based AR in precision farming. The system is designed for targeting two applications of plant disease detection and toxic content detection. In this system, there are two modules: (1) augmented reality

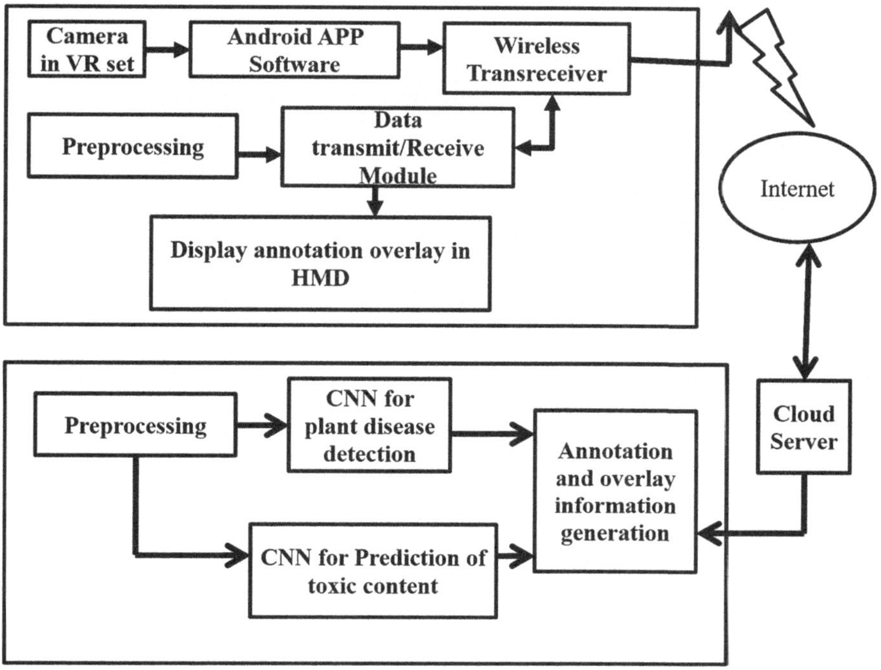

Fig. 5 AR-based plant disease detection and toxic content detection

head-mounted display (ARHMD) set and (2) the software module running in the cloud. There are many ARHMD sets available for AR application deployment. Among them, Moverio BT-350 is one of the leading AR platforms with more facility of programming which can be used in the smart farming application [22].

The system specifications for Moverio BT-350 are listed:

- Lifetime is approximately 6 h (2950 mAh).
- Operating voltage—5 V DC (when it is connected with USB port).
- Sensors—GPS/geomagnetic sensor/accelerometer/gyroscopic sensor/illumination sensor.
- Panel size—0.43″ wide panel (16.9 flip-up style).
- Screen size—40″ support (virtual viewing distance 2.5 m).
- Memory capacity—2 GB internal (16 GB external extended to 32 GB).
- Camera resolution—five million pixels.

Moverio BT-350 device enables android-based programming using Java. A python program supporting low-cost smart class is developed by us for augmented reality application. Figure 6 shows the picture of our design of smart glass which can be used as ARHMD device. The deign uses Raspberry Pi Zero board and touch HD display which enables navigation of augmented images. It also has a memory card and a USB charger.

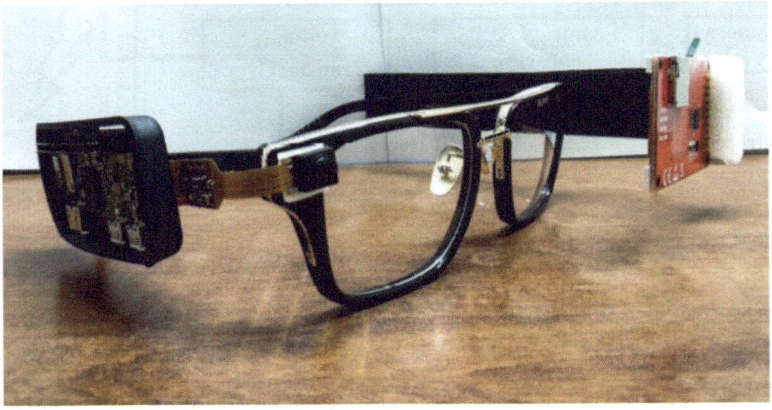

Fig. 6 Raspberry Pi Zero board-based smart glass

Under this system, the farmer has to wear the ARHMD set and watch the field through the glass. It gives a virtual view of the specific information about plants that are received from the cloud server. The camera in the HMD set captures the images and delivers them to a cloud server for analytics. The android software in the ARHMD set is responsible for the preprocessing of the captured image like extraction of a specified field present in the image, i.e., ROI (region of interest). After preprocessing, it is forwarded to the wireless transceiver for the transmission to the cloud server. Once the annotated information is received from the cloud server, the HMD set will display on the see-through glass display.

In the software processing, block running in the cloud uses two kinds of convolutional neural networks for analyses of plant disease and the identification of food toxic content. Based on the output of convolutional neural networks, the software module generates the annotation and overlay information to be displayed on the HMD and communicated to it.

This cloud-based AR system makes the farmer to quickly identify the plant disease and toxic content at each plant level more accurately and efficiently compared with the naked-eye inspection.

4.3 Soil Sampling Using AR

Soil sampling is an essential tool to gather information for making proper decisions regarding the fertilization of fields. The soil sampling process involves the collection of soil samples from the specified regions and evaluating nutrients (potassium, sodium, phosphorus, etc.) present in that region. It will be helpful to the farmers to plan crop cultivation accordingly.

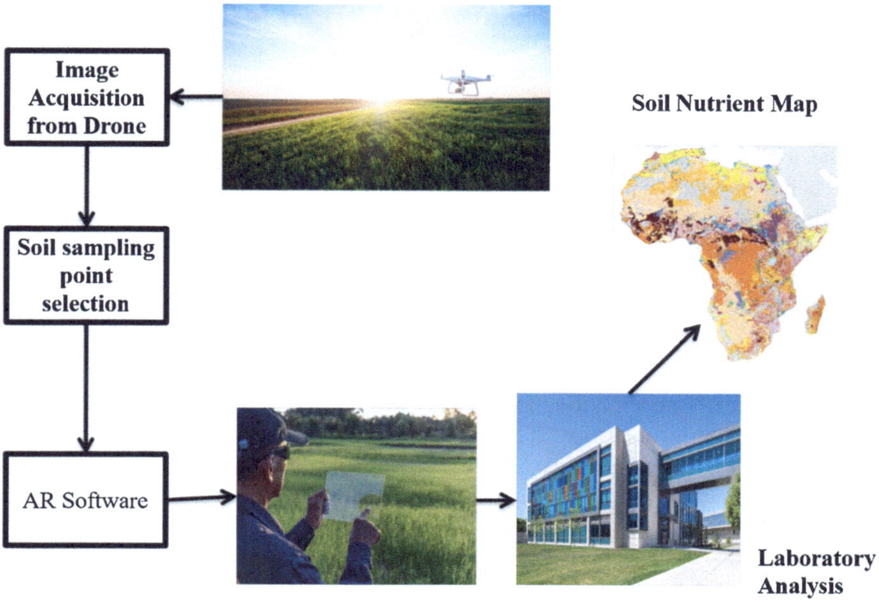

Fig. 7 Soil sampling with drones and AR in precision agriculture

In soil sampling, selecting the sampling point in a large agricultural field is a challenging one. On the field, augmentation is required for the farmer to choose the sampling point. A wearable headset-based augmentation system will solve the above challenge. The drone can provide sky view-based mapping information about the soil and can create augmentation for the real-time sample collection. Figure 7 shows soil sample processes with AR and drones. The prerequisite needs soil sampling to be carefully examined before taking the samples. The first step is to select the region of farmland, where the survey to be done. In the second step, the field must be plowed precisely without any stagnation of water. This prerequisite requires to avoid error in the soil analysis. The region must be primary tillage with crop residue covered. Then, the drone will take the images in the tillage area, and it is processed to know about the nature of the region. Since drones can collect images and cover a large area, it will be helpful for the researchers to identify the peculiar area where the soil sampling analysis can be done. Management zones are formed by marking the sample regions and sample points with the help of the image captured by the drone. The AR set is designed with the required algorithm with information about management zones. The person wearing the head-mounted set can collect the samples in the mentioned soil sampling point of management zones. Later, samples are produced for evaluation in laboratories, and the resulting soil mapping can be obtained. Through this soil sampling, it is possible to get information about its nutrient [19].

4.4 Integrating Augmented Reality (AR) and IoT: Crop Monitoring

IoT provides data visualization in off-site and mostly textual, which is not enough for the user's sensorial perception and interaction. Integrating AR with IoT can make better interaction and data visualization. The AR-IoT kit [23] overlays the IoT data directly onto real-world objects and enhances object interaction. Crop monitoring by the AR-IoT system can be implemented successfully for on-field monitoring [23]. By utilizing this system, the farmers can interact with IoT data directly from the real-world environment. It helps farmers to ensure more accurately about the quality of the crop and also helps in reducing the planting costs.

In the AR set, the software is deployed to map the IoT data to physical crops with a virtual object representation. The virtual text is overlaid on the physical crop images to show information such as the date of planting, growth level of the plant, and the type of fertilizer utilized. A virtual cube with color-coding in the text is used to convey the message of crop growth stages and its readiness for harvesting [23].

4.5 Tractor Navigation System Using AR

The tillage of the agriculture field is not uniform when the navigation of the tractor is carried out by the naked eye of the driver. A tractor navigation system using augmented reality (AR) can produce a superimposed image of a computer-generated virtual three-dimensional (3D) image and the area captured camera image [24]. The 3D model can be generated using the tractor position and direction determined by two real-time kinematic GPS and an inertial measurement unit. This AR system will guide the tractor driver to cover all the field areas uniformly, which prepares the soil for productive cultivation. The positioning accuracy of the AR navigation system was examined experimentally by changing the roll, pitch, and yaw angles of a tractor at the actual field. The AR-based navigation system will be able to work with not greater than 3 cm positioning error within 3 m from the front end of the tractor.

5 Machine Learning in Precision Agriculture

Machine learning is defined as the scientific field that gives machines the ability to learn without being strictly programmed [25]. Machine learning is a concept that allows the machine to learn from examples and experience to perform a task [25]. Under machine learning, machine built a logic automatically to understand the given data with the help of an algorithm. Applying machine learning in agriculture improves the productivity and quality of the crops. We can use agriculture robots with machine learning to carry out agriculture-related tasks like harvesting the crops. The robots with machine learning can execute the task without any tiredness and fatigue.

The following section presents first an overview of the machine learning. Then, machine learning applications of predicting the yield of the crop, detecting weed and disease, livestock management, soil management, automated irrigation, and smart robot for picking fruits are presented.

5.1 An Overview of ML Algorithm

The ML algorithms are classified into supervised and unsupervised algorithms based on the data used for training is labeled or not. ML algorithm is further classified based on the nature of the task it accomplishes as a classification algorithm, regression algorithm, clustering algorithm, and dimensionality reduction algorithm. The general process in the machine learning algorithm is presented in Fig. 8. The training data may be labeled or unlabeled depending on its nature which is fed to the machine learning algorithm. The learning process learns the pattern in the data and builds a logic model. The trained model can deliver its results as a class label or predicted value or clusters using the experience gained on the training phase [26].

5.2 Yield Prediction Using ML

Yield prediction is one of the essential elements of precision agriculture. ML algorithms can be used for yield prediction and mapping. It helps to match the crop supply with the demand of the consumer market with low cost and nondestructive manner [27].

A deep learning method called convolutional neural networks (CNN) is applied to predict the yield using NDVI data and RGB image data, which is acquired from a UAV [28]. The selection of deep learning algorithm parameters is an important one for accurate prediction. In this work, the impact of deep neural network parameters on the prediction accuracy is analyzed. The training algorithm, number of layers, regularization method, and tuning of the hyperparameters are considered as

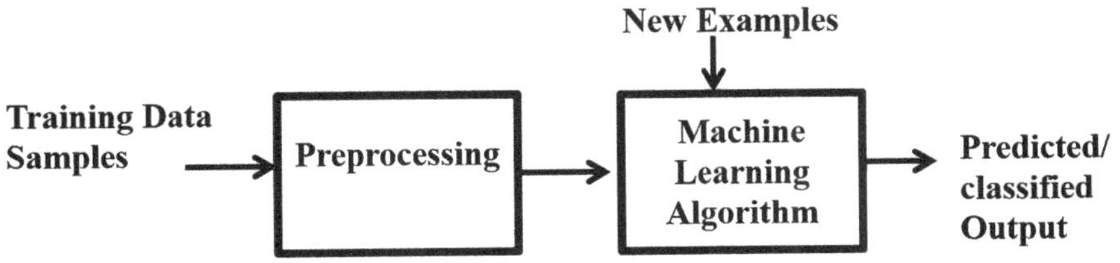

Fig. 8 The general flow of ML tasks

the parameter to study the impacts of them on the prediction accuracy. Wheat and barley crop yield prediction is achieved with a mean absolute percentage error (MAPE) of 8.8% in the early period of the growing season and with 12.6% on a later stage. The work also proved that the RGB data perform better than the NDVI data.

5.3 Plant Disease Detection

Plant disease is a significant threat to global food security. The statistic shows plant diseases produce around 10–16% losses in the worldwide harvest of crops each year. Research on pest and disease control in the agricultural field is required to overcome the above issue. The earlier detection of infected plants using machine learning increases yield. It also reduces time consumed compared with a naked-eye examination. The hyperspectral images are widely used to discriminate between healthy and diseased plants.

The spectral signatures of a diseased plant are challenging to analyze by parametric approaches such as simple or multiple regression. The non-parametric methods of principal component analysis (PCA), fuzzy logic, support vector machine (SVM), and neural networks (NNs) are used for this purpose [29].

5.4 Weed Detection

Weed detection is a challenging one because crop and weed have more similarities, and discriminating between them is difficult. There are several challenges for real-time weed detection on the field, such as occlusion and overlap of leaves, varying lighting conditions, and different growth conditions. Computer vision with ML algorithms can detect the weed at a low cost. In the future, these technologies will drive robots that will destroy weeds, minimizing the need for herbicides [30]. To handle the challenges of weed detection in real time, we need four robust mechanisms called preprocessing, segmentation, feature extraction, and classification [31].

In preprocessing, color space transformation, normalization, resizing, contrast enhancement, and denoising are carried out. The segmentation is used to separate plants from background elements like soil and resides. This vegetation segmentation involves three approaches: (1) color index-based segmentation, (2) threshold-based segmentation, and (3) learning-based segmentation. Biological morphology, spectral features, visual textures, and spatial contexts can be used as a feature set for the weed classification. Support vector machine (SVM) and artificial neural networks (ANN) are machine learning methods that can be used for weed detection [32].

A novel method that uses hyperspectral images for weed classification is presented [33]. This method proposes counter propagation (CP)-ANN and multispectral images captured by UAV for weed identification. The mixture of Gaussian active learning and self-organizing map provides better accuracy compared to any other methods [34].

5.5 Livestock Management

Livestock farming is a concern about domestic animals raised in an agricultural field to produce labor and commodities such as milk and egg. The livestock management consists of two subcategories, namely, animal welfare and livestock production. Animal welfare deals with the health and well-being of animals with the main application of ML in monitoring animal behavior for the early detection of diseases. The second category describes the issues in a production system, where the main scope of ML is to estimate accurately economic balances for the producers based on production line monitoring [30].

In precision livestock farming, computer vision with ML is used to monitor the health information of animals. A cattle health monitoring system using deep learning, which is able to analyze body condition score, live weight, and activity behaviors, is presented [35]. Segmenting the cattle or any interested animal and extraction of the contour of them in the form field are challenging tasks which can be solved by R-CNN (region-convolutional neural network). The R-CNN is a modified CNN architecture for object detection with selective search to reduce the computational time. The R-CNN is able to achieve segmentation accuracy of 0.92 mean pixel accuracy (MPA). The performance of contour extraction is made with an average distance error (ADE) of 33.56 pixels.

5.6 ML in Soil Management

Prediction of soil parameters or their nutrient content helps to reduce cost on fertilizer and increase profits. It also saves time and improves soil health. Machine learning techniques can be applied to the prediction of soil properties. ML algorithm gives the estimation of dryness of soil conditions, soil temperature, and moisture level in the soil. The predicted soil characteristics allow farmers to understand the dynamic change of the ecosystem and the impact of them in the agriculture process [26]. An application using the Microsoft Cortana Intelligence Suite, which monitors the condition of the soil and recommends to select the necessary fertilizers, is reported [30]. As a result of the recommendation, the farmers can harvest 30–40% more than the yield of the usual cultivation process.

The soil fertility is measured using the content of five nutrient components, namely, phosphorus (P), potassium (K), organic carbon (OC), and boron (B). The pH value measurements indicate the soil reaction. Extreme learning machine (ELM) method is employed to predict the five nutrient contents [36]. ELMs with the

Gaussian radial basis function as kernel classify the soil fertility content with an accuracy of above 80%. The pH classification is achieved near 90% accuracy with the hyperbolic tangent as the kernel function. The accuracy level achieved for the prediction of the nutrient content of the soil is not sufficient for practical deployment. There is a need for further research to improve the accuracy of prediction.

5.7 Automated Irrigation Systems

The scarcity of water resources demands the development of smart technology for optimum usage of water. Predicting the required amount of water for irrigation will make optimum usage of water in the agriculture field. Predicting the required amount of irrigation depends on the real-time sensing of parameters like the current soil moisture level, real-time soil temperature, environmental field conditions, and the climate forecast data.

An ML algorithm, which is a combination of supervised support vector regression (SVR) algorithm and *k*-means clustering unsupervised machine learning techniques, is presented [37] for predicting the amount of change in soil moisture because of weather condition changes. SVR is trained by the dataset of temperature present in the air, relative humidity in the air, radiation in the environment, soil moisture differences, and soil temperature. Once the training model is ready, it will be used to predict future soil moisture differences. The prediction accuracy of soil moisture difference is improved by k-means clustering.

This system is used to maintain the required soil conditions continuously, significantly requires less human labor, also has the potential to reduce production costs. This kind of technology will subsequently have a global impact on global water supply processes [30].

5.8 Smart Robots for Picking Fruit

The smart tractors are built up with ML software with "ready" intellectual technologies of sensors, radars, and GPS systems. The smart tractors go around the fields, cultivating the land and harvesting, without the need for driver company. This autonomous crop handling system enables to cultivate much more areas for more extended periods. The aim of the system is mainly to reduce the labor cost in harvest [38].

A strawberry harvesting robot is designed using mask region convolutional neural network (Mask R-CNN). ResNet-50 is used for feature extraction [39]. The region proposal network (RPN) was trained to get the region of interest (RoI). This RoI is used as input for two fully convolutional neural networks (FCN) to generate an image mask for ripe fruits and the location of coordinates to pick the strawberries. The performance of the ML-based picking robots is measured with a precision rate of 95.78% and a recall rate of 95.41% on testing with 100 test images.

6 Conclusion

The production and yield of the agriculture sector are reducing every year because of the improper farm management system. The latest technology adaptation in the agriculture field will improve the outcome of it. Even though the technologies of IoT, drone, augmented reality, and machine learning transform the agriculture process and increase productivity, the affordability of them by small farmland farmers is still in question. Even in the presence of some significant work to make those technologies affordable, the illiteracy of the farmer acts as another barrier to utilize those technologies.

The mobile application-based solution may make more usability of those technologies since smartphone has widespread use even among the illiterate farmer. The industry and government have to put effort into developing low-cost products and services with a user-centric approach for ease of use for the farmer community.

References

1. Mogili, U.R., Deepak, B.B.V.L.: Review on application of drone systems in precision agriculture. Procedia Comput. Sci. **133**, 502–509 (2018). https://doi.org/10.1016/j.procs.2018.07.063
2. https://www.biz4intellia.com/blog/5-applications-of-iot-in-agriculture/
3. Mokaya, V.: Future of precision agriculture in india using machine learning and artificial intelligence. Int. J. Comput. Sci. Eng. **7**(2), 1020–1023 (2019)
4. Chen, W.L., Lin, Y.B., Lin, Y.W., Chen, R., Liao, J.K., Ng, F.L., Chan, Y.Y., Liu, Y.C., Wang, C.C., Chiu, C.H., Yen, T.H.: AgriTalk: IoT for precision soil farming of turmeric cultivation. IEEE Internet Things J. **6**, 5209 (2019)
5. Elijah, O., Rahman, T.A., Orikumhi, I., Leow, C.Y., Hindia, M.N.: An overview of Internet of Things (IoT) and data analytics in agriculture: benefits and challenges. IEEE Internet Things J. **5**(5), 3758–3773 (2018). https://doi.org/10.1109/jiot.2018.2844296
6. https://data-flair.training/blogs/iot-applications-in-agriculture/
7. Khattab, A., Abdelgawad, A., Yelmarthi, K.: Design and implementation of a cloud-based IoT scheme for precision agriculture. In: 28th International Conference on Microelectronics (ICM), pp. 201–204 (2016)
8. https://www.microdrones.com/en/content/drones-and-precision-agriculture-the-future-of-farming/
9. https://uavcoach.com/agricultural-drones/
10. http://www.droneguru.net/the-pros-and-cons-of-drones-in-agriculture/
11. Ji-Hua, M., Bing-Fang, W.: Study on the crop condition monitoring methods with remote sensing. Int. Arch. Photogramm. Remote Sens. Spat. Inf. Sci. **37**(B8), 945–950 (2008)
12. Rascher, E.: Monitoring spatio-temporal dynamics of photosynthesis with a portable hyperspectral imaging system. Photogramm. Eng. Remote Sens. **73**, 45–56 (2007)
13. Sofonia, J., Shendryk, Y., Phinn, S., Roelfsema, C., Kendoul, F., Skocaj, D.: Monitoring sugarcane growth response to varying nitrogen application rates: a comparison of UAV SLAM

LiDAR and photogrammetry. Int. J. Appl. Earth Obs. Geoinf. **82**, 101878–101878 (2019). https://doi.org/10.1016/j.jag.2019.05.011
14. Kavoosi, Z., Raoufat, M.H., Dehghani, M., Abdolabbas, J., Kazemeini, S.A., Nazemossadat, M.J.: Feasibility of satellite and drone images for monitoring soil residue cover. J. Saudi Soc. Agric. Sci. **19**(1), 56–64 (2020). https://doi.org/10.1016/j.jssas.2018.06
15. https://blog.dronedeploy.com/soil-sampling-drone-mapping-combine-to-deliver-better-prescriptions-e2aac3307e46
16. Kharim, M.N.A., Wayayok, A., Shariff, A.R.M., Abdullah, A.F., Husin, E.M.: Droplet deposition density of organic liquid fertilizer at low altitude UAV aerial spraying in rice cultivation. Comput. Electron. Agric. **167**, 105045–105045 (2019). https://doi.org/10.1016/j.compag.2019.105045
17. Yu, F., Zhang, J.F., Zhao, Y., Zhao, J.C., Tan, C., Luan, R.P.: The research and application of virtual reality (VR) technology in agriculture science. In: International Conference on Computer and Computing Technologies in Agriculture, pp. 546–550. Springer, New York (2009)
18. Wu, H.K., Lee, S.W.Y., Chang, H.Y., Liang, J.C.: Current status, opportunities and challenges of augmented reality in education. Comput. Educ. **62**, 41–49 (2013)
19. Huuskonen, J., Oksanen, T.: Soil sampling with drones and augmented reality in precision agriculture. Comput. Electron. Agric. **154**, 25 (2018). https://doi.org/10.1016/j.compag.2018.08.039
20. Klerkx, L., Jakku, E., Labarthe, P.: A review of social science on digital agriculture, smart farming and agriculture 4.0: new contributions and a future research agenda. NJAS-Wagen. J. Life Sci. **90–91**, 100315–100315 (2019). https://doi.org/10.1016/j.njas.2019.100315
21. https://telanganatoday.com/augmented-reality-for-smart-farming
22. https://tech.moverio.epson.com/en/bt-300/
23. Phupattanasilp, P., Tong, S.R.: Augmented reality in the integrative Internet of Things (AR-IoT): application for precision farming. Sustainability. **11**(9), 2658–2658 (2019). https://doi.org/10.3390/su11092658
24. Kaizu, Y., Choi, J.: Development of a tractor navigation system using augmented reality. Eng. Agric. Environ. Food. **5**(3), 96–101 (2012). https://doi.org/10.1016/s1881-8366(12)80021-8
25. Chlingaryan, A., Sukkarieh, S., Whelan, B.: Machine learning approaches for crop yield prediction and nitrogen status estimation in precision agriculture: a review. Comput. Electron. Agric. **151**, 61–69 (2018). https://doi.org/10.1016/j.compag.2018.05.012
26. Liakos, K., Busato, P., Moshou, D., Pearson, S., Bochtis, D.: Machine learning in agriculture: a review. Sensors. **18**(8), 2674–2674 (2018). https://doi.org/10.3390/s18082674
27. Rauf, H.T., Saleem, B.A., Lali, M.I.U., Khan, M.A., Sharif, M., Bukhari, S.A.C.: A citrus fruits and leaves dataset for detection and classification of citrus diseases through machine learning. Data Brief. **26**, 104340–104340 (2019). https://doi.org/10.1016/j.dib.2019.104340
28. Nevavuori, P., Narra, N., Lipping, T.: Crop yield prediction with deep convolutional neural networks. Comput. Electron. Agric. **163**, 104859–104859 (2019). https://doi.org/10.1016/j.compag.2019.104859
29. Golhani, K., Balasundram, S.K., Vadamalai, G., Pradhan, B.: A review of neural networks in plant disease detection using hyperspectral data. Inf. Process. Agric. **5**(3), 354–371 (2018). https://doi.org/10.1016/j.inpa.2018.05.002
30. https://towardsdatascience.com/7-reasons-why-machine-learning-is-a-game-changer-for-agriculture-1753dc56e310
31. Wang, A., Zhang, W., Wei, X.: A review on weed detection using ground-based machine vision and image processing techniques. Comput. Electron. Agric. **158**, 226–240 (2019)
32. Behmann, J., Mahlein, A.K., Rumpf, T., Römer, C., Plümer, L.: A review of advanced machine learning methods for the detection of biotic stress in precision crop protection. Precis. Agric. **16**, 239 (2015)
33. Pantazi, X.E., Tamouridou, A.A., Alexandridis, T.K., Lagopodi, A.L., Kashefi, J., Moshou, D.: Evaluation of hierarchical self-organising maps for weed mapping using UAS multi-

spectral imagery. Comput. Electron. Agric. **139**, 224–230 (2017). https://doi.org/10.1016/j.compag.2017.05.026
34. Pantazi, X.E., Moshou, D., Bravo, C.: Active learning system for weed species recognition based on hyperspectral sensing. Biosyst. Eng. **146**, 193–202 (2016). https://doi.org/10.1016/j.biosystemseng.2016.01.014
35. Qiao, Y., Truman, M., Sukkarieh, S.: Cattle segmentation and contour extraction based on Mask R-CNN for precision livestock farming. Comput. Electron. Agric. **165**, 104958–104958 (2019). https://doi.org/10.1016/j.compag.2019.104958
36. Suchithra, M.S., Pai, M.L.: Improving the prediction accuracy of soil nutrient classification by optimizing extreme learning machine parameters. Inf. Process. Agric. **7**, 72–82 (2020)
37. Goap, A., Sharma, D., Shukla, A.K., Krishna, C.R.: An IoT based smart irrigation management system using machine learning and open source technologies. Comput. Electron. Agric. **155**, 41–49 (2018). https://doi.org/10.1016/j.compag.2018.09.040
38. Barbedo, J.G.A.: Detection of nutrition deficiencies in plants using proximal images and machine learning: a review. Comput. Electron. Agric. **162**, 482–492 (2019). https://doi.org/10.1016/j.compag.2019.04.035
39. Yu, Y., Zhang, K., Yang, L., Zhang, D.: Fruit detection for strawberry harvesting robot in non- structural environment based on Mask-RCNN. Comput. Electron. Agric. **163**, 104846–104846 (2019)

IoT-Based Brinjal Crop Monitoring System

Navdeep Kaur and Gaurav Deep

Abstract Many real-time applications perceive the various advances made in numerous domains using new technologies. Still, the use of these new innovations in the agricultural sector remains a challenging task. This paper uses a remote crop monitoring mechanism using LoRaWAN in the greenhouse. The motive of this paper is to enhance the traditional way of agriculture in rural areas with the help of a wireless sensor network using LoRaWAN protocol. This paper proposes a design and implementation of a system having an IoT-based sensor that senses the temperature of the environment inside the greenhouse for brinjal with the aim to control the devices like fan and heater according to the optimal range for food production and proper growth of the crop. The proposed model is implemented and validated using the CupCarbon simulator. The results show that our approach is fast and that this model provides shorter simulation time with greater efficiency.

Keywords IoT (Internet of Things) · WSN (wireless sensor network) · LoRaWAN (long-range wireless area network) · TFP (total factor productivity rate) · UAVs (unmanned aerial vehicles)

1 Introduction

Climate change is the most challenging issue of our time. Human activities and natural variability both are the reasons behind it. Excess use of land and greenhouse gas is the impact due to humans [1]. Climate change includes a raise in temperature, frequent rainfall, wind, and shorter growing seasons, and understanding its adaptation is a challenge in itself and raises the complexity of decision-making [2].

Solar radiations, net radiation, heat condition, evapotranspiration, etc. Are the factors of land surface radiation that are responsible for changing the environment [1]. Climate change has generated severe impacts on the security of food. A frequent change in weather and drought for a long duration is having a great effect on

N. Kaur · G. Deep (✉)
Department of CSE, Punjabi University, Patiala, India
e-mail: deepgaurav48@pbi.ac.in

D. Gupta et al. (eds.), *Smart Sensors for Industrial Internet of Things*, Internet of Things, https://doi.org/10.1007/978-3-030-52624-5_15

food production. Therefore, farmers realized a need for approaching different techniques so that they will be prepared in advance for climate change [2].

Due to these climatic changes and to fulfill the demands of an increasing population, a new technique is needed to be generated. Internet of Things (IoT) comes as an emerging field in agriculture to sense the change in climate in advance so that appropriate action can be taken at the right time. This also helps in the fulfillment of food for thousands of people. In IoT, many things are connected to the internet forming a new infrastructure in the network system using IPv6 protocol [3].

A wide range of potential applications like smart cities, smart agriculture, smart home, smart health care, monitoring system, etc. Are provided under IoT technology [3]. The basic idea of IoT is to make physical objects as smart objects by embedding electronics, software, sensors, actuators, and network connectivity to them [4].

IoT benefits offer farmers special technological skills with regard to yield development, assessment, manure management, and detailed cure strategy to be used. The advantages of using IoT in agriculture include a reduction in generation cost, sustainability, and food well-being. The IoT-based sensor system is useful for agriculture improvements. IoT offers various innovations such as innovation in the management of food, innovation in environmental control, innovation in clean water systems, etc. [5]. IoT-based solutions reduce the transport cost, prediction of prices, the current market conditions, and the number of middle agents between the farmer and the end consumer [6].

Internet of Things has given a model for keeping records of seedlings and tracking high-quality agriculture products. It allows controlling the greenhouse system by controlling its temperature and irrigation system and indoor environment automatically. IoT is used in a traceability system that helps to track seeds from their selection to the final stage. In it, there is an effective use of land and no chemicals and pesticides were used. Nutrients and fertilizers are supplied to crop using an irrigation system [7].

Agriculture is the area where there are significant impacts of change in climatic conditions, and the accurate assessment of impacts helps to surge production rate in agriculture, and thus it is necessary for sustainability and security of food in the country. Crop production is greatly influenced by temperature, precipitation, or by the management of fertilizers, pesticides, and irrigation [8].

The temperature has a great effect on crops, agriculture fields, consumption of energy, morbidity, trade, and productivity of labor. An increase in variability in climate decreases crop yield. It has been observed that countries having a cold zone had an increase in per capita income, whereas there is a decline in the per capita income of hot climatic zones. Climate change affects the total factor productivity rate (TFP) that is the growth rate of production efficiency and reduces the productivity of agricultural land and labor [9].

Agriculture is greatly affected due to crop sensitivity to climate changes like shortage of water, temperature, and humidity. Negative effects of climate subsequently plunged the quality as well as the yield of crops. It also increases the pests and diseases of plants which results in a downgrade in the production of crop level.

Farmers holding small farms are totally dependent on the rainfall. They have to suffer a lot due to climatic conditions as they are economically not in good condition and not able to access or purchase any technical support from outside [10].

Changing climate has a greater impact on food security and the production of food grains in India. Environmental changes have greater impacts on the fertility of the soil, pest incidence, and frequency of droughts and floods. Also, there are biophysical and socioeconomic impacts of climate change like impact on crop growth, crop development, crop yield, future food demand, land usage, etc. [11].

IoT collects real-time data from various sensors deployed at different places in wireless sensor network (WSN) and sends that data to application via cloud storage. There are many challenges farmers have to face, which are solved by using IOT like monitoring of land, the scarcity of water, cost managing, etc. Sensors used to detect ultraviolet (UV) rays, the water level of crops, and pH range in the soil are of small size that they can be integrated into any application of agriculture [12].

Growers can determine gross profit, daily prices, the quality of the product, and need of crops. Different sensors are used in the field of agriculture like humidity sensor, soil temperature, light intensity sensor, and air temperature. These sensors are deployed in the field and Pennellii-1 is used to collect data. It is an environmental monitoring system, and then data is transmitted through GPRS (General Packet Radio Service) to the cloud servers and viewed by the user using mobile phones and personal computers (PC) [13].

1.1 Motivation

Due to the vast variation of temperature in the Indian subcontinent, there is a need to provide a constant temperature for increasing the productivity of the brinjal crop with the use of a greenhouse that can be monitored remotely. The research paper uses a remote crop monitoring mechanism in the greenhouse. It uses wireless sensor network and uses long-range wireless area network. This approach is applicable to a wide range. This mechanism can be easily implemented to resolve the problems in the agriculture field due to climatic changes. This can help end users, including farmers, to better understand farming techniques that need to be implemented for crop management.

1.2 Research Contribution

The proposed work makes the following contribution:

- The proposed mechanism is able to work under vast temperature variation in the Indian subcontinent with the help of IoT devices.

- The proposed mechanism allows monitoring of temperature in the greenhouse with the use of the temperature sensor.
- Proposed system controls the environment of a greenhouse using various IoT devices remotely by using LoRaWAN.

The research paper is organized as follows: Section 2 presents the literature review of a comparison of features achieved by CupCarbon simulator in different scenarios and various approaches related to crop features. Section 3 highlights the proposed remote crop monitoring mechanism using LoRaWAN. Section 4 includes the testing of the suggested technique, using the CupCarbon simulator as the verification tool, and finally, the paper concludes in Sect. 5.

2 Related Works

The practical implementation of CupCarbon simulator is highlighted by Bounceur et al. and proposed a new interface known as CupCarbon-Lab which is based on existing simulator CupCarbon. Codes implemented in CupCarbon simulator can be directly uploaded to hardware platforms like Arduino, Raspberry Pi cards, etc. This proposed platform is helpful to test the limitations of the remote controlling, scalability, mobility, and the feasibility of an algorithm in real conditions of the environment [14].

Previous researchers have proposed various applications and work done by CupCarbon simulator in various fields and work by IoT in the agriculture field, but still, there is a lot to do. Before proposing a new application by CupCarbon, a literature survey is done on CupCarbon agriculture in IoT that shows how this simulator works in different fields and how in the future it can be implied in the agriculture field.

Lopez-Pavon et al. [15] present a paper that describes a network simulator called CupCarbon and its characteristics with examples like how to detect the mobile node. CupCarbon integrates LORAWAN, Zigbee, and Wi-Fi protocols. CupCarbon is having the ability to draw a graph of battery level and energy consumption of nodes. A modified version of the routing algorithm such as the Dijkstra algorithm was proposed which allows changing routes dynamically on the basis of the remaining battery level of node compared to the adjacent neighbor node battery level to evaluate the best route with minimal cost. The proposed algorithm provides a more stable network and increases a 20% lifetime of the network. Comparison of CupCarbon with other simulators has been also proposed in this paper to evaluate its use in academics and the research field.

Vaidya and Vishwakarma [16] discussed a comparative analysis on automated home systems to monitor, control, and provide security in homes. Comparison is done on the basis of features and disadvantages using different technologies like IoT, GSM (Global System for Mobile communication), Bluetooth, and PIC microcontroller using Zigbee modulation. In the smart home system, one can control appliances of home using smart mobiles sitting far in remote areas. The main objective of home automation is to reduce human efforts and electricity consumption and to provide a comfortable and easy life for people. Zigbee is a short-range wireless communication module. The popularity of home automation is increasing day by day as it is highly secure and reduces stress and comfort.

Lourenço et al. [17] proposed a scenario for early detection of fire in forest areas using surface plasmon resonance phenomenon, which senses increasing refractive index that is burning of wood and uses multimode interference consisting of 3 dB coupler which provides an optical path to send information of detected fire to sensing nodes. To perform the simulation for monitoring fire event, CupCarbon simulator is used. One hundred seventy-eight nodes with a 50 m sensing range are deployed in the forest. When the fire is detected by sensor nodes, the alarm is sent by nodes to neighbors, and after 61 received messages, alarm reaches to the gateway, and then message is sent to the remote control center for further action (Table 1).

Sumi and Ranga [18] have given a framework for the management of the traffic system in emergencies based on IoT, wireless sensor network, and vehicular ad hoc network (VANET) using CupCarbon simulator. This system helps emergency vehicles like an ambulance to find the nearest path to a destination on the basis of real-time information on traffic. In the proposed system, RSUs (roadside unit), traffic management server, sensors, and vehicles are present and all these can communicate with each other. When an ambulance has to bypass heavy traffic, then it sends a message to RSU, which directs all vehicles and automatically detects the shortest path to reach a destination and clears that path by managing and controlling traffic light for the smooth passage of emergency vehicle, and after the crossing of the ambulance, the whole system resumes to its normal functioning.

Table 1 A comparison of features achieved by the CupCarbon simulator in different scenarios

Features available	Lopez-Pavon et al. [15]	Vaidya and Vishwakarma [16]	Lourenço et al. [17]	Sumi and Ranga [18]	Saoudi et al. [19]
Simulation	No	No	Yes	Yes	Yes
Model used	Dijkstra algorithm	No	Plasmon resonance phenomenon	Vehicular ad hoc network	Data mining technique
Improve energy consumption	No	Yes	No	No	Yes
Improve lifetime of sensor	Yes	No	No	No	Yes
Test scalability and feasibility	No	No	No	Yes	No
Security	No	Yes	No	No	No
Remote controlling	No	No	Yes	Yes	No
Remote crop monitoring	No	No	No	No	No

Saoudi et al. [19] proposed an approach to detect fire in forests by the deployment of sensors which uses WSN and data mining techniques, and there is a clustering of each node to detect fire using classification techniques. After that, an alert is sent to firefighters through the sink by using gateways. Clustered network approach, route discovery to sink, and routing alerts to the sink are three main phases of this approach. CupCarbon simulator is used to implement the whole approach. Measuring real data for fire detection from various sensors is done through the artificial neural network. Results obtained show that this approach is fast and achieves 93.3% precision and reduces energy consumption.

Many researchers have worked in the field of agriculture. Different models are used in the field of agriculture to sense the sudden climatic changes. Table 2 presents a comparison of various approaches based on crop features.

Ayaz et al. [20] discussed wireless sensor networks and IOT in agriculture and challenges faced while the collaboration of IoT in previous agriculture techniques. Optimization of crop yield from sowing to harvesting and then packaging and transportation are analyzed in this technology. Different IoT-based technologies like geo-mapping, variable rate technology, Crop Water Stress Index (CWSI), robots, and UAVs are used for fertilization, irrigation, pest management, and yield monitoring. Advance agriculture practices like a greenhouse, vertical farming, phenotyping, etc. Are used. Communication techniques, sensors, and UAVs are used to measure temperature, humidity, soil moisture, etc. and are explained in detail. The overall conclusion induced is that automating crops is important to maximize the production of crops using IOT-based sensors and technologies.

Table 2 Comparison of various approaches related to crop features

Features available	Ayaz et al. [20]	Núñez V et al. [21]	Núñez V et al. [22]	Sharma et al. [23]	Weinberger and Srinivasan [24]
Model used	Crop water stress index, communication techniques, etc.	Internet of Things architecture	Smart node platform	High-performance liquid chromatography	Pest management strategy
Real-time information	Yes	No	No	No	No
Shortest path to destination	No	No	No	No	No
Reduction in energy consumption	No	Yes	No	No	No
Network distribution	No	No	No	No	No
Increase in production	Yes	Yes	Yes	No	No
Sensing of data	No	Yes	No	No	No

Núñez V et al. [21] presented the design and implementation for the tomato crop in Colombia for its better yield using wireless sensor network and the Internet of Things. Impacts of different climatic changes like temperature, humidity, luminosity, and soil moisture on the tomato crop have been observed. Also, given an IOT architecture for the traceability and storage of the crop, the proposed system is at low cost, transmits and collects data fast, considers all weather conditions, and consumes less energy, and this system basically consists of three parts, software, hardware, and WSN, and obtained the proper range of variables for the good yield production of tomato crop.

Núñez V et al. [22] proposed a system for the implementation and design of a WSN for the proper management of cabbage crop by analyzing frequent climatic changes with the help of temperature, humidity, and soil moisture sensors to take the appropriate action for the good production of crop and to enhance its yield by obtaining optimal ranges for the good production of cabbage crop in China and Panama. A smart node platform is used to obtain a range of variables. In its hardware and software, agroclimatic variables for the precise crop production are monitored.

Sharma et al. [23] proposed the high-performance liquid chromatography (HPLC) technique to analyze the persistence of spinosad in cabbage and cauliflower in the soil under changing weather conditions in India. Cabbage and cauliflower are one of the important vegetables in India, and they have to suffer from various problems related to pests. HPLC is used to obtain the optimal range of residue of spinosad to control pests and time periods for which pesticides retain in cabbage and cauliflower.

Weinberger and Srinivasan [24] discussed excessive use of pesticides in india by the farmers to enhance the cabbage and cauliflower production and to protect them from pests. There is a tremendous effect of pesticides on environment degradation, human health, insecticide resistance, and resource loss. Due to these hazardous impacts, a need for pest management strategy arises. The proposed work identified pests and diseases, strategy to control them, and the economy used on pesticides and also obtained that 74% of health problems occur due to pesticides.

It is evident from Tables 1 and 2 that there is no work done in the monitoring of brinjal crops with various features. There is a need to introduce the new technologies and automate the crop production in the field of agriculture with the aim of improving the quality and quantity of crops in the Indian subcontinent. It will also help in better management of crops in terms of damage control that can occur due to frequent climatic changes. Thus it will help to standardize the life of both consumers and growers of crops. Humidity and temperature influence the growth of the plant so the need for monitoring them arises for proper management of nutrients, weeds, pests, and diseases.

There is a need for remote monitoring of temperature and humidity of the environment for a different crop in the Indian subcontinent to improve the productivity of those crops and protect them from diseases and pests. Temperature and humidity play a vital role in increasing the yield of any crop. They help to keep a crop healthy and protect it from the frequent climate changes which can destroy the crop. Monitoring of humidity and temperature can be done using a different kind of sensors which are deployed in the remote areas using wireless network and the data collected can be monitored using IoT devices.

3 Proposed Remote Crop Monitoring Mechanism Using LoRaWAN

This section explains the proposed model architecture that is interactive software that will act as a helping hand to the farmer and algorithm and flow chart of remote crop monitoring.

3.1 The Architecture of Remote Crop Monitoring Using LoRaWAN

The greenhouse is partitioned into four parts P1, P2, P3, and P4. It consists of a temperature sensor that is embedded in the greenhouse to sense the temperature of the greenhouse. There is one main base station used which is connected with the temperature sensing node. After sensing the temperature sensor node sends the value of temperature to the main base station, and a further main base station is connected to the multiple base stations that are sub-base stations that are embedded at some distance in the greenhouse. Devices like a fan and heater are connected to the sub-base stations. All sub-base stations receive the value of temperature from the main base station and send the obtained values to the devices connected to the multiple base stations as shown in Fig. 1. The T1 node in yellow color in the figure is the sensor node, heaters are indicated by red color, fans are of brown color, the main base station B0 is of green color, and the sub-base stations (B1, B2, B3, and B4) are of blue color. Values sent by the main base station to the sub-base stations are shown by using red arrows.

3.2 Algorithm Used in Remote Monitoring of Brinjal Crop

Input: Sensed temperature value.
Output: Monitoring and controlling IoT devices available in the greenhouse.
T1: Sensed temperature value; $\boldsymbol{\alpha} = 21$(high); $\boldsymbol{\beta} = 13$(low); $\boldsymbol{\gamma}$ = recommended range; *B0*: main base station.
B1, B2, B3, B4: Multiple base stations.

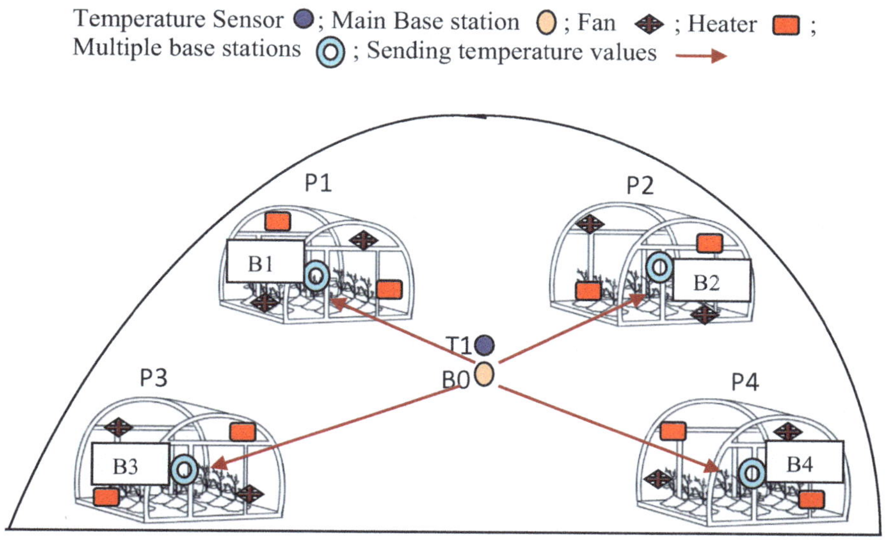

Fig. 1 A proposed architecture for remote monitoring of brinjal crop

Step1: The B0 fetches T1 from the temperature sensor installed in the field.
Step2: Send the value of T1 from the B0 to the B1, B2, B3, and B4.
Step3: B1, B2, B3, and B4 check values of T1 = γ, i.e., value of multiple base stations in range.
Step4: If the values of T1 = γ, then go to step8.
Step5: If T1 > α, then switch on the fans in the greenhouse.
Step6: If T1 < β, then switch on the heaters in the greenhouse.
Step7: If the value of T1! = γ, then repeat steps 2 to 7.
Step8: Exit.

3.3 Flowchart of Proposed Mechanism (Fig. 2)

Working details of various base stations and IoT sensors are shown in the flowchart in Fig. 2. In the beginning values of temperature are fetched then sent to the main base station. B1, B2, B3,... B*n* are multiple base stations that receive the value of temperature from the main base station. F1, F2,... F*n* are fans attached with sub-base stations. H1, H2,... H*n* are the heaters connected with sub-base stations. Fans and heaters receive sensed values of temperature from their base station in the greenhouse and get on and off according to their recommended range.

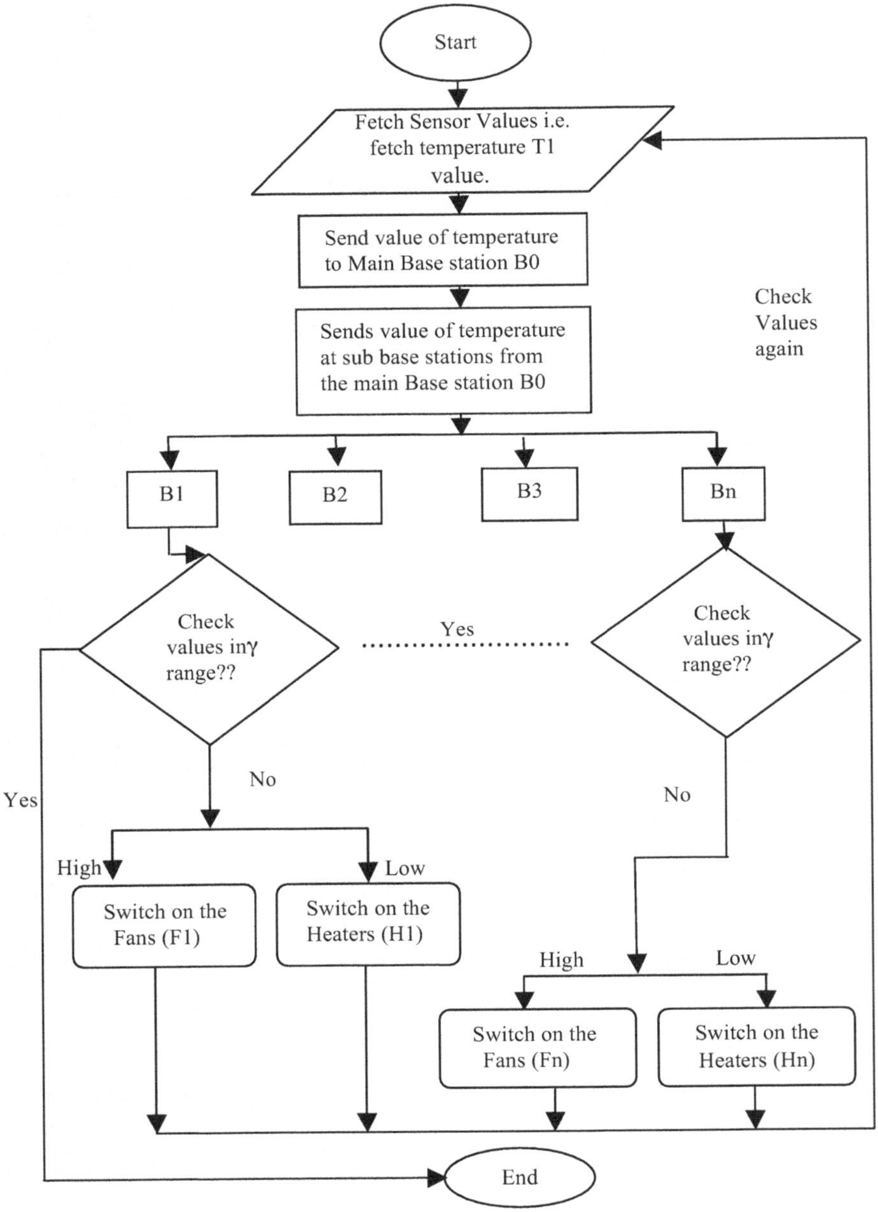

Fig. 2 Flowchart of the proposed mechanism

4 Experimental Results

For evaluating the proposed approach, we have carried out extensive simulation experiments. In this section, we will explain the simulator used and then present our experimental results and discussions.

4.1 CupCarbon Simulator

CupCarbon is an open source framework for modeling and simulating wireless sensor networks for smart city and the Internet of Things. In the Java language, it is programmed [25]. CupCarbon is multiple agents and a discrete event-based wireless sensor network simulator driven on geolocation. This allows for the modeling and simulation of sensor networks on the OpenStreetMap platform [26]. The aim is to design, model, and validate environment monitoring algorithms, data collection, etc. It is able to simulate the Zigbee, LoRaWAN, and Wi-Fi protocols [27]. CupCarbon also consists of a window called "Natural Event Generator" which generates values that are detected by sensor nodes such as temperature, humidity, etc. This comes with a script called SenScript which enables programming and configuration for an individual sensor node. CupCarbon provides a multistep simulation of algorithms and scenarios. Before its actual deployment, it helps to describe the design, feasibility, and practical implementation of a network [28].

Figure 3 shows an example of our approach using the CupCarbon simulator.

In this scenario, there is one temperature sensor embedded on the greenhouse which is sensing the value of temperature and sending values to the main base station. There are another four base stations used named base station1, base station2, base station3, and base station4 which are placed at different distances from the main base station and get the value of temperature. Further devices like fans and heaters are connected to them. Scripts are assigned to every sensor node and base station. Scripts contain a program for each node according to how every node works. All nodes are communicating with each other using LoRaWAN (long-range wireless area network) protocol. Smart cities use LoRaWAN as its range is 0–20 km, and it can be used in remote areas where there is no access to the Internet. In it, the data rate is also low and applicable for a long range. Millions of low-power devices use it due to its efficiency. It works on radio frequency. It is less expensive.

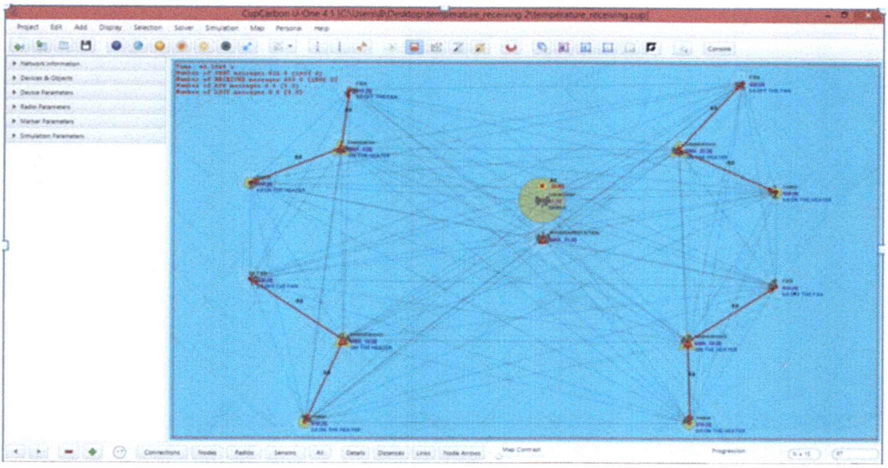

Fig. 3 CupCarbon simulation

Figure 3 shows that if the temperature obtained by the devices from the base station is not in range as shown in Table 3, for the good production of brinjal crop, then devices get off and on automatically.

Natural event generator is used to generate the values of temperature which are further sensed by the temperature sensor. It uses Gaussian distribution to generate the values of temperature. We feed the value of mean, standard deviation, period, and size and then click on the generate button to generate the values of temperature as shown in Fig. 4.

As shown in Fig. 5, the mean is 22, the standard deviation is 4, the period is 60, and the size is 1000. Period refers to the gap of time after which values are generated. It is in seconds. Size refers to the count of values generated. After clicking on generate button, values of temperature are generated. Further, this file containing generated values is assigned to the sensor node.

According to the recommend range in Table 3, if the temperature gets low that is less than 13 °C, then amount of cooling is increased in the greenhouse which will affect the growth of brinjal plant so there is need to control the temperature so heaters get on by themselves if the temperature is below its range as shown in Fig. 6.

Table 3 Ranges of variables for a good production [29]

Type of sensor	Crop	Range of temperature for proper growth
Temperature	Brinjal	13–21 °C

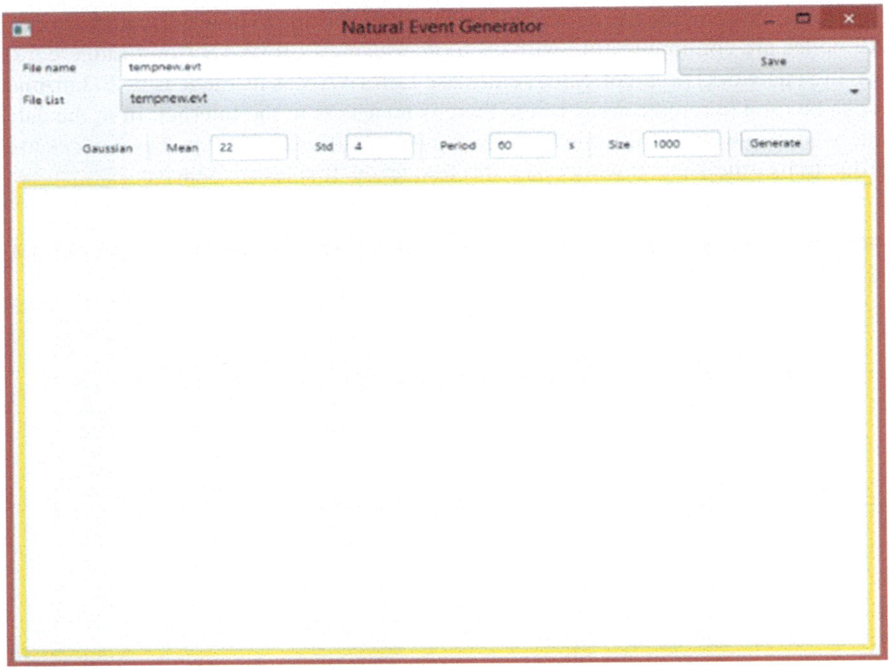

Fig. 4 Generation of temperature values by Gaussian distribution

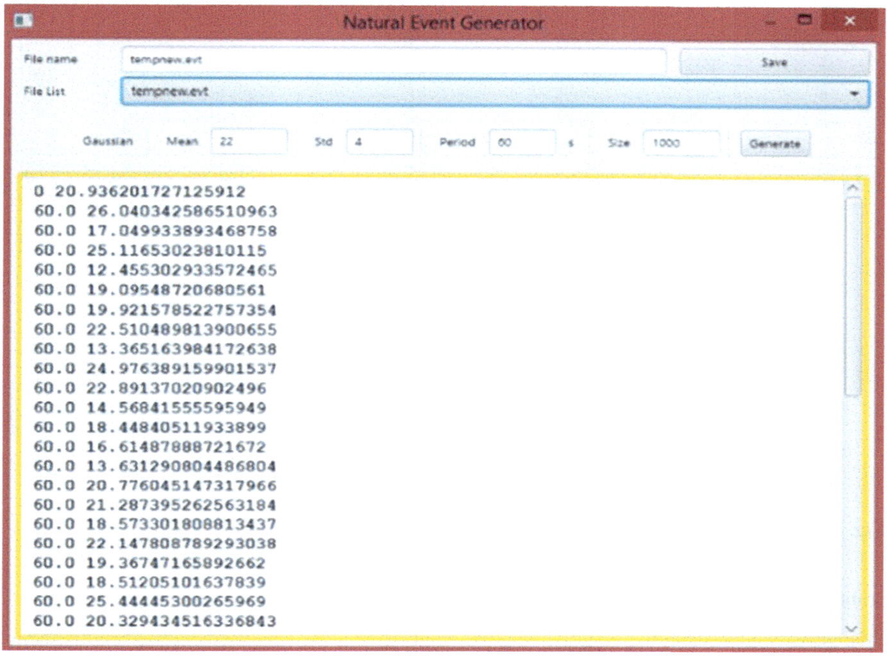

Fig. 5 Temperature values generated by Gaussian distribution

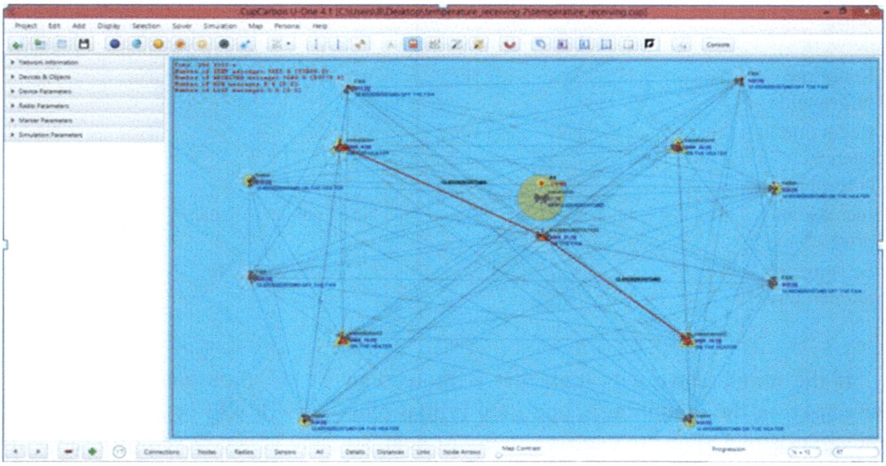

Fig. 6 Effect of decreased temperature

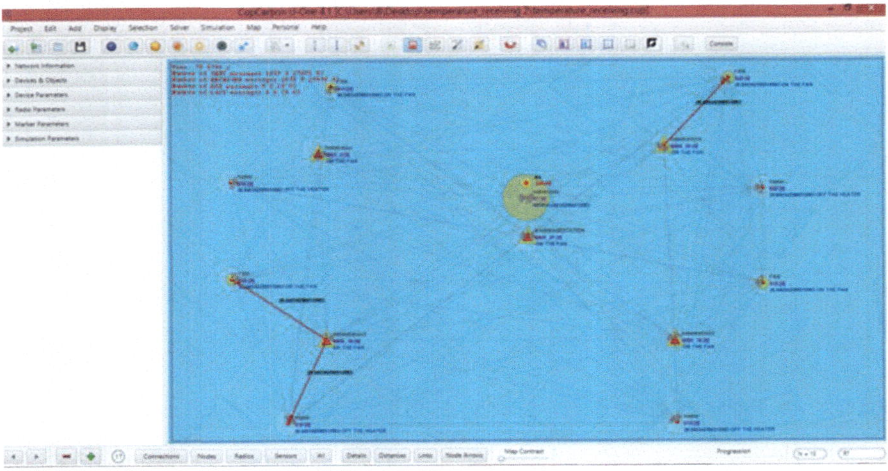

Fig. 7 Effect of increased temperature

Table 4 The comparison of the proposed scheme and other related scenarios for brinjal crop

Features available	Proposed remote crop monitoring mechanism in LoRaWAN	Shinde and Siddiqui [30]	Nandhini et al. [31]	Das et al. [32]	Verma et al. [33]	Verma et al. [34]	Rasdi and Salmah [35]	Affrin et al. [36]
Wireless sensor network	Yes	Yes	No	No	Yes	Yes	No	Yes
Real-time information	Yes	Yes	Yes	Yes	Yes	No	No	No
Reduction in energy consumption	Yes	No	Yes	No	No	No	No	No
Increase in production	Yes	Yes	No	No	No	Yes	No	Yes
Sensing of temperature	Yes	Yes	No	No	Yes	No	No	No

If the temperature gets high that is more than 21 °C, then amount of heat is increased in the greenhouse, and then brinjal plant growth will be highly affected they get damaged due to excess amount of heat, so there is need to control the temperature so fans get on by themselves if the temperature exceeds its range as shown in Fig. 7.

From Figs. 6 and 7, it is concluded that brinjal crops can be remotely monitored with the use of LoRaWAN and various IoT devices such as IoT-based fans and heater which are installed in the greenhouse increasing crop productivity remotely.

The benefits of the proposed mechanism can be seen from Table 4. It has more features as compared to the implemented existing techniques for the brinjal crop. It contributes to enhancing the production of the brinjal crop. It uses a temperature sensor that further helps to control the temperature in the greenhouse using IoT devices. The proposed mechanism provides real-time information.

5 Conclusion

Using the Internet of Things technology will solve the greater part of the difficulties of tracking a brinjal crop toward its production and makes the agriculture smarter. Agriculture provides the main means of subsistence for the large Indian population. So, there is a need to increase the productivity of food. Increasing technology can help farmers to find and plan the most efficient farming practices to expand their understanding of which crops to produce at their best. It could improve their production and benefit more from what they already have. The proposed model will get real-time data from the temperature sensor and control the devices according to sensed temperature so that there will be no bad effect on the growth of brinjal. In a simulation environment, our work has been evaluated, and the tests have shown good performance. As future work, this can be extended for multiple crops in remote areas in a long range.

References

1. Liang, S., Kustas, W., Schaepman-Strub, G., Li, X.: Impacts of climate change and land-use changes on land surface radiation and energy budgets. IEEE J. Sel. Top. Appl. Earth Obs. Remote Sens. **3**(3), 219–224 (2010)
2. Waldman, K.B., Attari, S.Z., Gower, D.B., Giroux, S.A., Caylor, K.K., Evans, T.P.: The salience of climate change in farmer decision-making within smallholder semi-arid agroecosystems. Clim. Chang. **156**, 527–543 (2019)
3. Deshmukh-Bhosale, S., Sonavane, S.S.: A real-time intrusion detection system for wormhole attack in the RPL based internet of things. Procedia Manuf. **32**, 840–847 (2019)
4. Malik, M., Dutta, M.: Contiki-based mitigation of UDP flooding attacks on the internet of things. In: 2017 International Conference on Computing, Communication, and Automation (ICCCA), pp. 1296–1300. IEEE, Washington, DC (2017)
5. Jaiganesh, S., Gunaseelan, K., Ellappan, V.: IOT agriculture to improve food and farming technology. In: In 2017 Conference on Emerging Devices and Smart Systems (ICEDSS), pp. 260–266. IEEE, Washington, DC (2017)
6. Shenoy, J., Pingle, Y.: IOT in agriculture. In: 2016 Third International Conference on Computing for Sustainable Global Development (INDIACom), pp. 1456–1458. IEEE, Washington, DC (2016)
7. González-Amarillo, C.A., Corrales-Muñoz, J.C., Mendoza-Moreno, M.Á., Hussein, A.F., Arunkumar, N., Ramirez-González, G.: An IoT-based traceability system for greenhouse seedling crops. IEEE Access. **6**, 67528–67535 (2018)

8. Zhuo, Z., Gao, C., Liu, Y.: Regional grain yield response to climate change in China: a statistic modeling approach. IEEE J. Sel. Top. Appl. Earth Obs. Remote Sens. **7**(11), 4472–4479 (2014)
9. Kumar, S., Khanna, M.: Temperature and production efficiency growth: empirical evidence. Clim. Chang. **156**(1–2), 209–229 (2019)
10. Pham, Y., Reardon-Smith, K., Mushtaq, S., Cockfield, G.: The impact of climate change and variability on coffee production: a systematic review. Clim. Chang. **156**(4), 609–630 (2019)
11. Mall, R.K., Singh, R., Gupta, A., Srinivasan, G., Rathore, L.S.: Impact of climate change on Indian agriculture: a review. Clim. Chang. **78**(2–4), 445–478 (2006)
12. Pandithurai, O., Aishwarya, S., Aparna, B., Kavitha, K.: Agro-tech: a digital model for monitoring soil and crops using the internet of things (IoT). In: In 2017 Third International Conference on Science Technology Engineering & Management (ICONSTEM), pp. 342–346. IEEE, Washington, DC (2017)
13. Kang, M., Fan, X.R., Hua, J., Wang, H., Wang, X., Wang, F.Y.: Managing traditional solar greenhouse with CPSS: a just-for-fit philosophy. IEEE Trans. Cybern. **48**(12), 3371–3380 (2018)
14. Bounceur, A., Marc, O., Lounis, M., Soler, J., Clavier, L., Combeau, P., et al.: Cupcarbon-lab: An IoT emulator. In: 2018 15th IEEE Annual Consumer Communications & Networking Conference (CCNC), pp. 1–2. IEEE, Washington, DC (2018)
15. Lopez-Pavon, C., Sendra, S., Valenzuela-Valdés, J.F.: Evaluation of CupCarbon network simulator for wireless sensor networks. Netw. Protoc. Algor. **10**(2), 1–27 (2018)
16. Vaidya, V.D., Vishwakarma, P.: A comparative analysis on smart home system to control, monitor and secure home, based on technologies like GSM, IoT, Bluetooth and PIC microcontroller with ZigBee modulation. In: 2018 International Conference on Smart City and Emerging Technology (ICSCET), pp. 1–4. IEEE, Washington, DC (2018)
17. Lourenço, P., Fantoni, A., Vieira, M.: Simulation of an early warning fire system. In: Doctoral Conference on Computing, Electrical and Industrial Systems, pp. 305–317. Springer, Cham (2019)
18. Sumi, L., Ranga, V.: An IoT-VANET-based traffic management system for emergency vehicles in a smart city. In: Recent Findings in Intelligent Computing Techniques, pp. 23–31. Springer, Singapore (2018)
19. Saoudi, M., Bounceur, A., Euler, R., Kechadi, T., Cuzzocrea, A.: Energy-efficient data mining techniques for emergency detection in wireless sensor networks. In: 2016 International IEEE Conferences on Ubiquitous Intelligence & Computing, Advanced and Trusted Computing, Scalable Computing and Communications, Cloud and Big Data Computing, Internet of People, and Smart World Congress (UIC/ATC/ScalCom/CBDCom/IoP/SmartWorld), pp. 766–771. IEEE, Washington, DC (2016)
20. Ayaz, M., Ammad-Uddin, M., Sharif, Z., Mansour, A., Aggoune, E.H.M.: Internet-of-things (IoT)-based smart agriculture: toward making the fields talk. IEEE Access. **7**, 129551–129583 (2019)
21. Núñez V, J.M., et al.: Design and implementation of WSN and IoT for precision agriculture in tomato crops. In: 2018 IEEE ANDESCON, pp. 1–5. IEEE, Washington, DC (2018)
22. Núñez V, J.M., et al.: Design and implementation of WSN for precision agriculture in white cabbage crops. In: 2017 IEEE XXIV International Conference on Electronics, Electrical Engineering and Computing (INTERCON), pp. 1–4. IEEE, Washington, DC (2017)
23. Sharma, A., Srivastava, A., Ram, B., Srivastava, P.C.: Dissipation behavior of spinosad insecticide in soil, cabbage, and cauliflower under subtropical conditions. Pest Manag. Sci. **63**(11), 1141–1145 (2007)
24. Weinberger, K., Srinivasan, R.: Farmers' management of cabbage and cauliflower pests in India and their approaches to crop protection. J. Asia Pac. Entomol. **12**(4), 253–259 (2009)
25. Kadjouh, N., Bounceur, A., Tari, A., Euler, R., Lagadec, L., Laouid, A.: Temperature-based models of batteries for the simulation of wireless sensor networks. In: Proceedings of the Third International Conference on Future Networks and Distributed Systems, p. 22. ACM, New York (2019)

26. Mehdi, K., Lounis, M., Bounceur, A., Kechadi, T.: Cupcarbon: a multi-agent and discrete event wireless sensor network design and simulation tool. In: Seventh International ICST Conference on Simulation Tools and Techniques, Lisbon, Portugal, 17–19 Mar 2014, pp. 126–131. Institute for Computer Science, Social Informatics and Telecommunications Engineering (ICST), Belgium (2014)
27. Akkurt, M., Küçük, K.: Simulation of smart city applications based on IoT technologies with CupCarbon. In: 2018 Third International Conference on Computer Science and Engineering (UBMK), pp. 179–184. IEEE, Washington, DC (2018)
28. Lounis, M., Bounceur, A., Euler, R., Pottier, B.: Estimation of energy consumption through parallel computing in wireless sensor networks. J. Ambient Intell. Human. Comput. 1–13 (2017). https://doi.org/10.1007/s12652-017-0582-5
29. No, P.B., Jakhini, P.O.: E-Manual on Improved Production Technologies in Vegetable Crops
30. Shinde, D., Siddiqui, N.: IOT based environment change monitoring & controlling in greenhouse using WSN. In: 2018 International Conference on Information, Communication, Engineering and Technology (ICICET), pp. 1–5. IEEE, Washington, DC (2018)
31. Nandhini, S.A., Hemalatha, R., Radha, S., Indumathi, K.: Web enabled plant disease detection system for agricultural applications using WMSN. Wirel. Pers. Commun. **102**(2), 725–740 (2018)
32. Das, S., Nayak, S., Chakraborty, B., Mitra, S.: Continuous radial growth rate monitoring of horticultural crops using an optical mouse. Sensors Actuators A Phys. **297**, 111526 (2019)
33. Verma, S., Chug, N., Gadre, D.V.: Wireless sensor network for crop field monitoring. In: 2010 International Conference on Recent Trends in Information, Telecommunication and Computing, pp. 207–211. IEEE, Washington, DC (2010)
34. Verma, S., Bhatia, A., Chug, A., Singh, A.P.: Recent advancements in multimedia big data computing for IoT applications in precision agriculture: opportunities, issues, and challenges. In: Multimedia Big Data Computing for IoT Applications, pp. 391–416. Springer, Singapore (2020)
35. Rasdi, Z.M., Salmah, M.C.: Multitrophic system: effect of different concentration of nutrient and pre-infested brinjal (*Solanum melongena*) on whitefly (*Bemisia tabaci*) population. In: 2011 Third International Symposium & Exhibition in Sustainable Energy & Environment (ISESEE), pp. 236–242. IEEE, Washington, DC (2011)
36. Affrin, K., Reshma, P., Kumar, G.N.: Control and management of *L. orbonalis* using WSNs. In: 2017 IEEE Technological Innovations in ICT for Agriculture and Rural Development (TIAR), pp. 8–13. IEEE, Washington, DC (2017)

Internet of Drones: An Engaging Platform for IIoT-Oriented Airborne Sensors

Ambuj Kumar and Purnima Lala Mehta

Abstract Internet of things, since its conception in the year 2000, has been steadily permeated in almost every sector that humans have created till now, which includes industry, business, entrepreneurs, academia, and so on. The impact is so high that IoT is now creating a paradigm shift for an overall ecosystem. Besides any other form, in which IoT may be playing its part, Internet of drones (IoD) is a recent yet greatly appreciated and rapidly incorporated form. Being an aerial vehicle, IoD provides extreme reachability for any IoT devices under operation, and for that reasons, various stakeholders and business consider it as pivotal for their needs. IoD is now being explored for its feasibility and operability scenarios, where it can cascade to existing drone operations. This chapter shall take a walk on describing IoD and proposes state-of-the-art architecture and it's applications, especially those that are oriented to industrial IoT.

Keywords Internet of drones · Internet of things · HANET · SIIARA · Place time coverage and capacity

1 Introduction

Ever since Kevin Ashton coined the term Internet of things (IoT) in the year 1999, a system of interrelated machines that have the capability to be connected over the internet, this technology has never looked back [1]. The Internet of things, also known as machine-to-machine (M2M) or machine-type (MT) communications without human interventions, have penetrated in our daily routine of tasks, helping in comfort and convenience through automation. By 2020, it is expected that there

A. Kumar (✉)
CTIF Global Capsule, Department of Business Development and Technology, Aarhus BSS, Aarhus University, Herning, Denmark
e-mail: ambuj@btech.au.dk

P. L. Mehta
IILM Academy of Higher Learning-College of Engineering & Technology, Greater Noida, India
e-mail: purnima.mehta@iilmcet.ac.in

D. Gupta et al. (eds.), *Smart Sensors for Industrial Internet of Things*, Internet of Things, https://doi.org/10.1007/978-3-030-52624-5_16

would be billions of IoT connected devices. This would require a low-powered wide-area network. The IoT depends on proper connectivity through reliable and robust wireless sensors, including low-latency data transfer. IoT incorporates various platforms, applications, and technologies, which will be executed under several radiocommunication services. There is a fast increase in the deployment of MTC infrastructure or IoT networks. IoT is an infrastructure for providing services by interconnecting (physical and virtual) objects. This is a vision that has technological and societal implications. IoT is poised to integrate technologies, namely, M2M communications, autonomic networking, data mining and decision-making, security and privacy protection, and cloud computing, with advanced sensing mechanisms. The IoT's capabilities for devices to interact on a common communication platform have added a new dimension to the drone's technology.

IoT applications include smart grid, smart home, e-health, agriculture, support consumer, water management, waste management, intelligent transport, wireless industry automation, deep indoor coverage, traffic efficiency and safety, government applications, etc. with a variety of services. These all will help in boosting the exponentially economic development of the country benefitting society at large. For communication purposes, terrestrial International Mobile Telecommunications (IMT) network including unlicensed bands play a significant role for IoT or MTC applications. On a case-by-case basis, IoT/MTC has ranges of narrowband and broadband applications. Sensors say for health monitoring, etc. fall in the category of broadband, while narrowband connection needs low-power consumption. It is expected that the wireless sensors and/or actuator telecommunications will cater to wide-coverage area and variety of things on a cell-by-cell basis considering the traffic requirements. Radio spectrum below 1 GHz, currently for IMT applications, will be for narrowband, while spectrum between 1 and 6 GHz offering higher bandwidth will support broadband applications.

The drones have existed for more than three decades and been earlier used mainly for detection and surveillance purposes by security agencies. Drones can be categorized according to their aerodynamic structure and level of automation. Talking about structure, drones, depending on the applications, are available in different sizes, viz. very small, small, medium and large. These sizes may either be multiblade or fixed-wing types. The fixed-wing drones with faster speed have longer duration covering wider areas. Though they need more space as a runway for landing and taking off, they are most suitable for aerial surveys, taking high-resolution photos and mapping, etc. Multi-wing drones have lower speed with less flight duration and limited capacity. Their ability to move over a limited area makes them quite useful for attributing IoT devices with controlled itinerancy. Further, a drone may be autonomous or manually controlled. As per the definition of the International Civil Aviation Organization (ICAO), those who perform manoeuvring operations without human intervention are often referred as the unmanned aerial vehicle (UAV) [2] also known as unmanned aerial system (UAS). UAS can receive or transmit signal from the ground simultaneously at multiple locations, through a dedicated transceiver known as a ground station. Drones that are controlled by a human are termed as remotely piloted aircraft systems (RPAS) [3]. UAVs, which are majorly fixed-wing drones, are often used for military purposes and are often designed to carry large weights [4, 5]. RPAS are primarily small-scale drones that are controlled by a human from a distance through remote control.

Mobility is the biggest challenge for the efficient utilization of IoT devices. It has been observed that the usability of an IoT device can be enhanced significantly if mobility is incorporated. Also, the altitude makes sensor-based IoT devices more potent and suitable for certain applications (surveillance, monitoring, seeding, etc.). Recently, attempts are being made to amalgamate drones and IoT leading to a combinatorial technology known as Internet of drones (IoD) [6], which is also defined as the Internet of flying things [7]. These attempts of combining the two technological streams are still at an evolving stage, and a lot of developments are yet to come. Nonetheless, the technological duo seems very promising for both present and future. Drone adds mobility to the IoT device that is mounted as a payload. The 'thing' of IoT, which is the functional equipment such as a sensor or an actuator, can now be placed at a vantage point, where its efficiency is enhanced manifolds. Meanwhile, the wireless communication module or the 'Internet' part of IoT helps these 'things' to communicate with the rest of the world. As technology progresses, instead of placing IoT device on the payload, the entire system can be integrated on a common board as a single functional device. IoD converges the electromechanical systems and wireless technologies, where in this case electromechanical systems also account for the drone's functional systems.

Combing the two frameworks is a challenging task; UAV, being an aerial vehicle, has many obligations to follow, most of which are regulatory-related. As an example, drone operation beyond the visible line of sight (also known as BVLoS) is prohibited in many countries. Also, many countries do not allow *swarm of drones*, which means that a single user cannot control multiple drones at a time. On the other hand, both streams have to deal with the fusion of technological frameworks. The mobility of UAV-mounted IoT devices adds significant complexity in establishing seamless wireless communication channels.

Pertaining to aforementioned discussions, this chapter broadly covers topics on drones; IoD architecture; harmonized radio spectrum for IoD applications and their navigation techniques, allocation, and management of airspace; and systems and technologies for various services/applications. *This chapter focalizes on the professional applications of IoD, with a considerable level of automation so that human intervention can be minimized.* Therefore, we are discussing the swarm architecture, where multiple drones can be used to complete an operation collectively, or each drone can be dedicated to mutually independent tasks.

2 Internet of Drones (IoD): Swarm Architecture

Presently, IoD has no definite architecture and is normally considered as drones with mounted IoT devices that have some connectivity with the ground station. However, this is not sufficient for the future of IoD, and therefore, we propose an

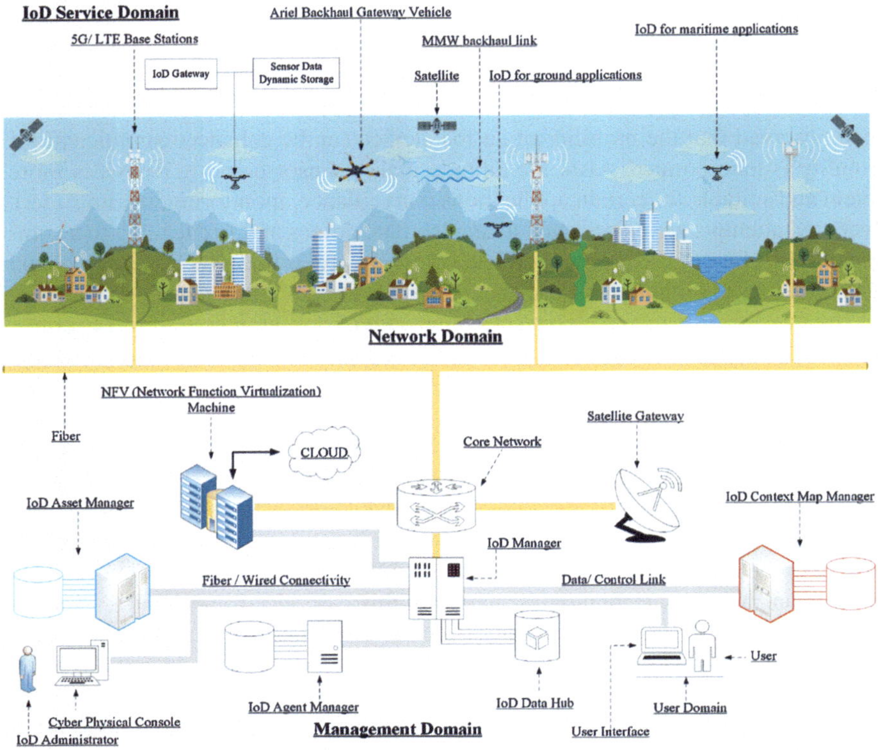

Fig. 1 Perceived IoD architecture

architecture that defines a generic architecture of IoD that can be extended from single use to swarm of flying IoT. Figure 1 shows that generic and proposed IoD swarm architecture is an amalgamation of IBM-connected vehicle concept [8] and communication architecture, where the IoT domain and network domain are shown separately [9], while the core elements are converging at the network core [10]. For end-to-end description, we have extrapolated the individual drone operation to an architecture that enables multiple drones to operate in coherence.

As shown in Fig. 1, the IoD architecture has mainly three components: the drone, the IoT devices which are the sensors and the attached communication modules that are mounted on the drone as a payload to be carried aerially, and the underlay communication architecture.

2.1 Architecture Description

Each of these components is functional in their respective domains; the drone (carrier vehicle) and attached IoT devices are functional in the service domain, the serving communication network in the network domain, and IoD management functions belonging to the management domain. The three domains cohere for a comprehensive IoD activity, where the service domain includes functions of airborne motion and functions of the IoT devices and their storage, management domain takes care of all the functions related to drone-IoT system, and network domain performs functions that facilitate the exchange of control and data between service and management domains.

We perceive IoD to be built around the existing communication systems. Instead of a dedicated ground station, we recommend conventional mobile communication network to share access to the drone-mounted IoT devices. This will not only promote infrastructure reuse but also provide seamless coverage to the flying IoT devices. The IoD architecture connects to the network core as a point of interconnect for the IoD management domain and the flying IoT devices. The architecture of the terrestrial communication network, shown in Fig. 1, accommodates 4G/LTE and 5G characteristics for its present and futuristic propositions. The communication network is primarily a cellular network that has network core, which connects with the geographically distributed BSs via fibre-optic or wireless links, often recognized as the backhaul links.

The figure shows that the flying IoT system has two major components, one being the IoD gateway (IDG) and another as the sensor dynamic data storage (SDDS). SDDS collects all the real-time data from the IoD sensors and stores temporarily to be sent to the IoD data hub (IDDH), whereas the IoD gateway connects to the terrestrial communication network to communicate with the IoD manager (IM) via the network core. IDG is a micro-backhaul for the individual flying system, and each of these systems has their gateways. However, for certain cases, this micro-backhaul link may not be feasible enough to push the data through the communication system. The reasons are enormous; path loss, signal blocking, and low battery are some of them. In such cases, we propose to make use of the concept of SIIARA [11], where a sturdy drone can serve as the repeater/ range extender system for one or multiple IoD flying devices, which is shown in Fig. 1 as the gateway aerial vehicle for hovering ad hoc network (GAVH). As discussed in [11], GAVH is a futuristic concept, where the authors of the concept proposed to provide improvised backhaul link for aerial radios (see SIIARA), and we used the same concept to connect the flying IoT devices. We can see in the figure that GAVH is connected to the ground station (terrestrial base station) via a millimetre wave channel, which shall be elaborated more in the later section.

The drone-mounted IoT devices are connected to this base station via dedicated channels through IDG. However, the environment in which the drones are traversing may not be feasible enough for continuous streaming of the sensor data. In such cases, SDDS provides redundancy for data protection and saves the data until the system reaches to the feasible location from where the flying IoT devices' data can be backhauled to the destination. When a drone reaches to a suitable position, where the signal propagation is feasible enough, the SDDS data is pushed to the IoD management system via network domain. In this way, all the data observed by sensors of all the flying IoT devices, mounted on different drones that are itinerating in a common zone, are stacked in the IDDH.

IM is responsible for managing the overall IoD devices in a particular zone. In an IoD architecture, IMs are connected to the core of a local network. IM parallelizes with the cellular architecture; an area can have multiple regions, and each region may have its network and its core, to which individual IMs are connected. The IM manages the IDH data and assigns them to IoD Agent Manager (IAgM), IoD Asset Management (IAsM), and IoD Context Mapping Manager (ICMM). Concepts of IAgM, IAsM, and ICMM are motivated by IBM's connected vehicles' concept [8].

In the IoD context, we define the aerial vehicle as to the 'agent' and the mounted IoT devices as 'asset'. The aerial drones serve as the vehicles in the air, as the cars serve on the ground. They have the on-board positioning, navigation, and timing (PNT) systems and ambience sensor (AS) systems to itinerate from position to position. The PNT systems use satellite signals to identify locations and waypoints of the carrier aerial vehicle. Besides these on-board devices, the drone in IoD has the attached IoT devices designed for a dedicated operation. When a drone is a part of an IoD architecture, it's PNT and AS systems use the network domain to communicate with the IAgM. Similarly, the IoT devices that are mounted on the drone use the network domain to interact with (IAsM).

The IAgM manages the airborne vehicles, which includes identifying, storing, analysing, and deriving the sensor information of the flying vehicle. The IAgM performs the mission plan, which is to take the inputs from the user/clients, plan the voyage, and activate the sensors as when needed and return home. This also incorporates contingencies and redundancies of the operations. IAgM also defines the mounted IoT devices and their characteristics. Once the mounted devices and their characteristics are identified, IAsM manages the data that is generated by these IoT devices, which includes data storage, analysis, interpretation, and actuation. IAgM.

ICMM is responsible for context-based functions and maps the IDDH data with geospatial frameworks, such as U-space [12], while incorporating other environment-based parameters such as temperature, humidity, wind gusts, payload weight, etc. ICMM assists IAgM in monitoring and guiding the drone movements during any operation or service.

Another significant point that needs to be mentioned here is the role of satellites in the IoD architecture. As shown in Fig. 1, besides the entire IoD architecture, there is a significant presence of satellites. These satellites are not the integral part of IoD architecture; however, they are essential co-partner in the entire framework. Satellites are necessary for autonomous vehicles, including drones, as it is used for position navigation and timing (PNT) services. A vehicle with a significant level of automation uses the PNT service to identify its own coordinates and also the coordinates of the destination point. Based on the difference in the coordinates, the vehicle decides its mobility parameters, such as direction, speed, and acceleration. In the future, a drone shall adhere to the U-space frameworks, where PNT service shall also be used in geofencing and waypoint locus from initial and destination points. These satellite-based PNT services are provided by many service providers, out of which the significant ones are Global Positioning System (GPS) [13] by the USA, GLObal NAvigation Satellite System (GLONASS) [14] by Russia, GALILEO [15] by Europe, and upcoming satellite system NAVIC [16] by India.

2.2 Architecture-Assisted IoD Operation

An IoD user, who wants to utilize IoD for operations or services, launches a request through a web-based portal, specifically designed by IoD service provider. The IAgM identifies the agent characteristics, in terms of specifying requirements (user requirements, technological requirements, regulatory obligations, and their frameworks) and assigning agents to cater to these requirements (mounted sensors' specifications and drone characteristics). ICMM, on the other hand, tallies the requirements with context mapping, which includes generating the waypoints and marking the geofenced locations.

Once the drone is airborne, IAgM manages its movements based on the context mapping by ICMM. IAgM also takes care of the safety and security of the aerial vehicle. IAgM has features to ensure collision avoidance of the multiple drones in an area, for both collaborator [17] drones and intruders [18]. As most of the battery/fuel power of the IoD system is consumed by drones, IAgM takes care of the energy efficiency of the flying vehicle so that the longevity of the operation is prioritized. Energy efficiency optimization includes optimization of drone communication [19], the possibility of optimizing communication using relays (GAVH) [20]. When a drone reaches to the desired location, IAsM manages the mounted equipment, such as sensors and actuators, by activating and deactivating them as on when required. IAsM is also responsible for the management of data that this equipment generated during any operation.

This is a visualized architecture that enables swarm of drones to work in coordination and coherence and realizes progressive approach towards full automation of IoD. The architecture also looks forward to the universal drone registration that is anticipated to be incorporated by 2025 [12]. Thus, the architecture shown in Fig. 1 is an evolving concept that accommodates IoD features, which is soon to be realized in the coming future.

2.3 Possibility of Integration of Drones of Other Domains

The proposed and discussed architecture mentioned above is not just limited to airborne drones. Since airborne vehicles are the pioneer in defining and evolving the drone concept, often the term drone is misunderstood as aerial vehicles. However, the drone, which is a system/robot/vehicle that can operate without the human presence within the vehicle, applies to any domain that satisfies the definition. Nowadays, apart from the conventional airborne drones, there are waterborne drones that are submerged in water, water-buoyant drones that float on the water surface, and surface vehicles that walk on the ground. Figure 2 shows the three scenarios coexisting in the IoD architectural framework.

Fig. 2 Multimodal drone scenario

3 Harmonized Radio Spectrum for the Internet of Drones (IoD) Applications

Aerial drones residing in airspace have a wide range of applications [21] from terrestrial to airspace to water bodies in combination of terrestrial to airspace, water bodies to terrestrial, and water bodies to airspace or in several permutations of all the three spaces. Many times, for the establishment of these links, wireless plays a dominant role. In IoT, and sensing of objects, the tasks are only performed through wireless communication technologies. IoT is a network consisting of computing devices connected with sensors and software. These devices may be in billons that collect real-time data and share via wireless communications. It is well known that any wireless connectivity is established deploying radio spectrum in the appropriately earmarked frequency bands for that services/applications. *The International Telecommunications Union (ITU), a United Nations specialized agency for information and communication technologies (ICTs), among other mandates, allocates* radio spectrum globally for terrestrial, sea, and space operations. The tasks of allocation of radio spectrum for various radiocommunication services/applications are performed by the *World Radiocommunication Conferences* (WRCs) held every 3 to 4 years. The function of WRC is to revise the Radio Regulations, an international treaty for the usage of the radio-frequency spectrum. These are made based on agenda items decided by the previous WRCs. The *Conference Preparatory Meeting (CPM)*, for facilitating work of a WRC, prepares a consolidated report. This report is based on input contributions from member countries/stakeholders, ITU's study groups, etc. relating to regulatory, technical, and operational aspects. In this section, a brief on radio spectrum requirements for MTC/IoT, safety and security in the sea, and wireless sensing is described.

3.1 Radio Spectrum for MTC/IoT Narrowband and Broadband Applications

In this perspective, WRC-2015 resolved [22] that urgent studies supporting the implementation of narrowband and broadband machine-type communications/IoT infrastructures under WRC-19 agenda item are needed. The resolution states 'to carry out studies on the technical and operational aspects of radio networks and systems, as well as spectrum needed, including possible harmonised use of spectrum supporting the implementation of narrowband and broadband machine-type communication (MTC) infrastructures'.

There is another resolution of ITU, which directs for studies related to wireless systems and applications for the development of IoT. This resolution considered that the (a) globally connected IoT is built on connectivity and functionality, being realized by telecommunications networks; (b) growing numbers of IoT applications require higher transmission speed for supporting a huge amount of data among a large number of devices; (c) wireless mobile systems and inland mobile service provide telecommunications network to a number of sensors and actuators; and (d) different frequency bands used by many radiocommunication services provide the capacity needed by IoT's deployment. This deployment ensures efficient utilization of the radio spectrum.

The narrowband and broadband MTC/IoT utilizes radio-frequency spectrum within the band earmarked/assigned for IMT applications. Narrowband shall use frequency band below 1 GHz while broadband between 1 and 6 GHz. Possible examples of frequency bands for use by MTC/IoT are enunciated in Table 1.

3.2 Radio Spectrum for Global Maritime Distress and Safety System (GMDSS)

One of the possible applications of drones is for package delivery from port to ship parked 10–15 km away in the sea. Situations do arise during the case of emergency when distress signals to coast stations are required to be sent. Today, approximately 80% of the world's export and import is via sea. Hence, more safety precautions are required for safe navigation. When the ships are in the sea, terrestrial communication systems do not function. Therefore, an exclusive robust, seamless communication system including in distress situations in the sea is required. The medium

Table 1 Possible examples of frequency bands that could be considered for use by MTC and IoT

Frequency band in MHz		
Mobile station transmitter	Base station transmitter	Remarks
452.5–457.5	462.5–467.5	This corresponds to an arrangement in which IMT-based MTC/IoT networks are operating or planned to operate within these bands in certain countries
703–748	758–803	
733–736	788–791	This corresponds to an arrangement in which certain countries are considering this frequency band as a potential option for harmonized use for IMT-based narrowband MTC/IoT networks deployment
824–849 z	869–894	This corresponds to an arrangement, in which IMT-based MTC/IoT networks are operating or planned to operate within these bands in certain countries
880–915	925–960	
832–862	791–821	
1710–1785	1805–1880	
1920–1980	2110–2170	

frequency (MF, 300 kHz to 3 MHz) bands are used for maritime radio beacons. During the eighteenth century, for sending any distress signal to coast stations, Morse codes were used by the ships sailing in coastal waters. This kind of information was never very clear for taking remedial measures by the concerned authorities in the events of emergency because of all kind of losses. On first February 1999, an internationally agreed safety procedure was adopted, known as GMDSS or Global Maritime Distress Safety System. GMDSS is a set of standards for the use of communication protocol, procedures, and safety equipment at the time of distress situation by the ship. Under GMDSS, all the passenger ships and cargo ships sailing in international water are required to carry equipment for safety; it makes rescue operations of ships/boats easier.

GMDSS accomplishes jobs of (1) alerting (including position determination of the unit in distress) and search and rescue coordination, (2) locating, (3) maritime safety information broadcasts, (4) general communications, and (5) bridge-to-bridge communications. The ship using GMDSS sends a distress signal via a satellite or radio communication equipment. This is also used for sending or receiving maritime safety information and general communication channel. For the purposes of maritime communication, the range of 156–174 MHz is allocated. The frequency 156.800 MHz is for distress, urgency, and safety communications. However, the frequency 156.525 MHz is for routing VHF digital selective calling watch.

4 IoD Applications

Almost all drone-related application is analogous to the IoT paradigm. There will be a sensor or actuator or both mounted on a flying vehicle that connects to a communication network to transceive the data from the ground station, which is exactly what happens in case of an IoT device. Walking on this line, in this section, we shall mention some categories of applications with some relevant examples.

4.1 IoD as an Aerial Radio Architecture

One of the novels and futuristic applications of IoD is the aerial radio architecture (ARA) framework, as discussed in [23]. Figure 1 shows a proposed architecture, defined as SIIARA [23] by the authors, which explicitly describes all the major components for such an architecture to cover all the needs for an ad hoc aerial base station network. SIIARA complements the terrestrial network by providing a temporary aerial base station-based wireless communication network at locations where the place time coverage or capacity is desired, thereby eliminating the needs of having additional base stations from time to time [24, 25].

This architecture is designed to address the additional coverage and capacity requirement that wobbles up due to random accumulation of network users at an unprecedented location. These random accumulations deteriorate the overall network performance and create requirements of additional base stations, although no additional user is added in the network. The base stations from where the users have moved away to create congestion in other BSs create a void in the previous BS, which results in the underutilization of the capacity of the previous BS. Hence, wherever users accumulate, an additional capacity and/or coverage requirement is created irrespective of the original network status. This dynamic capacity demand due to mobility and accumulation of users is defined as the place time coverage and capacity (PTC^2) [25]. Thus, huge mobility of users and accumulating at random locations create a huge disbalance in the network capacity utilization. In the absence of a suitable solution, the network service provider (NSP) has to keep on adding additional network sites, thereby bearing additional costs on infrastructure and network elements.

The architecture shown in Fig. 3 is actually base stations mounted on IoD. The IAsM manages the sensors of the IoD to detect the user accumulations or PTC^2, in an area of the event (AoE). IAgM performs the functions to distribute the drones according to the user scenario underneath. Each drone is an ad hoc/temporary base station that serves the users until they dissipate. These IoD-based flying base stations are explicitly termed as Hovering Ad Hoc Network (HANET) member [26], for its exclusive use in the aerial communication network. HANET that serves the underneath users is defined as the HANET Serving Member (HSM). IAsM uses the base station parameters of an HSM to identify the physical placement of the member. The signal base placement is discussed in [27]. A user group may require one or multiple drone-mounted base station to satisfy the collective capacity of the group. For seamless serving, the drone follows the group of accumulated users until they are traversing together. This mobility of drone-mounted base stations eliminates the need for an additional aerial base station, thereby saving the infrastructure elements.

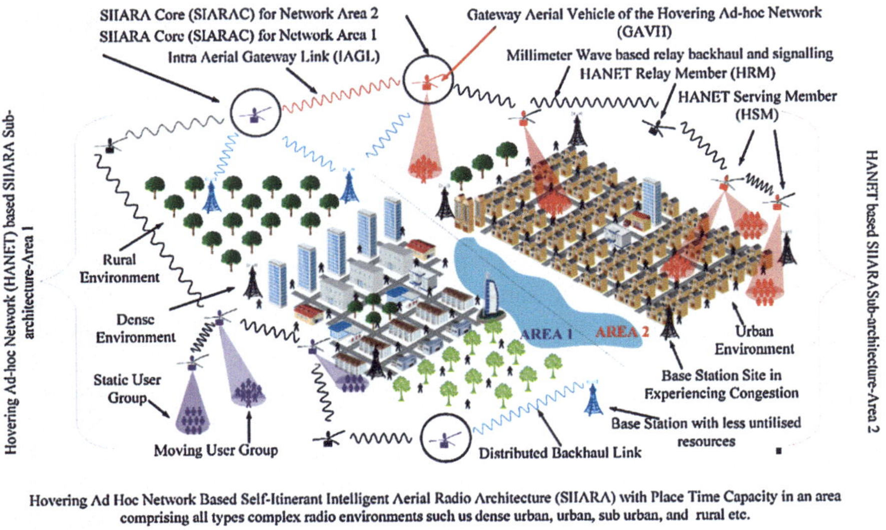

Fig. 3 Aerial radio architecture [23]

However, this airborne wireless communication network requires efficient backhauling. SIIRA has another HANET element, SIARAC, to serve as network core of the local ARA. SIRAC uses millimetre waves to create an efficient backhaul link that connects local ARA with an underutilized terrestrial base station that has low traffic in its serving area. The terrestrial BS that participates in backhauling of the SIIARA is termed as HANET Gateway BS (HGBS). This backhaul link is dynamically allocated as HSM proceeds on its way following the underneath group of users, as shown in Fig. 3. In case an HSM is far enough from the SIARAC, an IoD-based relay BSs are placed to pass on the backhaul link from serving member to the core member. This relay HANET member is defined by authors as the HANET Relay Member (HRM). IM follows the distributed dynamic backhauling [28] procedures to update IAsM, IAgM, and ICMM data.

There can be multiple local groups of these flying BSs in different areas of a target zone (shown as area 1 and area 2 in Fig. 3 as examples). The communication between users of these groups can be backhauled through a local gateway, GAVH, whereas internetwork communication (communication between users latched to SIIARA and users connected to the terrestrial network) is backhauled through HGBS.

SIIARA is a futuristic network, and considering the U-space and regulatory requirements, this kind of architecture is possible no earlier than 2030, when the drones are completely autonomous, and there is strong compatibility between aerial and terrestrial networks. As we can see in Fig. 3, SIIRA uses a swarm of drones for a complete airborne network deployment, which signifies the importance of swarm IoD architecture shown in Fig. 1.

4.2 IoD as Airborne IoT for Industrial Safety Applications: Monitoring, Surveillance, and Search and Rescue

One of the prominent uprising application of drones is in the monitoring, surveillance, and search and rescue sector. This includes single or coordinated drone applications. We have grouped these four functions, as this is related to human/commodity safety and requires intense technologies, computations, procedures, and regulatory obligations to perform such activities. Whatever the case it may be, the IoD architecture mentioned in Fig. 1 shall be the efficient framework to suffice this operation. In the surveillance/monitoring-related operations, IoD has the pertinent sensors to read physical parameters. Referring to Fig. 4, the following are the specific cases, where IoD can be of great use:

4.2.1 IoD for Environmental Monitoring

The environment is an increasingly prime concern for us, critical enough to challenge our existence. Several measures are being taken to control the factors that affect the environment in some or the other way. Besides taking controlling actions, the significant endeavour is being put on observing and monitoring the environment components and keeping an eye on the responsible sources. Drones are already being used to observe the air and water quality, especially near the industrial areas and high-density urban areas, where human activities have affected the environment the most. Drones are being used to promptly reach the difficult locations, such as above the chimney or near life-threatening areas such as toxic chemical containers, which are beyond human safety boundaries. Just like these dedicated drones, IoD can play significant role activities, and environmental monitoring can be done by mounting appropriate sensors such as CO_2, SO_2, temperature, humidity, etc. Figure 4a shows how IoD can replace the dedicated drones with a full-fledged self-reliant automated system. The figure shows two kinds of IoDs, one being airborne to monitor the air quality of an industrial area and another being waterborne to measure the water quality of the nearby river. Both of these drones can be centrally controlled by IoD administrator, and the combined data, managed by IAsM, can reveal holistic information of the area more holistically and precisely. The system can allow more IoDs of various sensors as per requirements, giving flexibility and scalability to the entire environment monitoring systems.

Besides monitoring the industrial area, IoDs can be used to cater to the ocean environment by monitoring ships and marine vessel containers for their law tolerance and for any unnoticed/ignored leakages. Both air- and waterborne drones can measure the effect on air and water quality due to a malicious ship passing by a port. Based on the comprehensive information, the port and border control authorities can block and/or penalize the offending vessel. This is an important area of IoD application as many ships do not follow international regulation of using efficient methods to control CO_2, oil spillage, and dumping the garbage.

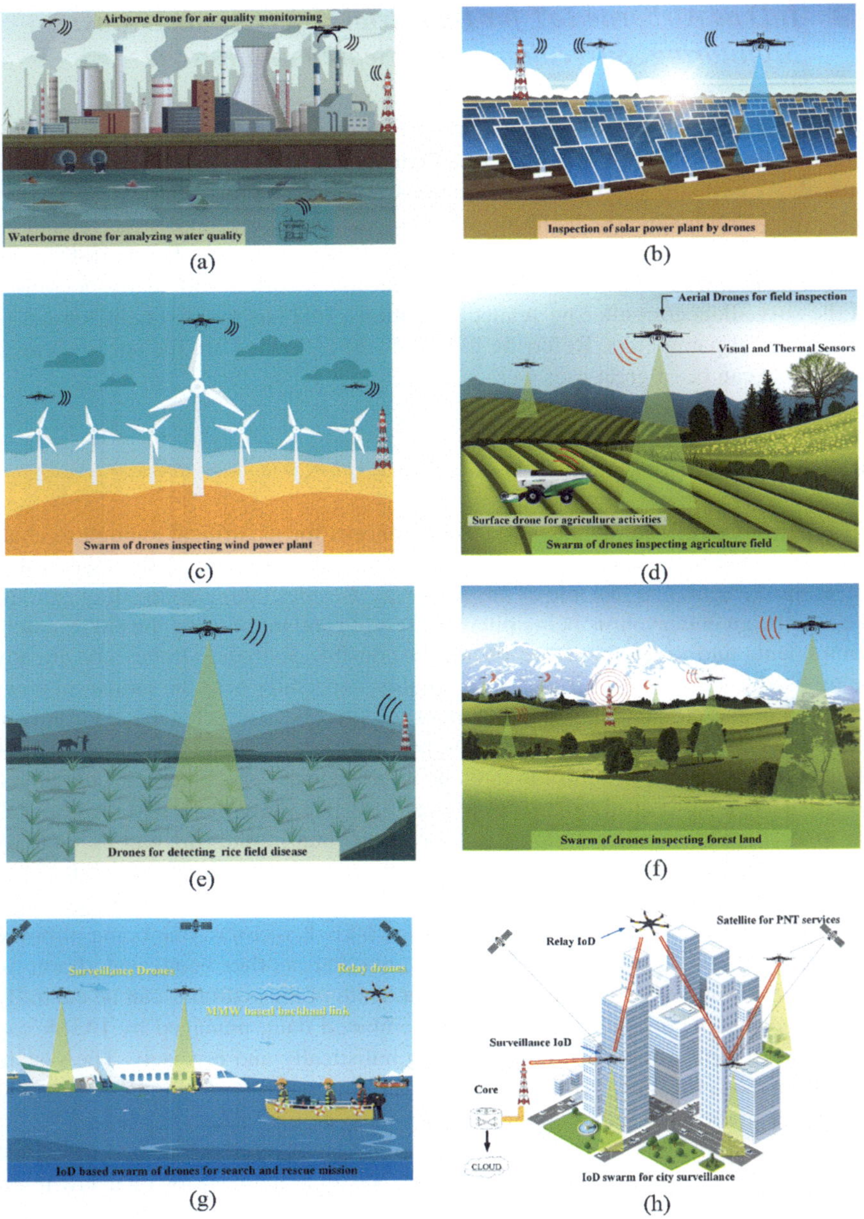

Fig. 4 IoD applications for industrial safety

4.2.2 IoD for Inspection of Power Plants

Just like environment inspection and monitoring, IoDs can also be used to monitor the power plants, particularly solar photovoltaic (PV) power plants, as shown in Fig. 4b, and wind turbine power plants, as shown in Fig. 4d. It has been observed that PV solar panels develop faults and eventually cease to operate, which can happen due to several factors such as time-based deterioration, mishandling, weather adversities, bird droppings, corrosions, etc.

In a PV plant, there are thousands of solar panel installed, which makes for the service provider to monitor each and every panel manually and to identify the faulty panels. Recently, a project known as Easy PV [29] was successfully completed, which uses drones with a mounted thermal camera, to identify the health of the PV panels and to identify the faulty panels. This is an eminent example of the usefulness of IoDs in the inspection of PV plants. Integrating these dedicated drones with IoD architecture shall not only make it centrally controllable but also raises possibilities to integrate with other drones to operate simultaneously to cover area fastly and more precisely.

On the same line, wind turbines also pose needs of motoring and maintenance. Unlike solar panels, a monitoring wind turbine is far more tedious and a threat to human safety. Severe accidents have been reported that happened during the monitoring of the wind turbines. Recently, drones have been introduced to monitor the health status of the turbine blades, which may get damaged due to passage of time, adverse weather, bird strikes, etc. Such drones can be raised to integrate with IoD architecture to perform a monitoring campaign. Such solutions shall also cater the maritime wind turbines, which are even more difficult and safety threat to humans.

IoD shall make maintenance of wind turbines and PV plants more efficient and less time-consuming and reduce the threat to human safety at the same time.

4.2.3 Inspection of Agricultural Fields

The use of *drones* for management that includes monitoring of crops, farm animals, fisheries, forests, etc. opens new vistas for technological inventions. Drones have great potential in addressing problems of the agriculture sector. This sector is going to be one of the largest users of IoD.

For improving farm production, the essential parameters are good soil for the crop; proper irrigation at appropriate timings; timely use and application of fertilisers; use of right kind of pesticides for treating crop diseases; etc. IoDs shall be deployed to collect raw data, including local weather information that can be analysed for developing simulation models for agriculture, as shown in Fig. 4d. The information so collected shall be used for taking appropriate actions and suggesting solutions. IoDs fitted with infrared and multisensors are used for monitoring crop health. They are also deployed for spraying fertilisers and pesticides for growth of crop including detection of diseases. These factors have a direct link with healthy crop yield. The insurance companies in central and western states of India are taking services of drones for agriculture surveys. This is required for accurate insurance claims. In the event of disaster risk, for its reduction, the drones are used. The useful data so collected are analysed for the agencies to effectively organize reliefs. The drones equipped with photogrammetric and navigation systems detect water stress, etc. helping to predict landslides, damage due to storm, erosion in agriculture land, etc. This information is provided to agriculture landowners and action-taking agencies, who can use it in mitigating the risks.

Figure 4d also shows the integration of multimode IoDs. Airborne IoDs shall also serve the relay station for the ground-based IoD, which is an agrobot. This agrobot can be controlled by the user sitting somewhere very far from the agricultural land, and still, he/she can instruct and monitor the agrobot to perform several tasks.

4.2.4 Detection of Crop (Rice) Disease

Rice, in many Asian countries, is a dominant crop with high yield, and the majority of the population is rice eaters. This crop is grown in waterborne areas, and many a time paddy stems, leaves, and unripe produce suffer from a quite common disease appearing as bacterial stain and brown spot [30]. These diseases are visibly seen and severely affect production. As soon as these diseases are detected, remedial actions would need to be addressed. Normally, humans are deployed for spraying of pesticides, but they cover neither monitoring of larger crop areas nor uniformly spraying of pesticides. IoDs can be utilized for monitoring the crops and timely spraying of pesticides to control the disease, as shown in Fig. 4e. The monitoring can be performed by sensor-based machine-to-machine (M2M) communications/IoT with cloud computing, and in this way, drones replace human intervention.

4.2.5 Monitoring Illegal Fishing

The drones can be deployed for detecting illegal fishing that may help take legal actions against lawbreakers. This can also help in enforcing fishing regulations in water areas (sea, river, or pond) and conduct mapping [31, 32].

4.2.6 Forestry

The forests use drones for landscape mapping, helping to evaluate and monitor the activities. The pictures taken by the drones are used to prepare ortho maps useful for creating and updating the database for geographical information system (GIS) [33]. The landscaping helps in knowing the illegal and encroachment activities in a forest for taking remedial actions and arresting them. As shown in Fig. 4f, IoDs also help in understanding individual tree's geometry and ecological features and preparing and updating forest records on a local and wider area basis. Drones fitted with HD and high-resolution cameras are used for remote tracking and monitoring of animals in the forest/wildlife. This monitoring with the help of sensors also helps to know about any illegal human's dwellings and thieves hiding in a deep forest. The data collected by the drones are processed and analysed.

4.2.7 IoDs for Search and Rescue Missions

Figure 4g shows the wreckage of an aeroplane in the deep sea. In such cases, identifying the drowning people in the scattered mess and bumpy waves is extremely difficult. Recently a project named SARA [34] has commenced with an objective to use drones to help Italian coast guards to detect people lost in the sea, especially during low light conditions. IoD can walk on the same line to support comprehensive and prompt solution to such conditions. Figure 4g shows the surveillance IoDs trying to detect the people between the wreckage pieces and relay IoD that shall connect the distant ground station with the serving IoDs.

4.2.8 IoDs for Search City Surveillance

Nowadays, drones are being used to know city whereabouts. Some smart cities are progressively incorporating drone abilities for surveillance and monitoring. Figure 4h shows the possibility of using IoD for such applications, where individually operating drones can be integrated with other subordinates for better actions. The figure also shows how the HRM, or relay drone, can be used to connect the IoD operating in cut-off areas with the ground station.

4.3 IoD Applications as a Delivery Vehicle

There has been a recent upsurge in a wide range of commercial applications of drones like package delivery of goods. The IoD provides synchronized access to the controlled airspace for the drones. These can be ushered with an effective and on-time management of navigation and airspace. Air-package delivery of large goods from location 'A' to location(s) 'B' poses a greater challenge, because of availability of limited 'airspace' in each geographic location, in which many drones may be planned to navigate. For this, the allocation of airspace needs to be precise causing any conflicts during the duration navigation of a number of drones. The accurate allocation of airspace for movement of drones is equally significant for any other applications.

4.3.1 Port to Ship

On-demand delivery of essential items of low weight and small-sized packets containing medicines, repair parts, documents, etc. from port to ship entering the sea and mostly parked 10–15 km away from the port is a challenging task. Presently, these packets are loaded in a small carrier boat managed by persons sailing all the way to the ship parked in seawater. Most of the time, these boats face rough weather making its delivery difficult, involving risks and human safety. Sometimes, even for a few packets, same infrastructures of a boat with a person are required leading to less efficient deployment of resources or wastage. The diesel consumption by the carrier boat(s) also is substantial, leading to carbon footprint. Drones can efficiently come to rescue by automating such operations. Although, the shipping industry have undergone semi or full automation and a lot of efforts have been made continuously in this direction. However, well-organized automatic on-demand supply chain from port to ship is waiting for operations. The purpose is for complete replacement of human intervention from the tasks and to put in place a semi or fully automated system. The on-demand supply will be performed by integrating operations of semi-automatic carrier boats and aerial drones to be coordinated by a control centre stationed at the port. The advent of multidisciplinary technologies in semi-automatic water-buoyant and aerial unmanned drones, sensors, and communication systems, MTC/IoT/IoD, with safety and security features will effectively solve issues of on-demand supply of goods from ports to ships [35].

4.3.2 Ground to Ground

Drones are now everywhere. They are becoming the best pal for the ground-to-ground delivery, whether it is a mere item as domino's pizza delivery [36] or as critical as kidney transplant [37]. With IoD architecture, having these services shall be more organized and accessible. A user can place an order using the IoD-dedicated portal, and IAgM and ICMM shall make the voyage plan, based on the criticality and environmental factors, and the IM shall manoeuvre to conduct the operation. Anticipating the future business prospects, companies like Amazon are now coming up with delivery drones [38]. Drone shall be the adorable means of transport for small package delivery, for its promptness and time efficiency.

5 Challenges of IoD

Until now, we have discussed the benefits of IoD as a futuristic technology. However, as rightly said, there are no pros without cons, and IoDs are no exception. Below are some anticipated challenges that IoD shall be facing:

5.1 Deliberate Misuse of Technology

Unintegrated/unsecured IoDs can be used for cross-border roguish activities and crimes. As an example, recently some attempts have been made to disrupt airport security by intruder drones. Also, there are cases when drones are being used for illegal transportation of weapons and prohibited drugs. IoDs being abler entities pose a severe threat to human safety and security. Voices have been raised towards the potential of IoDs to disrupt security paradigm and the vulnerabilities that it poses on the society. IoD shall enable criminals to perform actions without being actually present at the crime location, thereby leaving a clean slate of pieces of evidence. This shall encourage criminals to incline towards this evolving technology.

5.2 Technological Barriers

Both drone and IoT are evolving technologies, and the integration of both is very recent. Until now the use of IoD is almost negligible, and therefore, identification of gaps in the technologies is yet to commence. A significant technological barrier is a wireless communication, for both vehicle and the mounted IoT device. It is still hard to say how to manage IoDs in situations where the communication ceases to exist, especially for the vehicle. Also, since the use of IoDs is at a nascent stage, factors such as interference, shadowing, congestion of channels, coverage holes, and low throughputs are yet to show their impacts.

The second barrier is the poor progress in battery technology, due to which the drones' operational longevity is severely compromised. Despite the wonderful designs and low power electronics, most of the drones can fly between 20 and 40 min, without any significant payload. Long-performing drones can last up to 45 min by using multiple battery packs that not only increase the weight of the drones significantly but also take a lot of time to recharge all of them. With this limitation prevailing, IoD can never be in the mainstream business. Due to this limitation, most of the battery-driven drones cannot lift beyond 10 kg of weight. Fuel-driven drones can be an alternative; however, considering the inflammable and pollution risks, they cannot be used for delicate missions like kidney or pizza delivery.

The aerodynamics of small-sized drones are also a technology challenge. Such drones are not very efficient to create sufficient thrust to lift payload weight.

5.3 Regulatory Barriers

As mentioned earlier, regulations like BVLoS and swarm of drones can be a huge jolt in the IoD realization. As per BVLoS, a drone cannot be piloted beyond the directly visible sight, which significantly reduces the possibilities of IoD applications. Also, the prohibition of a swarm of drones shall definitely limit the potential of any potent framework. Besides the existing bindings, it is possible that some undesirable regulations might be imposed to keep a check on IoD-related crimes.

5.4 Operational Challenges

Until now, aviation is a considered an abiding and professionally precise sector. Each flying vehicle, whether a plane, helicopter, etc., is operated by professionals who have undergone months of professional training and have spent hours and hours of flight time. They follow a strict path and are not allowed to deviate under any normal situations. For these precise procedures, the aviation industry is the safest of all modes of transport. However, with IoD coming into the limelight, the flying vehicles, although very small in size as compared to their prevailing counterparts, are now controlled mostly by amateurs or by people whose professional expertise cannot be quantified. In such a case, drone-related accidents are very much prone to occur. Also, with more drones airborne, the possibility of collisions is also very high. Although intense research is being carried out to make a drone a safer vehicle, however, probability of such accidents is still unavoidably high. With more drones entering into the business, the probabilities of such mishaps shall considerably increase. This might lead to public unacceptability of these emerging technologies.

6 Conclusion

In this chapter, a discussion on the Internet of drones has been made. In the beginning of the chapter, along with the background overview of the Internet of things and drones, a concept that amalgamates the attributes of both the technologies, formally known as the Internet of Drones (IoD) is presented. A futuristic IoD architecture is proposed that is derived from IoT, connected vehicles, and drone technologies. This chapter familiarizes with aerial radio architecture, SIIARA, which is one of the significant application of IoD architecture. In the background of the IoD and SIIARA, this chapter discusses the areas where IoD shall prove to be of significant use. An illustrative discussion is made to show IoDs' use for the safety and security of people and how it is a promising mode of transport for the delivery of small packets. At the end of this chapter, a brief discussion is made about the challenges of IoD technology. We mentioned how this wonderful technology can be misused and how the increasing numbers of non-professional users can hamper the public acceptance of this evolving technology.

References

1. What is internet of things (IoT)? – Definition from WhatIs.com, IoT Agenda [online]. https://internetofthingsagenda.techtarget.com/definition/Internet-of-Things-IoT. Accessed 05 Oct 2019
2. The UAV – unmanned aerial vehicle [online]. https://www.theuav.com/. Accessed 05 Oct 2019
3. IATA. Remotely Piloted Aircraft Systems (RPAS) – drones [online]. https://www.iata.org/whatwedo/safety/pages/drones.aspx. Accessed 05 Oct 2019
4. Unmanned aerial vehicle | Definition, history, types, & facts. Encyclopedia Britannica [online]. https://www.britannica.com/technology/unmanned-aerial-vehicle. Accessed 05 Oct 2019
5. Unmanned aerial vehicles – an overview | ScienceDirect topics [online]. https://www.sciencedirect.com/topics/engineering/unmanned-aerial-vehicles. Accessed 05 Oct 2019
6. Gharibi, M., Boutaba, R., Waslander, S.L.: Internet of drones. IEEE Access. **4**, 1148–1162 (2016)
7. Hassan, Q.F.: The internet of flying things. In: Internet of Things A to Z: Technologies and Applications. IEEE, Washington, DC (2018)
8. Product architecture, 24 Oct 2014 [online]. www.ibm.com/support/knowledgecenter/en/ssnq4v_bas/iot-automotive/overview/key_concepts.html. Accessed 02 Oct 2019
9. Figure 2: High-level IoT architecture. ResearchGate [online]. https://www.researchgate.net/figure/High-level-IoT-architecture_fig1_281896657. Accessed 02 Oct 2019
10. A converged architecture approach for 5G (Reader Forum). RCR Wireless News, 05 Jun 2017
11. PhD_dissertation_Purnima_Lala_Mehta.pdf
12. SESAR Joint Undertaking | U-space [online]. https://www.sesarju.eu/U-space. Accessed 03 Oct 2019
13. GPS: The Global Positioning System [online]. https://www.gps.gov/. Accessed 06 Oct 2019
14. Information analytical centre of GLONASS and GPS controlling [online]. https://www.glonass-iac.ru/en/. Accessed 06 Oct 2019
15. GALILEO | European Global Navigation Satellite System
16. Indian Regional Navigation Satellite System. Wikipedia, 02 Oct 2019
17. Yoo, C., Cho, A., Park, B., Kang, Y., Shim, S., Lee, I.: ADS-B HILS test for collision avoidance of Smart UAV. In: 2011 Tyrrhenian International Workshop on Digital Communications – Enhanced Surveillance of Aircraft and Vehicles, pp. 253–257 (2011)
18. Yoo, C., Cho, A., Park, B., Kang, Y., Shim, S., Lee, I.: Collision avoidance of Smart UAV in multiple intruders. In: 2012 12th International Conference on Control, Automation and Systems, pp. 443–447 (2012)
19. Zeng, Y., Zhang, R.: Energy-efficient UAV communication with trajectory optimization. IEEE Trans. Wirel. Commun. **16**(6), 3747–3760 (2017)
20. Hanyu, A., Kawamoto, Y., Kato, N.: On improving flight energy efficiency in simultaneous transmission and reception of relay using UAVs. In: 2019 15th International Wireless Communications Mobile Computing Conference (IWCMC), pp. 967–972 (2019)
21. Radio regulations [online]. https://www.itu.int/pub/R-REG-RR. Accessed 05 Oct 2019
22. World Radiocommunication Conference 2015 (WRC-15), Geneva, Switzerland, 2–27 Nov 2015 [online]. https://www.itu.int/en/ITU-R/conferences/wrc/2015/Pages/default.aspx. Accessed 05 Oct 2019
23. Mehta, P.L., Prasad, R.: Aerial-heterogeneous network: a case study analysis on the network performance under heavy user accumulations. Wirel. Pers. Commun. **96**(3), 3765–3784 (2017)
24. Kumar, A., Prasad, R.: Place-time responsive network: an approach to the perennial clandestine challenge. IEEE Aerosp. Electron. Syst. Mag. **33**(5–6), 66–73 (2018)
25. Kumar, A.: Active Probing Feedback based Self Configurable Intelligent Distributed Antenna System: For Relative and Intuitive Coverage and Capacity Predictions for Proactive Spectrum Sensing and Management. Aalborg Universitetsforlag, Denmark (2016)
26. Lala, P., Sørensen, T.B., Prasad, R.: HANET: millimeter wave based intelligent radio architecture for serving place time capacity issue. In: Proc. Fifth Int. Conf. Wirel. Commun. Veh. Technol. Inf. Theory Aerosp. Electron. Syst. Technol. Wirel. ViTAE (2015)

27. Mehta, P.L., Sørensen, T.B., Prasad, R.: SINR based capacity performance analysis of hovering Ad-Hoc network. In: 2016 19th International Symposium on Wireless Personal Multimedia Communications (WPMC), pp. 147–152 (2016)
28. Distributed dynamic backhauling in aerial heterogeneous networks | SpringerLink [online]. https://link.springer.com/article/10.1007%2Fs11277-019-06582-w. Accessed 05 Oct 2019
29. EasyPV – a complete solution for PV maintenance [online]. http://www.easy-pv.eu/. Accessed 06 Oct 2019
30. Kitpo, N., Inoue, M.: Early rice disease detection and position mapping system using drone and IoT architecture. In: 2018 12th South East Asian Technical University Consortium (SEATUC), vol. 1, pp. 1–5 (2018)
31. Global drone regulations database [online]. https://www.droneregulations.info/C%C3%B4te+d%27Ivoire/CI.html. Accessed 06 Oct 2019
32. Stöcker, C., Bennett, R., Nex, F., Gerke, M., Zevenbergen, J.: Review of the current state of UAV regulations. Remote Sens. **9**(5), 459 (2017)
33. Goodbody, T.R.H., Coops, N.C., Marshall, P.L., Tompalski, P., Crawford, P.: Unmanned aerial systems for precision forest inventory purposes: a review and case study. For. Chron. **93**(01), 71–81 (2017)
34. The Sara Project – search and rescue aid using high accuracy EGNSS
35. Muhammad, B., Kumar, A., Cianca, E., Lindgren, P.: Improving port operations through the application of robotics and automation within the framework of shipping 4.0. In: 2018 21st International Symposium on Wireless Personal Multimedia Communications (WPMC), pp. 387–392 (2018)
36. Murphy, M.: The future is here: drones are delivering Domino's pizzas to customers. Quartz [online]. https://qz.com/838254/dominos-is-delivering-pizza-with-autonomous-drones-to-customers-in-new-zealand/. Accessed 06 Oct 2019
37. A drone has been used to deliver a donor kidney for the first time. MIT Technology Review [online]. https://www.technologyreview.com/f/613449/a-drone-has-been-used-to-deliver-a-donor-kidney-for-the-first-time/. Accessed 06 Oct 2019
38. Lacy, L.: June 5 and 2019. Amazon announces 30-minute drone deliveries via prime air [online]. https://www.adweek.com/digital/amazon-announces-30-minute-drone-deliveries/. Accessed 06 Oct 2019

A Novel Approach on Renewable Energy Harvesting Using Internet of Things (IoT)

S. Chandragandhi, E. Udayakumar, and K. Srihari

Abstract The goal of this undertaking is to audit existing sustainable power source and imminent methodologies in vitality reaping procedure as methods for having an economical and low support activity of WSN (wireless sensor networks). The Internet of Things idea has been utilized to build up the undertaking. Sunlight-based vitality reapers are getting to be basic in home and structures to give considerable advantages to our atmosphere, well-being, and economy. It together with other inexhaustible sources will soon outdate the customary fossil vitality to turn into a column for economical improvement. An IoT-based system is implemented to distribute energy among the solar board. The power gathered by sunlight-based clusters, in any case, relies on different factors, for example, climate conditions, photovoltaic (PV) boards, and vitality transformation that need broad and constant examinations to boost its potential. In the primary bearing, fixed, single, and two-fold pivot trackers are created to follow the course of the sun. An electromechanical framework to pursue the situation of the sun was structured. It is coordinated by a shut circle servo framework that takes estimations of direct sun-based radiation as the criticism. The control methodology is to drive two little dc engines with the goal that the sun picture is kept at the focal point of the four-quadrant photograph indicator detecting the sun position.

Keywords Wireless sensor network · IoT · LCD · Current transformer

S. Chandragandhi (✉)
Department of CSE, JCT College of Engineering and Technology, Coimbatore, Tamil Nadu, India

E. Udayakumar
Department of ECE, KIT-Kalaignarkarunanidhi Institute of Technology, Coimbatore, Tamil Nadu, India

K. Srihari
Department of CSE, SNS College of Technology, Coimbatore, Tamil Nadu, India

D. Gupta et al. (eds.), *Smart Sensors for Industrial Internet of Things*, Internet of Things, https://doi.org/10.1007/978-3-030-52624-5_17

1 Introduction

Sunlight-based vitality reapers are getting to be basic in home and structures to give significant advantages to our atmosphere, well-being, and economy. It together with other inexhaustible sources will soon outdate the conventional fossil vitality to turn into a column for supportable improvement. The power collected by sun-based clusters, notwithstanding, relies on different factors, for example, climate conditions, photovoltaic (PV) boards, and vitality change that need broad and constant examinations to augment its potential. In the primary heading, fixed, single, and twofold hub trackers are created to follow the course of the sun. An electromechanical framework to pursue the situation of the sun was planned [1]. It is coordinated by a shut circle servo framework that takes estimations of direct sun-powered radiation as the criticism. The control methodology is to drive two little dc engines with the goal that the sun picture is kept at the focal point of the four-quadrant photograph identifier detecting the sun position. In this work, we propose another way to deal with the issue of the board of sunlight-based vitality in miniaturized scale lattices handling the sun following issue as well as the unwavering quality and versatility of the framework. We initially acquaint a system which deals with the vitality stream by incorporating different equipment parts in a homogeneous system. The associated equipment gadgets structure an IoT organize that dependent on it, implanted PC sheets can be utilized to control sun based trackers.

The trademark and mechanical checking because of a huge segment of such frameworks keep running on nonessential power source that offers WSN with a kept association lifetime. One of the principle commonsense resists in shrewd matrix arranging, control, and activity with DERs is the voltage guideline at the conveyance field level. This issue rouses the sending of sensors and actuators in power matrices with the goal that the voltage guideline can be controlled at the ideal level. To do that the estimations from the sustainable microgrid state data are transmitted to a vitality board through the web of things (IoT)-based correspondence arrangement [2].

The adjoining masses with clean supportable power source have changed into a benchmark in setting up another time of conceivable society that is normally savvy. An unending of research has been started with an inspiring temper in analyzing elective approaches to manage abuse as of late referenced wellspring of centrality [3]. The benefits of centrality social event have been routinely comprehended on a certainly fundamental scale the degree that the bringing down of ozone draining substance outpourings, importance security and the flood impacts, yet the impact of such criticalness structure for a humbler scale application, for instance, the WSN will be surveyed further here. As the computational furthest reaches of the WSN center is generally inadequate as for, load is reliably compelled to simply the basic figuring required [4].

The Internet of Things (IoT) is a wise framework of interestingly recognizable gadgets prepared to do remote speaking with each other, administrations, and individuals for a huge scope through the Internet [1]. IoT plans to make the Internet omnipresent and unavoidable and can possibly influence numerous parts of clients' personal satisfaction. The arranged heterogeneous gadgets associated with an IoT structure are commonly furnished with sensors, controlling processors, remote handsets, and a vitality source (e.g., a battery) to screen their condition and send/get information. Applications imagined for IoT cover a wide scope of fields including home mechanization, social insurance, observation, transportation, savvy situations, and some more. One of the predominant obstructions to executing such a pompous plan is providing satisfactory vitality to work the system in an independent way without bargaining QoS. Along these lines, it is basic to improve the vitality productivity and life span of gadgets in IoT.

Despite the fact that there are various techniques to accomplish vitality proficiency, for example, utilizing lightweight correspondence conventions or receiving low-power radio handsets, the ongoing innovation pattern in vitality collecting gives a central technique to draw out battery life span. Along these lines, vitality collecting is a promising methodology for the rising IoT [4]. For all intents and purposes, vitality can be collected from ecological sources: warm, sunlight based, vibration, and remote RF vitality sources [3]. While collecting from the previously mentioned natural sources is reliant on the nearness of the relating vitality source, RF vitality reaping gives key advantages regarding being remote, promptly accessible as transmitted vitality (TV/radio supporters, versatile base stations, and handheld radios), ease, and what's more, little structure factor execution. This article presents an outline of remote vitality collecting units with regard to remote vitality reaping IoT (WEH-IoT) frameworks. Right now, sensor hubs ordinarily transmit information to a typical sink hub. The sink hub, otherwise called a passage, is associated with the system and is open to the outside world over the Internet.

A WEH-empowered sensor gadget generally comprises a reception apparatus, a handset, a WEH unit, a force the board unit (PMU), a sensor/processor unit, and conceivably an installed battery. Among those parts, there are two basic units for vitality collecting: unit and PMU are the WEH units responsible for gathering the RF vitality and delivering a steady vitality hot spot for the remainder of the gadget. It additionally interfaces with the PMU, while the PMU controls the handset, detecting unit usefulness, and deals with the vitality utilization of every unit and additionally obliges battery charging utilizing the gathered vitality. Right now, likewise center around the empowering advances including high-productivity remote/RF vitality collecting rectifiers and low-power wake-up radio for WEH units. We propose a PMU engineering that obliges a battery charging plan utilizing the gathered vitality through a WEH unit.

2 Related Works

The display appeared differently in relation to customary MPPT approaches and the future comprehension of the development in WSN. One of the fundamental handy oppositions in savvy lattice arranging, control, and activity with DERs is the voltage guideline at the dispersion field level. This issue spurs the arrangement of sensors and actuators in power networks with the goal that the voltage guideline can be controlled at the ideal level. To do that the estimations from the sustainable microgrid state data are transmitted to a vitality board by means of the web of things (IoT)-based correspondence arrangement. As it were, the proposed IoT correspondence foundation gives a chance to address the voltage guideline challenge by offering the two-way correspondence for microgrid state data accumulation and estimation. In light of this keen lattice correspondence foundation, we propose a Kalman channel-based state estimation technique for voltage guideline of the microgrid. At long last, the viability of the Kalman channel-based state estimation technique is represented utilizing the direct state space model of a microgrid joining DER [5].

This thought of programmed water system technique and the accompanying examination supports this thought. The undertaking of programmed water system is done through the help of soil dampness sensors. In the task, aside from soil dampness sensor, moistness and temperature sensors are likewise used to make the procedure increasingly advance. The proposed structure likewise has the element of GSM which makes this framework remote. The power required by segments is given through sun-powered boards, and subsequently this frees us from interference with power supply because of burden shedding. The water substance is always judged, and at whatever point dampness level of soil gets low, the framework sends a sign to engines requesting that they turn on. The engines consequently stop after soil achieves its greatest upper limit esteem which is chosen by client. Each time the engine turns over or stops consequently, the client will get a SMS about the status of activity. The significant focal points of the venture incorporate evasion from water wastage, development of plants to their most extreme potential, less odds of mistake because of less work, and continuous supply of water because of sunlight-based vitality.

As the interest of inexhaustible wellspring of vitality has grabbed the eye of various industry players and government bodies far and wide, the exertion in diminishing contamination and consumption in multiplying end-client sand nearby populace with clean sustainable power source have turned into a benchmark in setting up another age of reasonable society that is naturally cognizant. A bounteous of research has been started with an inspirational demeanor in investigating elective approaches to misuse the aforementioned wellspring of vitality. The benefits of essentialness social event have been normally comprehended on an increasingly unmistakable scale the extent that the bringing down of ozone exhausting substance surges, imperativeness security and the flood impacts, yet the impact of such imperativeness structure for a smaller scale application, for instance. A significant rate philosophy will be discussed and took a gander further in the subsequent pages.

The Internet of Things (IoT) is a worldview which intends to propel broadcast communications on the whole circles of human life. Sooner rather than later, evenness advances for the Internet of Things (IoT), alongside evenness and asymmetry applications for IoT security and protection, will restructure the sociobiological human territory morphology. This headway prompts considerable improvement in nature of human life and the world's financial development on the loose. IoT is viewed as the foundation of rising applications, as development assumes a key job in the enormous advancement of machine correspondences. Machine-to-machine (M2M) traffic is evaluated to represent roughly 45% of the absolute Internet traffic by 2022 [2]. IoT makes a stage where physical articles can emulate certain human tactile capacities, for example, discernment, vision, hearing, smell, and thinking. Floated by these human tangible abilities and the developing material Internet, machines can speak with one another, share significant data, and settle on ongoing choices with less human info, particularly since we are relocating to the 5G period, wherein the normal remote system delay is 1 ms. In this situation, different sensors, for example, radio recurrence recognizable proof (RFID) and detecting innovation, can attempt the community oriented capacity of detecting, gathering, and transmitting sensor data through the Internet [2]. Specialists report that IoT holds staggering potential for keen homes, shrewd urban areas, and also, human service applications.

With the ascent in omnipresent IoT developments, it has unfolded on the telecom administrators that inventive and innovative arrangements are the basic devices expected to handle the difficulties and the capability of IoT. Today, the quantity of items with Internet network capacity is more prominent than that of the Earth's populace. As the abilities of the IoT extend, it is predicted that about 50 billion gadgets will have Internet network by 2020 [3]. For the most part, the IoT will be prepared to do interconnecting billions/trillions of heterogeneous items through the Internet by utilizing extraordinary kinds of sensors like RFID and detecting innovation. The two advancements will embrace community elements of detecting, gathering, and transmitting sensor data through the Internet.

In spite of the fact that dynamic RFID labels are low-vitality expending remote gadgets [6], when these RFID labels are associated with a mechanical scale including billions of them, they produce a large number of payload transmitted to the server farms for handling. To process these gigantic sensor payload, tremendous handling and examination abilities are required, which, thus, expends immense vitality assets. Thus, it is important to consider the issue of huge scope utilization of vitality assets by IoTs. Inspired by this, a generally novel research intrigue has developed known as "green IoTs." This activity will in general improve EE as well as diminish the discharge of CO_2 from ICT items [7]. It is commonly acknowledged that IoT will have extraordinary monetary and biological criticalness in years to come. Driven by these real factors, it has gotten basic to break down the cutting-edge strategies and approaches fit for handling the vitality needs of these billions of vitality-hungry gadgets. In this manner, the green IoT has become a significant point for scientists and merchants now like never before as the ordinary vitality assets are decreasing and the vitality utilization increasing.

3 System Design

The coming of assorted improvements prompted the insurgency of IoT in home computerization, wearables, and mechanical applications. This raised the interest to scale down and improve the conveyability of detecting what's more, correspondence gadgets in the field of IoT. The current innovation underpins either battery worked or coordinate force infused frameworks. Some low-vitality expending frameworks utilize 3.3-V microcontrollers related to sensors fueled by minimal effort coin cells as a well-known ease decision for power supply in a little structure factor. The major difficulty lies in its support if the sensors are put at a remote and an out of reach place. Further if there should be an occurrence of copious number of sensor being conveyed for a particular application like home computerization, at that point the things like changing battery and looking after them turns into a much more tedious task [6]. One significant viewpoint in giving a productive IoT sensor application is to gather vitality either by searching the squandered power or controlling the application by means of a normally accessible vitality source, for example, a photovoltaic cell [8].

Vitality collecting is in this manner considered to be a significant element of IoT applications. Here it is important to comprehend the need of vitality reaping and appropriately deal with the force source inside the electronic framework to extend its life also and diminish the general force necessities of the entire application. The field of vitality gathering has no solid proof in giving an unbending battery less or vitality gathering framework which makes a bottleneck in the improvements of a genuine IoT sensor framework. Having said that it is essential to characterize a model methodology that would fit in any IoT application particularly battery-less or vitality collected IoT-based sensor organization. Presently envision a perplexing system where there are a huge number of sensors supposed the edge gadgets or the hubs regarding the Internet of Things wording gathering plenteous measure of information by and large framing a major information source. Obviously this information needs to be handled by some preparing unit in the IoT engineering of the said application [9].

With a sharp perception its extremely clear that not all sensors sense also and push the information for handling low-voltage microcontroller, which mirrors our model of sparing vitality by choosing just a force source which would be sufficiently adequate to source just a required measure of hubs at a given measure of time. This makes us to thoroughly consider the work of the total structure of the application as far as the choice of the handling unit is concerned, power sourcing limit at a given snapshot of time where the application is expected to do the greatest out of it. So as to successfully plan a force collecting, we need to initially gauge the force requests of the remote sensor hub and distinguish whether the collected vitality is sufficiently adequate to source all the hinders in the vitality gathering square graph that appeared beneath. The square outline shows a commonplace vitality gathering framework for a remote IoT sensor hub. One of the fundamental structure squares of any power collecting framework is the transducer, which effectively catches the vitality being squandered in either case and changes over it into usable electrical signs.

These electrical signs are then passed onto the power the executives framework which further procedures what's more, directs the vitality for capacity. Here the Power. The executive unit (PMU) plays out an extremely pivotal errand of monitoring the vitality collected in the capacity until it is sufficiently adequate to source the third arrange (made out of sensors, microcontrollers, and actuators) of the square outline at any rate of one complete pattern of sense-process-upload movement. At that point contingent upon the force stockpiling accessibility, the PMU imparts fitting force signs to either stop or then again continue for the third stage once more. Accordingly the force the executives framework adds a type of insight to the total IoT sensor hub [10]. This brilliance is accomplished for the most part by utilizing explicit low force devouring force the board IC intended for the same proposed reason. On accepting a continuous signal the microcontroller is fueled on, and it just takes barely any milliseconds to boot up, sense the information from the sensor, and transfer it to the cloud capacity utilizing inner or outside RF handset.

The framework comprises a commonsense arrangement which is alluded as a proving ground for showing an RF vitality reaping framework. The proving ground is essentially made out of a RF receiving wire for accepting RF signals transmitted by any emanating gadget, for example, a phone. This transmitting vitality must be gotten remotely at close to separations in the scope of scarcely any centimeters. A RF radio wire planned explicitly to catch these high recurrence radiations is associated further with a sign sponsor which intensifies the sign level to a fitting vitality level. The utilization of battery-powered batteries or aggregators isn't reasonable since quite a while ago it runs and supports free foundation and there is a tremendous requirement for the organization stockpiling gadgets perhaps of electrostatic type component [11]. These intensified signs are put away in a super capacitor as vitality until adequate measure of vitality is created to drive the controller and the sensor hubs.

The utilization of intensity board unit turns out to be increasingly basic here where it needs to decisively screen the super capacitor vitality level and take a fast choice to drive the following stage. This entire procedure of reaping vitality from the transmitting reception apparatus from wireless and catching information bundle all the while to sending an affirmation signal back to the mobile phone takes 6–8 s on a normal on the proving ground. Frequently simple gadgets offer ultralow force plan answers for fabricate high-performance, low-power vitality collecting applications [12].

The vitality reaping sources are utilized to build the lifetime and effectiveness of the IoT framework. The vitality sources utilized for vitality reaping system are either natural sources like sun based, wind, and so on or other vitality sources like temperature contrast, movement, footfall, breathing, and so on. Table 1 shows the distinction between different vitality sources. A sun-powered and wind energy sun-powered vitality is the most proficient wellspring of ecological vitality, furnishing a large portion of the vitality with high productivity [13]. Daylight is gathered by utilizing sunlight based or photograph voltaic cells that are made of semiconductor material, for example, silicon. This innovation legitimately changes over daylight into usable power. For the area where accessibility of light is destined to be high, photograph voltaic cell is the advantageous arrangement. For indoor applications, specific PV materials are utilized which are more qualified for diminishing light. Enormous measure of vitality is lost when vitality is moved from collector. To conquer this issue, maximum force point tracker (MPPT) circuit has been proposed which capably moved collected sunlight-based vitality from collector to battery-powered battery. The support cost of sunlight-based cells is extremely low. Among all other sources, daylight is considered as the most suitable vitality hot spot for collecting. Warm vitality is the vitality that originates from heat. Warm vitality is likewise identified with varieties in temperature.

The systems utilized for warm vitality collecting are thermoelectric and pyroelectric. Thermoelectric system straightforwardly changes over temperature distinction into usable vitality structure by utilizing Seebeck effect. The collecting proficiency of thermoelectric is around 5–8%. The thermoelectric gadgets are utilized in space and earthbound applications. Pyroelectric system produces electrical vitality when there is change in temperature. Pyroelectric system utilizes pyroelectric precious stone that creates power when warmed or cooled. Pyro-power strategy ought not be mistaken for thermoelectric strategy in light of the fact that pyro-power results in transitory voltage over the gem and thermoelectric outcomes in perpetual voltage over the precious stone. There is consistently temperature contrast between condition temperature and human internal heat level, and this temperature distinction is a great wellspring of warm vitality. Warm vitality is hard to convert to usable electrical vitality. Vitality gathering is the procedure to change overencompassing vitality sources into electrical vitality. This influential idea has continually prepared with a few productive strategies for a number of sources. Inside any EH framework, there are discrete subframeworks working parallelly, and be that as it may, power creating components are key job players for overseeing control procedures.

Since a decade ago, numerous commitments have been given to address the low fueled gadgets by reaped vitality sources. The principle procedure for changing the vitality source into electrical vitality is through the transducers. All EH transducers are one of a kind because of their particular electrical qualities that are being tended to by both vitality transformation gadgets and framework overall. It incorporates photovoltaic cells (PV), piezoelectric generators, thermoelectric generators (TEG), radio recurrence, and some more. For structuring any IoT gadget, we have to design the power producing component dependent on the particular vitality source. It is urgent then to choose the proficient vitality reaping transducer that can capably collect the accessible vitality source however much as could reasonably be expected. In view of these stringent particulars, heaps of vitality collecting methods have been acquainted which give the most extreme vitality like MPPT calculations for extricating sun-based vitality source.

The strategies for augmenting the transformation effectiveness for thermoelectric generators, piezo-thermomechanical, and piezo-thermofluidic are being explored [6]. It is required to determine the estimation of intensity utilization measure of explicit IoT gadget for activity in a lengthened way. For this reason, planners and engineers must gauge the force utilization connection regarding the created power. There can be two situations: if the produced power is bigger than the devoured power or the produced power is lesser than the devoured power. Right now working modes have been sorted out cautiously for making the harmony among utilization and age. For generally speaking estimation, it is important to gauge the rest state, standby, and dynamic mode for the IoT gadget.

As generally known, the temperature distinction over the length of a thermoelectric material is a predominant factor in deciding the exhibition of a thermoelectric gadget. A bigger temperature distinction prompts a higher power yield, and along these lines, a suitable warmth sink is important to accomplish this huge temperature distinction. In that perspective, joining of warmth sinks with a thermoelectric gadget is basic in improving its exhibition yet is poorly examined. Right now, sharp plans of warmth sinks, not explicitly intended for TEGs, however conceivable to be incorporated with thermoelectric frameworks are illustrated. Among the different structures of warmth sinks that can be coordinated with thermoelectric gadgets is one with PCMs.

It has demonstrated incredible reason in diminishing the temperature of a warmth source. PCMs are known to have high inactive warmth, and in this way, such property has been explored generally for its latent capacity use in heat sinks. The instrument behind PCMs is using its high idle warmth when it changes stage from strong to fluid to retain the warmth from its warmth source. To improve the exhibition of the warmth sink dependent on PCMs, investigates to expand warm conductivity have been generally directed. Wang et al. [14] examine the impact of utilizing a porous metal fiber sintered felt (PMFSF) to expand the warm conductivity of PCMs. It has heat sink by including a low level of nanoparticles. They just numerically recreate the impacts of nanoparticles because of complexities in nanoparticle manufacture and had the option to acquire improved warm conductivities. In contrast to their desires, including just 2% of nanoparticles shows the best execution, and the exhibition continuously diminishes with further expansion of nanoparticles.

The structure a uninvolved warm administration framework by using a metal froth immersed with PCM in a warmth sink to diminish the temperature of a warmth source. Like the work, copper is utilized as the metal froth and contrasted with an unadulterated PCM and unadulterated copper storm cellar, and the froth PCM composite was the most proficient at forestalling the expansion in temperature of the warmth source. Regardless of every one of these endeavors, the most serious issue of the PCM heat sink despite everything stands even with a modification latent warm administration. Without question, as it changes stage, it can retain more warmth than in its strong structure because of its high idle warmth; notwithstanding, when the PCM arrives at its most extreme warmth limit after some time, the surface temperature of the warmth source starts to increment quickly [10]. Accordingly, thermally dealing with an electronic gadget with a PCM heat sink for a significant stretch of time is not down to earth and is just suitable for a brief span of time.

Another notable kind of warmth sinks is a framework made out of smaller-scale channels with a coursing liquid. First proposed by Tuckerman and Pease [7], a fluid cooled heat-exchanger configuration by utilizing liquid directs in minute measurements permits the activity of circuits with power densities higher than 1000 W cm^{-2}. The thought behind miniaturized scale channels is to utilize various separate channels, as opposed to a solitary channel to stream over the surface of the warmth source and to build the surface territory of the warmth sink that the liquid reaches. It includes an extra channel top of the base channel, eventually structuring a twofold layered counter-stream miniaturized scale channel with nanofluids and PCM slurries as the coolants as appeared.

With various kinds of liquid utilized in the upper and the lower channel, the warmth motion consumed by the warmth sink is the best when the nanofluids and the PCM slurries are conveyed in the upper and lower channel, individually. Not quite the same as warmth sinks with PCMs or smaller-scale channels referenced above, very little research has been finished on adaptable warmth sinks as their interest has been low. There have been a few works, which receive an adaptable warmth sink on a thermoelectric generator for body heat collecting. In spite of the fact that the presentation of body heat collecting has been improved with the presence of an adaptable warmth sink, those exploration have pointed more on the adaptable thermoelectric framework as opposed to adaptable warmth sink [15]. The interest for adaptable warmth sinks with supportability and elite is relied upon to be expanded for more extensive applications.

4 Proposed System

The proposed framework plans for sun-based vitality collectors are getting to be fundamental in home and industries giving considerable advantages to our atmosphere, well-being, and economy. It together with other sustainable sources will soon outdate both inexhaustible and non-inexhaustible vitality utilized for wireless sensor network. Arduino UNO ATmega 28 is a microcontroller utilized for working superiorly. The speed builds 20 MHz ranges in this framework. In this framework sunlight-based board can be utilized to store DC batteries. From DC to AC converter DC gives corrected and high voltage AC power is transmitted to another area at the less than desirable end, inverter used to static inverter plant changes over the power back to AC. Rather than utilizing WI-FI and Internet of Things (IoT) to get to information [6]. To accomplish our proposed framework, we need to utilize Arduino microcontroller ATmega328, current sensor, sunlight based, battery, and LCD. Typically burden takes power supply from matrix. DC battery is utilized for capacity reason. The square graph of vitality gathering utilizing IoT is shown in Fig. 1. DC to AC converter is utilized to change over sustainable power source to air conditioning vitality. IOT is utilized to send the determined measure of sustainable power source use. Voltage sensor is utilized to quantify the line voltage.

Here current sensor is utilized to shield the gadgets from deficiencies like over voltages and under voltage. On the off chance that any of the above issue happens, the gadget is consequently turned off. Controller status and everything is shown in

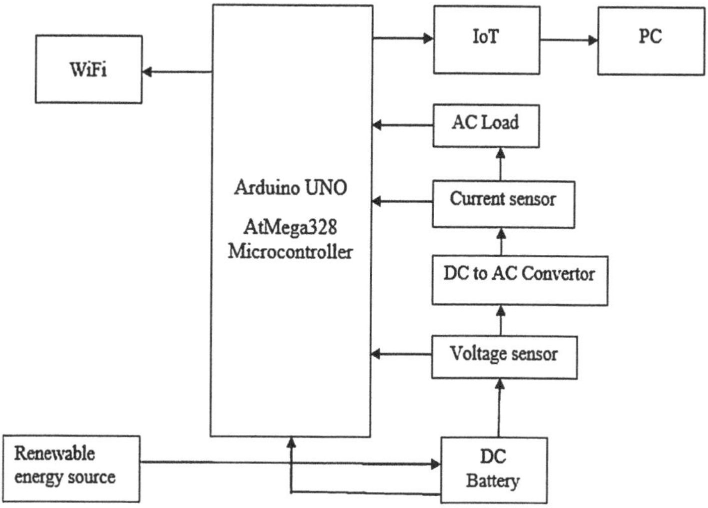

Fig. 1 Block diagram of energy harvesting using IoT

LCD. The entire procedure is constrained by a microcontroller. The information is sent to the cloud consistently and we can get to the site at any place [8].

4.1 Arduino Microcontroller

The Arduino microcontroller is a simple, yet amazing single-board PC that has increased significant footing in the side interest and expert market. The Arduino is open source, which means equipment is sensibly valued and improvement programming is free. This guide is for understudies in ME 2011, or understudies anyplace who are standing up to the Arduino just because [5]. For cutting edge Arduino clients, lurk the web; there are loads of assets. The most notable characteristic of Arduino isn't its capacity to crunch code, yet rather its capacity to interface with the outside world through its info yield (I/O) pins. The Arduino has 14 advanced I/O pins marked 0–13 that can be utilized to turn engines and lights on and off and read the condition of switches. Each computerized stick can sink or source around 40 mA of current. This is more than sufficient for interfacing to most gadgets, however means that interface circuits are expected to control gadgets [7] other than straightforward LEDs.

4.2 Current Sensor

Current sensor is a present identifying implies age of the voltage signal which is related to the present going in the circuit. A standard strategy for identifying current is to insert a resistor in the method for current to be sensed. By then we can put the distinguished resistor at wherever in game plan with the circuit may it be weight or switch. Along these lines current identifying contraptions are to be considered as present to voltage converter device which recognizes and changes over current to get a yield voltage, which is genuinely in respect to the current in the organized manner. Exactly when current is experiencing the circuit, a voltage drops over the way where the current is spilling. Also an alluring field is made near the present passing on conductor. These above miracles are used in the present sensor plan method [16].

4.3 Current Transformer (CT)

Current transformer (CT) is a transformer that quantifies electric flows and the present high current strong state vitality meters. It can make the grade regarding very high present and devours little power. It is additionally extremely valuable in estimating or checking high current, high voltage, and high power circuits. These are utilized in power arrangement of different types, for example, control supplies, engine controls, and lighting controls.

5 Results and Discussion

In this work sustainable power source can be utilized for mechanical reason to created, when this strategy to observing the machine working and any high voltage or drop current conditions in everyday distinguishing proof effectively and furthermore incorporate temperature. We use web of things (IoT) to transfer information base in cloud that is intended to store information. The hardware description is shown in Fig. 2.

The stage is worked to take in the gigantic volumes of information produced by gadgets, sensors, sites, applications, clients, and accomplices and start activities for continuous reactions. The IoT database is shown in Fig. 3. In this framework IoT-based sustainable power source collecting can be created. Also, utilizing most recent update can be refreshed naturally. The client can get to any place for these updates. Like the issue in modern machine running framework, in this work in general sustainable power source gathering, utilizing IoT-based correspondence system was effectively created [16].

Fig. 2 Hardware description

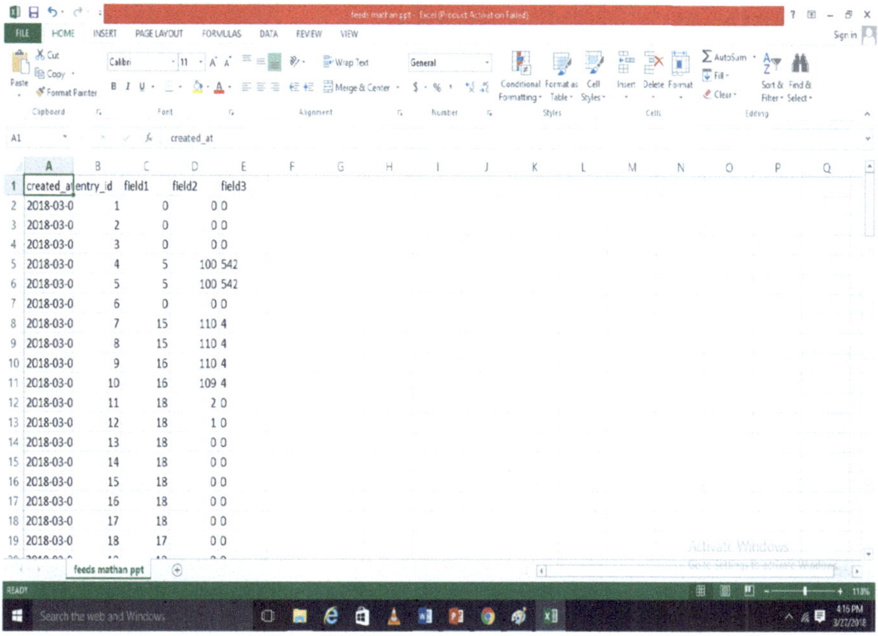

	A	B	C	D	E
1	created_a	entry_id	field1	field2	field3
2	2018-03-0	1	0	0	0
3	2018-03-0	2	0	0	0
4	2018-03-0	3	0	0	0
5	2018-03-0	4	5	100	542
6	2018-03-0	5	5	100	542
7	2018-03-0	6	0	0	0
8	2018-03-0	7	15	110	4
9	2018-03-0	8	15	110	4
10	2018-03-0	9	16	110	4
11	2018-03-0	10	16	109	4
12	2018-03-0	11	18	2	0
13	2018-03-0	12	18	1	0
14	2018-03-0	13	18	0	0
15	2018-03-0	14	18	0	0
16	2018-03-0	15	18	0	0
17	2018-03-0	16	18	0	0
18	2018-03-0	17	18	0	0
19	2018-03-0	18	17	0	0

Fig. 3 IoT database

6 Conclusion

In this work, we have acquainted a framework which deals with the sustainable power source reaped by utilizing sun-oriented vitality. Equipment and system structure have been effectively actualized with genuine sun-based vitality. The IoT innovation has been utilized refreshing current procedure, wireless sensor network (WSN) having erratic and restricted wellspring of intensity, the appropriation of harvestable vitality in the power the executives framework. In this framework IoT-based sustainable power source collecting can be created. Also, utilizing most recent update can be refreshed consequently. The client can get to any place for these updates. Like the issue in mechanical machine running framework, in this work in general sustainable power source reaping, utilizing IoT-based correspondence system was effectively created.

References

1. Chavali, P., et al.: Distributed power system state estimation using factor graphs. IEEE Trans. Signal. Process. **63**(11), 2864–2876 (2015)
2. Kayastha, N., et al.: Smart grid sensor data collection, communications, and networking: a tutorial. Wireless Comm. Mobile Comput. **14**(11), 1055–1087 (2014)
3. Li, Y.: Fully distributed state estimation of smart grids. In: Proc. of the Int. Conf. Comm., pp. 6580–6585 (2012)
4. Liang, H., Abdrabou, A., et al.: Stochastic information management for voltage regulation in smart distribution systems. In: Proc. of the INFOCOM, pp. 2652–2660 (2014)
5. Wang, X., Liang, Q.: Stabilizing the power sin microgrid using sensor selection. In: Proc. of the Global Comm. Conf., pp. 3513–3518 (2012)
6. Meliopoulos, A.P., et al.: Smart grid technologies for autonomous operation and control. IEEE Trans. Smart Grid. **2**(1), 1–10 (2011)
7. Vetrivelan, P., et al.: Chapter 13: A neural network based automatic crop monitoring robot for agriculture. In: The IoT and the Next Revolutions Automating the World, pp. 203–212. IGI Global, Hershey, PA (2019)
8. Weng, Y., Negi, R., Ilic, M.D.: Graphical model for state estimation in electric power systems. In: Proc. of the Int. Conf. on Smart Grid Comm., p. 108 (2013)
9. Kamalinejad, P., et al.: Wireless energy harvesting for the Internet of Things. IEEE Commun. Mag. **53**(6), 102–108 (2015)
10. Adila, A.S., Husam, A., Husi, G.: Towards the self-powered Internet of Things (IoT) by energy harvesting: trends and technologies for green IoT. In: Second International Symposium on Small-Scale Intelligent Manufacturing Systems (SIMS), Cavan, pp. 1–5 (2018)
11. Park, H., et al.: Energy harvesting using thermoelectricity for IoT (Internet of Things) and E-skin sensors. J. Phys. Energy. **1**(4), 042001 (2019)
12. Yogeshwaran, K., et al.: Design and Analysis of Efficient Special Modern Adder, pp. 1–77. Lambert Academic Publishing, Germany (2019)
13. Liang, H., et al.: Stochastic modeling & optimization in a microgrid: a survey. Energies. **7**(4), 2027–2050 (2014)
14. Wang, Y., et al.: Decentralized communication and control systems for power system operation. IEEE Trans. Smart Grid. **6**(2), 885–893 (2015)

15. Garg, N., Garg, R.: Energy harvesting in IoT devices: a survey. In: 2017 International Conference on Intelligent Sustainable Systems (ICISS), Palladam, pp. 127–131 (2017)
16. Udayakumar, E., et al.: An event based neural network architecture with content addressable memory. Int. J. Embedd. Real-Time Commun. Syst. **11**(1), 23–40 (2020)
17. Vetrivelan, P., et al.: Design of smart surveillance security system based on wireless sensor network. Int. J. Res. Stud. Sci. Eng. Technol. **4**(5), 23–26 (2017)
18. Kanagaraj, T., et al.: Foot pressure measurement by using ATMEGA 164 microcontroller. Adv. Nat. Appl. Sci. **10**(13), 224–228 (2016)
19. Srihari, K., et al.: Automatic battery replacement of robot. Adv. Nat. Appl. Sci. **9**(7), 33–38 (2015)
20. Vetrivelan, P., et al.: PAPR reduction for OQAM/OFDM signals using optimized iterative clipping and filtering technique. In: Proceedings of IEEE International Conference on Soft-Computing and Network Security (ICSNS'15), p. 72. SNS College of Technology, Coimbatore (2015)
21. Santhi, S., et al.: SoS emergency ad-hoc wireless network. In: Computational Intelligence and Sustainable Systems (CISS), EAI/Springer Innovations in Communications and Computing, pp. 227–234 (2019)
22. Santhi, S. et al.: Design and implementation of area and delay efficient FXLMS filter for active noise cancellation. In: Computational Intelligence and Sustainable Systems (CISS), EAI/Springer Innovations in Communications and Computing, pp. 115–129 (2019)
23. Alsharif, M.H., et al.: Energy harvesting techniques for wireless sensor networks/radio-frequency identification: a review. Symmetry. **11**, 865 (2019)
24. Patnaikuni, D., Patnaikuni, S.: A model approach towards energy harvesting and battery less self powered Internet of Things (IoT) applications. Int. J. Res. Advent Technol. **6**(6), 1038–1041 (2018)

Security and Surveillance at Smart Homes in a Smart City Through Internet of Things

Rinky Dwivedi, Koyel Datta Gupta, and Deepak Sharma

Abstract Smart home infrastructure consists of intelligent networking devices, infrastructure, and seamless integration of various devices using wired/wireless technologies. It allows the ease of use for household systems and creates a highly personalized and safe home space. There are many corporates that are seriously indulging in smart home systems which include GE, Cisco, Google, Microsoft, and others. Smart homes provide a productive and cost-efficient environment and maximize the effectiveness of the occupants. Smart homes provide efficient management with minimum life-time costs of hardware and facilities. They optimize the things such as structures, systems, and services and also manage the interrelationships between these three. The present research focuses on how security and surveillance are achieved in smart homes.

Keywords Smart home · Surveillance sensors · Raspberry Pi

1 Introduction

Very soon all the different things we are seeing around need to be connected. Global network is a connection of different networks and computing devices. The scope of the Internet is expanding beyond computers and computing devices; it is going to interconnect different things and physical objects that the people are seeing around. The different electrical appliances such as lighting system in a room, lights, fans, air conditioners microwave oven and refrigerator etc. in our homes along with the different machines at buiseness will be interconnected, Internet working different equipment and so on [1]. So each and every thing that is around us that the people

R. Dwivedi (✉) · K. D. Gupta
Department of Computer Science and Engineering, Maharaja Surajmal Institute of Technology, New Delhi, India
e-mail: rinkydwivedi@msit.in; koyel.dg@msit.in

D. Sharma
Division of Information Technology, Netaji Subhas University of Technology (Formerly Netaji Subhas Institute of Technology), New Delhi, India

D. Gupta et al. (eds.), *Smart Sensors for Industrial Internet of Things*, Internet of Things, https://doi.org/10.1007/978-3-030-52624-5_18

are using in home and in their business can be connected as a whole Internet of Things (IoT).

> "The internet of things (IoT) is the network of physical objects that contain embedded technology to communicate and sense or interact with their internal states of the external environment" [2].

In IoT unification of technologies is a big challenge since the devices can be of various types having different configurations and different specifications each of these supporting through other systems such as low-power embedded systems, cloud computing, big data, machine learning and networking, computer vision, etc.; a number of technologies from electrical and electronic sciences and even from mechanic sciences are required in order to build an IoT [3].

Manufacturing/business, healthcare, retail, and security are the major application domains of IoT; among all of these, the largest market share of IoT is in manufacturing and business which goes up to 40.2% approximately. Recently IoT is gaining importance in security and surveillance sectors; about 7.7% of market share goes to the same. IoT helps in integrating various technologies like biometrics and facial recognition locks, remote sensors, IRIS-based recognition, etc. to form efficient security mechanisms.

IoT is also one of the building blocks to be considered for use of developing smart homes and smart cities [4]. So IoT is one of the enabling technologies to make the home smart [5]. Let us consider the example of a school in a smart city; school basically works in isolation, but in a smart city we have different other things connected. For example, when a schoolchild wants to go to school, he must have information about different modes of transportation that he can take, and in the case of an emergency on road, the police, hospital, and traffic control are informed so that the child can get immediate help. The organization of the paper is as follows: Section II deals with the importance and characteristics of IoT and Sect. III describes the concept of smart cities and smart homes; after this the research moves forward and explained the home area network, the network formed for smart homes. Section V gives the details on how security and surveillance are achieved in smart homes.

2 Why IoT Is Required

The reason is that IoT is emphasized to provide an advanced level of surface to the society and to the business. With the help of this technology the different types of things the chair, the table, the watch etc. all of these can be fitted with embedded systems, embedded electronics and information technology so that they have some basic computing platform attached to them and in them they are going to act as different nodes in that particular Internet of Things [6, 7].

So what is going to happen is these systems/things all equipped with embedded systems and these embedded systems along with embedded electronics, embedded processors, and embedded communication systems help them to connect with other things around them, and depending upon the application requirements on the specific course of business, a big Internet is going to form that is much much bigger than the current Internet and is called as IoT.

The IoT is going to expand in two dimensions, first constitute the expansion of current internet where all the existing things are going to be connected to the existing Internet. The other dimension is to build a separate Internet world of the physical objects from scratch. Irrespective of what approach is addressed, each has its own challenge.

2.1 Characteristics of IoT

- *Efficient, scalable, and associated architecture*: It has to serve efficiently the requirements of the application for which they are deployed. They have to be scalable, because in IoTs a large number of things are associated not just with millions but several billions or trillions. So even if the number of sensing devices and IoT devices increases, the overall network performance should not be compromised.
- *Unambiguous naming and addressing architecture*: Presently, IP IV and DNAs are used for naming in the present Internet system. But now when the Internet is scaling at such a huge rate, a bigger problem of naming the physical nodes needs to be addressed.
- *Abundance of sleeping nodes, mobile, and non-IP devices*: A large number of devices connected with the IoT will be in a sleeping mode. Some devices are moving from one place to another like a smart watch, making the nework mobile. Network addressing in IoT is a challenging task and various reearchers are working on this area.
- *Intermittent connectivity*: It is another characteristic of a typical IoT; when these devices move, the network and subnetwork get portioned. One device that is in connectivity with other device at a latter instant of time may not be connected now. So, one solution to address this problem between the different nodes in the network is opportunistic networks (Op-Nets).

3 Concept of Smart Homes and Smart Cities

A smart city is an urban system that uses information and communication technology (ICT) tools that make the infrastructure very interactive, efficient, and accessible in an easier manner than before [8].

The term "Smart" in smart city means the services offered to respective stakeholders and citizens will be provided in a better manner than usual, with the help of latest ICT technologies, that includes advanced technology like an embedded systems to integrate computer devices and electronic devices together.

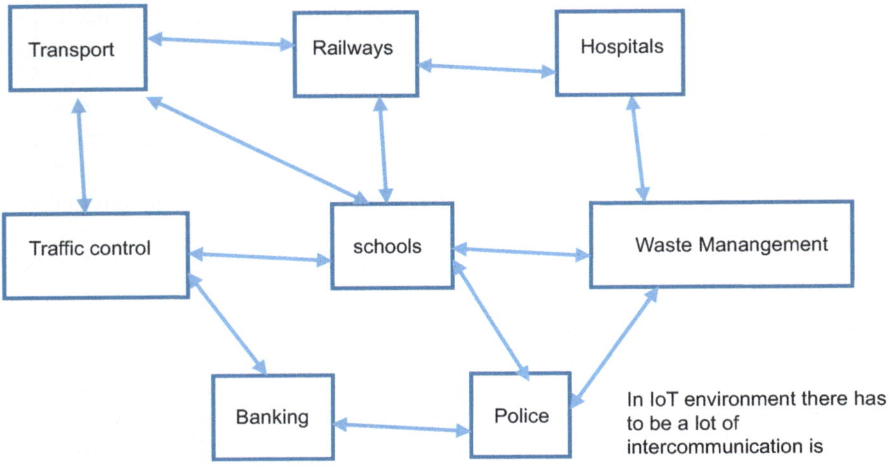

Fig. 1 Component level diagram of a smart city

This section describes how IoT can help in building smart cities and homes throughout the world. Nowadays there is a lot of scope in building smart cities.

Figure 1 shows the major components like school, railways, waste management, police, etc.; all these components play an important role in a city. So to transform all these components to the components of a smart city, there has to be a lot of intercommunication required among components. This one can be achieved with the help of ICT technologies like sensor networks, RFID, NFC, Z-Wave [9], etc.

In a smart city, if something wrong happens to a child at school, instead of the traditional way of calling, there would be smart messaging to the hospital in which the ambulance will come and throughout the way can be monitored, and parents would be informed immediately. So many things should be done seamlessly and automatically. Services can be offered to the subscribers only. So whether it is free or payment basis depends upon the smart city.

As an analogy between a prototype human and a regular city, the prototype has skeleton, organs, and skin, whereas the regular city has buildings, industries, people, transports, hospitals, police, banks, etc. Now extending the analogy further between a smart human and smart city, the smart human has brain, nerve, sensory organs, and cognitions, whereas the smart city has ubiquitously embedded intelligence, digital telecommunication networks, sensors, tags, etc. The smart city has following characteristics:

Smart economy: Smart cities emphasize on start-ups to improve economy, participation of global MNCs, worldwide buyers, involving government agencies, etc. And this all has to interconnect.

Smart governance: Core components of governance are government agencies, officials, and citizens. Around them there are other peripherals such as banking, finance, safety, surveillance, management, energy services, etc.; traditionally, in a regular city, they all work in isolation, and there is a minimum connectivity between them, but in a smart city, there is a smart connectivity between them.

Smart people: Education, healthcare and population are core components of a smart city from people's point of view and in the peripherals there can be banking, shopping, transport, social network, current affairs, safety and connectivity. From any component it should be possible to get information from another component; of course to preserve the privacy, there has to be some policy to float the data across. The citizens get information as per their need, and again that need has to be identified in a smart city with the help of cognition and intelligence embedded in the system.

Smart mobility: The core components for mobality are localities, cities and population. The peripherals are emergency responses, vehicular networks, electric vehicles, airways, railways, transports, and logistics. To move from point A to point B like a person books a train from point A in a city to point B in a city, the route is very congested and expensive or there may be any other reason to consider an alternative option. The person should get an alternative route or alternative option like bus, etc. The person can get the information with the help of smart tools.

3.1 Smart Homes in Smart Cities

Smart cities must have smart homes, smart parking lots, smart vehicles, smart health, pollution and calamity monitoring, smart energy, smart agriculture, etc. In a smart home, medical facilities like regular health monitoring should be done in a smart way. The medical data should be made available to the doctor whenever there is a critical situation; the corresponding physician should be called. Resources should be conserved (e.g., electricity, water, fuel); the tap water should be turned off with the help of ICT tools when not in use. For accomplishing these tasks, sensors need to be deployed at the various positions at home. The deployment of sensor networks in a smart home comes with different challenges—the choice of appropriate sensor for individual sensing task is crucial. For efficient utilization of energy, planning is very much required as different devices use different energy level, the scheduling of the duty cycle is a major issue. Device placement and network architecture are also very important for a reliable end-to-end IoT implementation [10].

3.2 How to Make the Home Smart Home

The objective of a smart home is to offer services to respective stakeholders in a better manner than usual, with the help of ICT technologies, for example, if one wants to cook in a smart manner at smart home. The procedure is, there may be

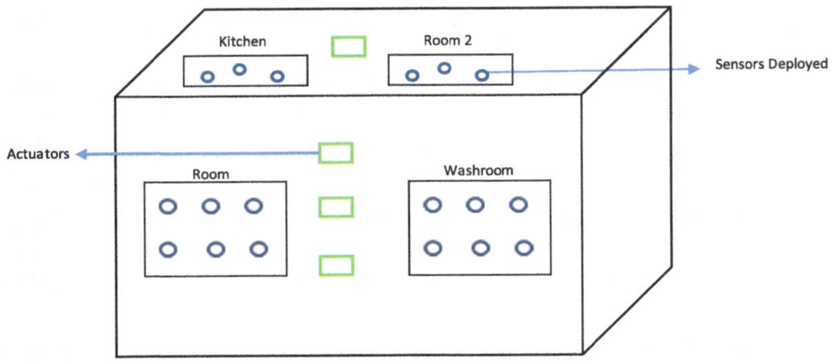

Fig. 2 Pictorial representation of a smart home

some devices at room 1 that go to the kitchen and open the refrigerator to take out something and then in kitchen there is a microwave, the device put the food taken from the refrigerator to microwave oven to warm it. After that the device serves the food at room 1 (a pictorial representaton of smart home is shown in Fig. 2). Similar to these other activities at home can be made smarter with the help of sensors, actuators, and other devices that need to set up an IoT.

4 Home Area Networks for Smart Homes

The network formed for a smart home is referred to as home area networks (HANs). The HAN is analogous to a local area network (LAN) in a building or partial area network (PAN) in a body. So, HAN is a network contained within a home; it enables remote access and control of devices and systems. HAN provides an amalgamation of various systems within a home such as security systems, home automation systems, personal media, communication media, etc. HAN supports multi-protocol gateway bridges for both IP network and non-IP network. HANs are wired and non-wired; a wired HAN is a low-cost network and easily integrates with pre-existing house infrastructure and can use powerlines, coaxial cables, telephone lines, optical fibers, and other such technologies for communication. At the same time, the limitation is mobility, i.e., it restricts the mobility of the user at home. Wireless HAN can use popular home Wi-Fi, Zigbee, and even new standards such as 6LoWPAN. Wireless networks improves the mobility and can be implemented easily thereby improving the satisfaction and utility of home networks.

5 Security and Surveillance at Smart Homes

The objective of smart home security is to provide security locks for door, comfort, convenience 24*7 security, and energy efficiency for the user, that further includes important factors. The factors that are necessary for the formation of a secured home are:

1. *Monitoring the door 24*7*: The very first motive of designing this project is providing the security at the door for 24 h. As it is impossible for a human being to sit all day at the outside of the door and keep a watch; it is basically suitable for the houses where at least a member is present inside the home. So if something happens, he/she will be able to call for the help. The same purpose can be solved by using a PI Cam [11] that will be installed at the door of the house and will constantly provide the surveillance.
2. *Get notified*: The second main objective is to alert the user or the owner of the house. The recording will be ON for 24 h, but as soon as someone comes near the door and gets in contact with the IR sensor [11] placed with the camera, the PI Cam [11] will immediately capture the image of the person at the door and will send an e-mail to the authorized user. In this way a user can keep a watch on the house even if there is no one present in the house.
3. *Auto door lock*: The other important objective is that it provides the access to the door. The mechanism will be set in such a way that as soon as sensor detects some intruder at the door, it will click the image and notify at the registered e-mail ID. The owner will see the image, and if the available person is his/her known, then they can simply provide access to the door.

Figure 3 presents a low-cost and flexible home control and monitoring system with IP connectivity for accessing and controlling devices and appliances remotely.

The main component of the system will be the Raspberry Pi board [12], i.e., the brain of the system. The Pi board consists of a processor and hosts a Linux Operating System named Raspbian Jessie made to suit the needs of IoT [13]. In our research, we mainly use Python for the automation system along with other suitable development tools. The development tools are used to create the web interface so that the system can be made accessible from anywhere in the globe. The proposed approach differs from the previous proposal in this field in a way that it provides high portability since the status of the electronic components of the house as all the information will be available right inside the web interface. Raspberry Pi, will be connected to an Internet-enabled router and a web server on its platform. The web server will host the web interface and will communicate with the control of devices.

The required functionality can be achieved in three steps:

Step 1: Sensors used for the purpose are temperature and humidity sensor (DHT11), light-dependent resistor (LDR), and passive infrared sensor (PIR).

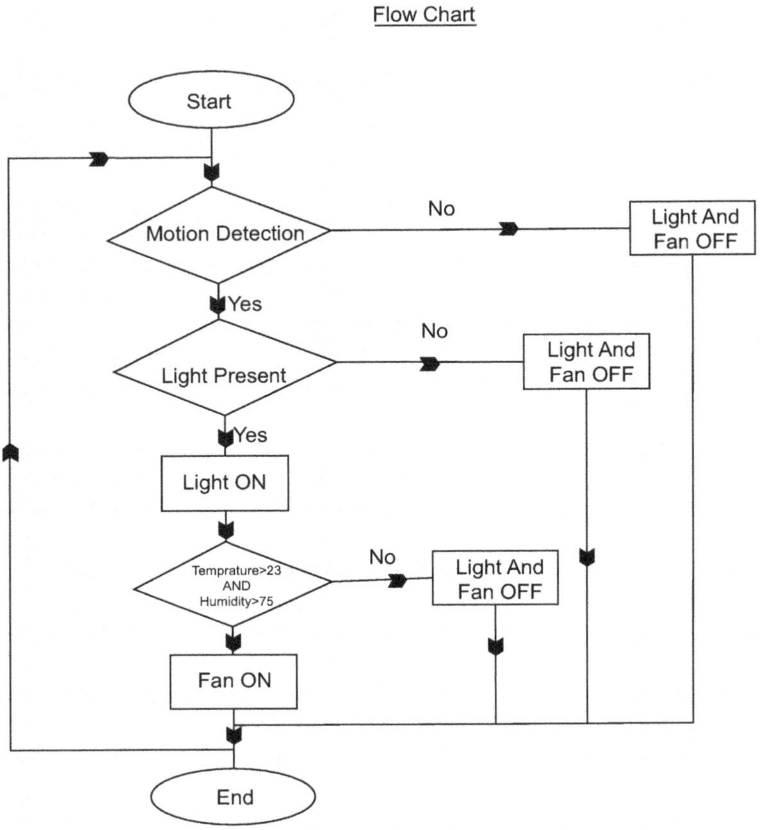

Fig. 3 Basic flow graph for home control and monitoring systems with IP connectivity

All the sensors were so connected with the GPIO pins of the Raspberry Pi to produce the desired output. The function is performed in a manner that supposes there is a PIR (motion sensor) is placed at the entrance of the door, it will detect the incoming and outgoing motion and the doors will be open accordingly, once someone entered the home, the next sensor that is DHT11 will start sensing the temperature of the home if the temperature is sufficient the other appliances present in the home will automatically turn ON or OFF (in the condition when there is no one in the home).

Step 2: Appliance control using web application—ON/OFF air conditioner, *ON/OFF* fridge, and *ON/OFF* water motor.

These three majorly used appliances were also controlled using the Raspberry Pi. The desired code was written in Python, and the controlling part was handled through Raspberry Pi.

Apart from the normal lights/fans, it became necessary to control some major heavy appliances which one often forgets to turn off or turn on. Here the controlling of the AC/fridge/water motor is done. The conditions are set on the Raspberry Pi, which further sends and receives the information according to the need and performs the task with the help of a web interface that is the website.

Step 3: Web application—The existing system in this field claims the project will work on Wi-Fi or Bluetooth. The use of Wi-Fi or Bluetooth resulted in a major drawback, i.e., the distance. Here the area of the use is being limited. The user and the hardware part should be in the network range of the system to work. But in our proposed system, this limitation is fully eliminated. We have designed a web Interface (website) to control all the appliances in a particular home. The linking of the website is done in such a manner that it was very user-friendly. Users can have their own user ID and password to log in their home control panel. After login, users can control their home from anywhere.

Basically, we created a text file on the server and store an ON/OFF value after selection ON or OFF options by a user, and every 3 seconds our python file stored in Raspberry Pi checks the text of this file using HTTP protocol and accordingly ON/OFF appliances.

6 Conclusion and Future Work

The major application domains of IoT are smart parking, structural health, noise urban maps, smartphone detection, traffic congestion, smart lighting, waste management, smart roads, river floods, smart grid, tank level, photovoltaic installations, water flow, stock calculation, perimeter access control, liquid presence, and hazard presence. Many IoT systems and prototypes are built to serve these applications. So, in the present day, what is required is trillions of sensors and billions of smart systems. Frequently used IoT enablers nowadays are RFID, sensors, actuators, and nanotechnology. Nanotechnology is an open area in IoT and is focused on building an IoT of nanosensor. Since it is going a lot of demand, many researchers are going in this area. The nanodevices are used for different purposes. For example, nanocapsules that can be consumed and can be spitted out after they performed the intended functions. These nanocapsules can talk to each other, these nanodevices are being conceptualized till date, and people are building nanodevices that can be used for IoT. So once we have an established nanodevice IoT, the IoT arises to a new horizon.

References

1. Sriskanthan, N., Karand, T.: Bluetooth based home automation system. J. Microprocess. Microsyst. **26**, 281–289 (2002)
2. http://www.gartner.com//it.glossary/internet of things/
3. Li, L.: Technology designed to combat fakes in the global supply chain. Bus. Horiz. **56**(2), 167–177 (2013)

4. Dohr, A., Modre-Osprian, R., Drobics, M., Hayn, D., Schreier, G.: The Internet of things for ambient assisted living. In: Proceedings Seventh International Conference of Information Technology, New Gener., pp. 804–809 (2010)
5. Toshi, G.M., Campos, L.B., Cugnasa, C.E.: Home automation networks: a survey. Comput. Stand. Interfaces. **50**, 42–54 (2017)
6. Atzori, L., Iera, A., Morabito, G.: The internet of things: a survey. Comput. Netw. **54**(15), 2787–2805 (2010)
7. Granjal, J., Monteiro, E., Sa Silva, J.: Security for the internet of things: a survey of existing protocols and open research issues. IEEE Commun. Surv. Tutorials. **17**(3), 1294–1312 (2015)
8. Ye, X., Huang, J.: A framework for cloud-based smart home. In: Computer Science and Network Technology (ICCSNT), vol. 2, pp. 894–897. IEEE, Washington, DC (2011)
9. Khattab, A., Jeddi, Z., Amini, E., Bayoumi, M.: RFID Security: A Lightweight Paradigm. Springer, Berlin (2016)
10. Arasteh, H., et al.: IoT based smart cities: a survey. In: IEEE 16th International Conference on Environment and Electrical Engineering (EEEIC) (2016)
11. Sukmana, F.M.G.K.D., Teja, H.: Wireless and mobile (apwimob). In: 2015 IEEE Asia pacific conference on, pp. 183–187 (2015)
12. Vujovic, V., Maksimovic, M.: Raspberry Pi as a Sensor Web node for home automation. Comput. Electr. Eng. **44**, 153–171 (2015)
13. Mercer, I.: Energy savings from home automation [online] (2010). https://blog.abodit.com/category/home-automation/environment-green/

Index

D. Gupta et al. (eds.), *Smart Sensors for Industrial Internet of Things*, Internet of Things, https://doi.org/10.1007/978-3-030-52624-5

The manufacturer's authorised representative in the EU is Springer Nature Customer Service Centre GmbH, Europaplatz 3, 69115 Heidelberg, Germany. If you have any concerns regarding our products, please contact ProductSafety@springernature.com

Printed and bound by CPI Group (UK) Ltd, Croydon, CR0 4YY
30/09/2026
02225694-0010